INVESTIGATIONS IN
YEAST FUNCTIONAL GENOMICS
AND MOLECULAR BIOLOGY

INVESTIGATIONS IN YEAST FUNCTIONAL GENOMICS AND MOLECULAR BIOLOGY

Edited by
Matthew Eckwahl, MPhil

Apple Academic Press

TORONTO   NEW JERSEY

Apple Academic Press Inc.	Apple Academic Press Inc.
3333 Mistwell Crescent	9 Spinnaker Way
Oakville, ON L6L 0A2	Waretown, NJ 08758
Canada	USA

©2014 by Apple Academic Press, Inc.

First issued in paperback 2021

Exclusive worldwide distribution by CRC Press, a member of Taylor & Francis Group
No claim to original U.S. Government works

ISBN 13: 978-1-77463-331-1 (pbk)
ISBN 13: 978-1-77188-010-7 (hbk)

Library of Congress Control Number: 2013955467

Library and Archives Canada Cataloguing in Publication

Investigations in yeast functional genomics and molecular biology/edited by Matthew Eckwahl, MPh.

Includes bibliographical references and index.
ISBN 978-1-77188-010-7
1. Saccharomyces cerevisiae--Molecular genetics. 2. Yeast fungi--Molecular genetics.
3. Yeast fungi--Biotechnology. 4. Molecular genetics--Technique. 5. Molecular biology.
I. Eckwahl, Matthew, editor of compilation

| QH470.S23I58 2014 | 571.6'29563 | C2013-907864-9 |

Apple Academic Press also publishes its books in a variety of electronic formats. Some content that appears in print may not be available in electronic format. For information about Apple Academic Press products, visit our website at **www.appleacademicpress.com** and the CRC Press website at **www.crcpress.com**

INVESTIGATIONS IN YEAST FUNCTIONAL GENOMICS AND MOLECULAR BIOLOGY

Edited by
Matthew Eckwahl, MPhil

Apple Academic Press

TORONTO   NEW JERSEY

Apple Academic Press Inc. | Apple Academic Press Inc.
3333 Mistwell Crescent | 9 Spinnaker Way
Oakville, ON L6L 0A2 | Waretown, NJ 08758
Canada | USA

©2014 by Apple Academic Press, Inc.

First issued in paperback 2021

Exclusive worldwide distribution by CRC Press, a member of Taylor & Francis Group
No claim to original U.S. Government works

ISBN 13: 978-1-77463-331-1 (pbk)
ISBN 13: 978-1-77188-010-7 (hbk)

Library of Congress Control Number: 2013955467

Library and Archives Canada Cataloguing in Publication

Investigations in yeast functional genomics and molecular biology/edited by Matthew Eckwahl, MPh.

Includes bibliographical references and index.
ISBN 978-1-77188-010-7
1. Saccharomyces cerevisiae--Molecular genetics. 2. Yeast fungi--Molecular genetics.
3. Yeast fungi--Biotechnology. 4. Molecular genetics--Technique. 5. Molecular biology.
I. Eckwahl, Matthew, editor of compilation

QH470.S23I58 2014   571.6'29563   C2013-907864-9

Apple Academic Press also publishes its books in a variety of electronic formats. Some content that appears in print may not be available in electronic format. For information about Apple Academic Press products, visit our website at **www.appleacademicpress.com** and the CRC Press website at **www.crcpress.com**

ABOUT THE EDITOR

MATTHEW ECKWAHL, MPhil

Matthew Eckwahl is a current graduate student at Yale University in the Department of Cell Biology, where he is a NSF Graduate Research Fellow. He received his BS in molecular biology and biotechnology at the University of Michigan-Flint. Eckwahl has research experience in regulatory networks of pseudohyphal growth response and is currently working on the packaging of non-coding RNA by retroviruses.

CONTENTS

ACKNOWLEDGMENT AND HOW TO CITE

The editor and publisher thank each of the authors who contributed to this book, whether by granting their permission individually or by releasing their research as open source articles. The chapters in this book were previously published in various places in various formats. To cite the work contained in this book and to view the individual permissions, please refer to the citations at the beginning of each chapter. Each chapter was read individually and carefully selected by the editor. The result is a book that looks at yeast functional genomics and molecular biology from many different angles and using many different methodologies. Specifically:

- Chapter 1 provides a very nice example of how differential regulation of the Start checkpoint is dependent on transcriptional programs that can interact with cell size control. Cell polarity is a nearly universal feature of eukaryotic cells. Cell division is asymmetrical in budding yeast, producing a larger mother and smaller daughter cell.
- Chapter 2 gives an in-depth look at the robustness tradeoffs organisms face in sensing and responding to chemical spacial gradients. Their model of yeast cell polarization resulting from mating pheromone gradients is especially interesting.
- The evolutionarily conserved Rho GTPase, Cdc42, is a small GTPase that is a regulator of cell polarization and is activated by Cdc24. The researchers in Chapter 3 examined whether phosphorylation of Cdc24 plays a role in its regulation. After thorough investigation, it ultimately seems that mutation of all currently identified phosphorylation sites does not cause observable defects in growth rate or morphology. Despite its non-confirmatory results, this article provides a nice example of hypothesis-driven research; moreover, the reporting of "negative data" is of critical value to scientific progress.
- Chapter 4 illustrates another technical advantage of yeast in its screening of an overexpression library of transcription factors to identify novel factors that have a role in hyphal development and virulence of *Candida albicans*.
- Chapter 5 reports the impressive power of yeast to better understand how human HSF1 (heat shock transcription factor 1) is regulated through post-translational modification, with the yeast assay also enabling a screen to

identify pharmacological activators of human HSF1 that could provide potential benefit in protein folding diseases.

- Yes, yeast also have prions. Intriguingly, they may even provide a survival benefit for yeast and other fungi. Chapter 6 provides a concise overview of this topic.

- Chapter 7 introduces an interesting approach to inhibit transmission of aggregate prions to the daughter cell via the sequestering of chaperones that would otherwise cut them into small, transmissible pieces.

- Chapter 8 provides a well-documented look at how gene expression is tightly coordinated with environmental changes. The authors focus on the example of ribosomal biogenesis, which utilizes a sizeable proportion of a cell's biosynthesis expenditure, and examine the contribution of internal versus external cues.

- RNA binding proteins play a critical role in post-transcriptional control of RNAs, such as splicing, mRNA stabilization, mRNA localization and translation. Chapter 9 uses genomic tools to identify RNA targets for RNA binding proteins and suggests that there may be an expansive post-transcriptional regulatory network between proteins that act in metabolism and RNA regulation that has previously been overlooked.

- Chapter 10 makes excellent use of the compact nature of the yeast genome and its comprehensive annotation to analyze the entire genetic architecture of chromatin accessibility regulation.

- Comparative genomics is a powerful method to look at how organisms are related to each other and to elucidate genomic function. In Chapter 11, the researchers discuss their *de novo* sequencing strategy to obtain chromosome-sized scaffolds for *Saccharomyces arboricolus* and achieve improved comparisons of genome organization with other strains.

- Chapter 12 gives an example of the analysis of functional divergence data and provides a global look at gene expression patterns between two yeast species.

- A uniform wild type yeast does not exist—extensive human manipulation has resulted in a genetically variable number of natural and domesticated strains, with the latter being specialized for specific industrial applications. Chapter 13 presents a fascinating look at the differences between brewing and winemaking strains compared to that of existing high-coverage strains. The authors also make a valid point about the limitations in the use of S288c as a yeast reference strain.

LIST OF CONTRIBUTORS

Jason P. Affourtit
454 Life Sciences, A Roche Company, Branford, Connecticut, United States of America

Alex N. Nguyen Ba
Department of Cell & Systems Biology, University of Toronto, M5S 2 J4, Toronto, Canada and Centre for the Analysis of Genome Evolution and Function, University of Toronto, M5S 3B2, Toronto, Ontario, Canada

Naama Barkai
Department of Molecular Genetics and Department of Physics of Complex Systems, Weizmann Institute of Science, Rehovot, Israel

Liliana Batista-Nascimento
Genomics and Stress Laboratory, Instituto de Tecnologia Química e Biológica, Oeiras, Portugal

Anders Bergström
Institute of Research on Cancer and Ageing of Nice (IRCAN), CNRS UMR 7284 - INSERM U1081, Université de Nice Sophia Antipolis, 06107, NICE Cedex 2, France

Martin Blythe
DeepSeq, Centre for Genetics and Genomics, Queen's Medical Centre, University of Nottingham, NG7 2UH, Nottingham, UK

Gireesh K. Bogu
Genome Institute of Singapore, Singapore, Singapore

Anthony R. Borneman
The Australian Wine Research Institute, Adelaide, Australia

Miri Carmi
Department of Molecular Genetics and Department of Physics of Complex Systems, Weizmann Institute of Science, Rehovot, Israel

Amy A. Caudy
Lewis-Sigler Institute for Integrative Genomics, Princeton University, Princeton, NJ 08544, USA and Donnelly Centre for Cellular and Biomolecular Research, University of Toronto, Toronto, Ontario M5S 3E1, Canada

Paul J. Chambers
The Australian Wine Research Institute, Adelaide, Australia

Jung Kyoon Choi
Department of Bio and Brain Engineering, Korea Advanced Institute of Science and Technology, Daejeon, Korea and Genome Institute of Singapore, Singapore, Singapore

Ching-Shan Chou
Department of Mathematics, Center for Mathematical and Computational Biology, Center for Complex Biological Systems, University of California Irvine, Irvine, California, United States of America

Frederick R. Cross
The Rockefeller University, New York, New York, United States of America

Francisco A. Cubillos
Centre for Genetics and Genomics, Queen's Medical Centre, University of Nottingham, NG7 2UH, Nottingham, UK. Current address: INRA, UMR1318, Institut Jean-Pierre Bourgin, F-78000, Versailles, France

Felix Dafhnis-Calas
Centre for Genetics and Genomics, Queen's Medical Centre, University of Nottingham, NG7 2UH, Nottingham, UK

Irina L. Derkatch
Department of Biological Sciences, University of Illinois at Chicago, Chicago, Illinois, United States of America and Department of Neuroscience, Columbia University, New York, New York, United States of America

Brian A. Desany
454 Life Sciences, A Roche Company, Branford, Connecticut, United States of America

Stefano Di Talia
The Rockefeller University, New York, New York, United States of America

Han Du
State Key Laboratory of Mycology, Institute of Microbiology, Chinese Academy of Sciences, Beijing, China

Maitreya J. Dunham
Lewis-Sigler Institute for Integrative Genomics, Princeton University, Princeton, NJ 08544, USA and Department of Genome Sciences, University of Washington, Seattle, WA 98195, USA

Michael Egholm
454 Life Sciences, A Roche Company, Branford, Connecticut, United States of America

Gilgi Friedlander
Department of Molecular Genetics and Department of Physics of Complex Systems, Weizmann Institute of Science, Rehovot, Israel

Angus H. Forgan
The Australian Wine Research Institute, Adelaide, Australia

Bruce Futcher
Department of Molecular Genetics and Microbiology, SUNY at Stony Brook, Stony Brook, New York, United States of America

André P. Gerber
Department of Chemistry and Applied Biosciences, Institute of Pharmaceutical Sciences, ETH Zurich, Zurich, Switzerland

Scott A. Gerber
Department of Cell Biology, Harvard Medical School, Boston, Massachusetts, United States of America

Guobo Guan
State Key Laboratory of Mycology, Institute of Microbiology, Chinese Academy of Sciences, Beijing, China

Yuanfang Guan
Lewis-Sigler Institute for Integrative Genomics, Princeton University, Princeton, NJ 08544, USA, Department of Molecular Biology, Princeton University, Princeton, NJ 08544, USA, and Current Address: Department of Computational Medicine & Bioinformatics, University of Michigan, Ann Arbor, MI 48109, USA

Joo Y. Hong
Department of Biological Sciences, University of Illinois at Chicago, Chicago, Illinois, United States of America and Department of Biochemistry and Molecular Biology, University of Nevada, Reno, Nevada, United States of America

Guanghua Huang
State Key Laboratory of Mycology, Institute of Microbiology, Chinese Academy of Sciences, Beijing, China

Jan Ihmels
Department of Molecular Genetics and Department of Physics of Complex Systems, Weizmann Institute of Science, Rehovot, Israel and Howard Hughes Medical Institute, Department of Cellular and Molecular Pharmacology, University of California, San Francisco, California, United States of America

Sarath Chandra Janga
Medical Research Council (MRC) Laboratory of Molecular Biology, Cambridge, United Kingdom

Inkyung Jung
Department of Bio and Brain Engineering, Korea Advanced Institute of Science and Technology, Daejeon, Korea

Shima Khoshraftar
Department of Cell & Systems Biology, University of Toronto, M5S 2 J4, Toronto, Canada

Dongsup Kim
Department of Bio and Brain Engineering, Korea Advanced Institute of Science and Technology, Daejeon, Korea

Kwoneel Kim
Department of Bio and Brain Engineering, Korea Advanced Institute of Science and Technology, Daejeon, Korea

Sang Cheol Kim
Korean Bioinformation Center, Korea Research Institute of Bioscience and Biotechnology, Daejeon, Korea

Anthony S. Kowal
Department of Molecular Pharmacology and Biological Chemistry, The Feinberg School of Medicine, Northwestern University, Chicago, Illinois, United States of America

Byungwook Lee
Korean Bioinformation Center, Korea Research Institute of Bioscience and Biotechnology, Daejeon, Korea

Heun-Sik Lee
Center for Genome Science, National Institute of Health, Cheongwon, Korea

Kibaick Lee
Department of Bio and Brain Engineering, Korea Advanced Institute of Science and Technology, Daejeon, Korea

Sanghyuk Lee
Korean Bioinformation Center, Korea Research Institute of Bioscience and Biotechnology, Daejeon, Korea

Sagi Levy
Department of Molecular Genetics and Department of Physics of Complex Systems, Weizmann Institute of Science, Rehovot, Israel

Liming Li
Department of Molecular Pharmacology and Biological Chemistry, The Feinberg School of Medicine, Northwestern University, Chicago, Illinois, United States of America

Rong Li
Stowers Institute for Medical Research, Kansas City, Missouri, United States of America and Department of Molecular and Integrative Physiology, University of Kansas Medical Center, Kansas City, Kansas, United States of America

Susan W. Liebman
Department of Biological Sciences, University of Illinois at Chicago, Chicago, Illinois, United States of America and Department of Biochemistry and Molecular Biology, University of Nevada, Reno, Nevada, United States of America

Gianni Liti
Institute of Research on Cancer and Ageing of Nice (IRCAN), CNRS UMR 7284 - INSERM U1081, Université de Nice Sophia Antipolis, 06107, NICE Cedex 2, France

Phillip C.C. Liu
Applied Technology Group, Incyte Corporation, Wilmington, Delaware, United States of America

Edward J. Louis
Centre for Genetics and Genomics, Queen's Medical Centre, University of Nottingham, NG7 2UH, Nottingham, UK

Sunir Malla
DeepSeq, Centre for Genetics and Genomics, Queen's Medical Centre, University of Nottingham, NG7 2UH, Nottingham, UK

Neel Mehta
Centre for Genetics and Genomics, Queen's Medical Centre, University of Nottingham, NG7 2UH, Nottingham, UK

Nitish Mittal
Medical Research Council (MRC) Laboratory of Molecular Biology, Cambridge, United Kingdom and Department of Biotechnology, National Institute of Pharmaceutical Education and Research, Punjab, India

Alan M. Moses
Department of Cell & Systems Biology, University of Toronto, M5S 2 J4, Toronto, Canada and Centre for the Analysis of Genome Evolution and Function, University of Toronto, M5S 3B2, Toronto, Ontario, Canada

Carolin A. Müller
Centre for Genetics and Genomics, Queen's Medical Centre, University of Nottingham, NG7 2UH, Nottingham, UK

Daniel W. Neef
Department of Pharmacology and Cancer Biology, Duke University School of Medicine, Durham, North Carolina, United States of America

Qing Nie
Department of Mathematics, Center for Mathematical and Computational Biology, Center for Complex Biological Systems, University of California Irvine, Irvine, California, United States of America

Conrad A. Nieduszynski
Centre for Genetics and Genomics, Queen's Medical Centre, University of Nottingham, NG7 2UH, Nottingham, UK

Isak S. Pretorius
The Australian Wine Research Institute, Adelaide, Australia

David Riches
454 Life Sciences, A Roche Company, Branford, Connecticut, United States of America

Claudina Rodrigues-Pousada
Genomics and Stress Laboratory, Instituto de Tecnologia Química e Biológica, Oeiras, Portugal

Adam P. Rosebrock
Department of Molecular Genetics and Microbiology, SUNY at Stony Brook, Stony Brook, New York, United States of America

Tanja Scherrer
Department of Chemistry and Applied Biosciences, Institute of Pharmaceutical Sciences, ETH Zurich, Zurich, Switzerland

Jungmin Seo
Research Institute of Bioinformatics, Omicsis, Daejeon, Korea

Cheuk C. Siow
Centre for Genetics and Genomics, Queen's Medical Centre, University of Nottingham, NG7 2UH, Nottingham, UK

Jan M. Skotheim
The Rockefeller University, New York, New York, United States of America

Yuan Sun
State Key Laboratory of Mycology, Institute of Microbiology, Chinese Academy of Sciences, Beijing, China

Dennis J. Thiele
Department of Pharmacology and Cancer Biology, Duke University School of Medicine, Durham, North Carolina, United States of America

Yaojun Tong
Graduate University of Chinese Academy of Sciences, Beijing, China and Chinese Academy of Sciences Key Laboratory of Pathogenic Microbiology and Immunology, Institute of Microbiology, Chinese Academy of Sciences, Beijing, China

Olga G. Troyanskaya
Lewis-Sigler Institute for Integrative Genomics, Princeton University, Princeton, NJ 08544, USA and Department of Computer Science, Princeton University, Princeton, NJ 08540, USA

Stephanie C. Wai
Stowers Institute for Medical Research, Kansas City, Missouri, United States of America and Biological and Biomedical Sciences Graduate Program, Harvard Medical School, Boston, Massachusetts, United States of America

Hongyin Wang
Department of Molecular Genetics and Microbiology, SUNY at Stony Brook, Stony Brook, New York, United States of America

Jonas Warringer
Department of Chemistry and Molecular Biology, University of Gothenburg, 41390, Gothenburg, Sweden

Adina Weinberger
Department of Molecular Genetics and Department of Physics of Complex Systems, Weizmann Institute of Science, Rehovot, Israel

Jing Xie
State Key Laboratory of Mycology, Institute of Microbiology, Chinese Academy of Sciences, Beijing, China and Graduate University of Chinese Academy of Sciences, Beijing, China

Zi Yang
Department of Biological Sciences, University of Illinois at Chicago, Chicago, Illinois, United States of America

Tau-Mu Yi
Department of Developmental and Cell Biology, Center for Complex Biological Systems, University of California Irvine, Irvine, California, United States of America

Lixin Zhang
Chinese Academy of Sciences Key Laboratory of Pathogenic Microbiology and Immunology, Institute of Microbiology, Chinese Academy of Sciences, Beijing, China

INTRODUCTION

The "awesome power of yeast genetics" is well known. Many features make *Saccharomyces cerevisiae* an ideal eukaryotic organism for research. This book intends to present readers with a range of topics that showcase some of their advantages in allowing us to understand complex biological phenomena. The first section deals with symmetry breaking, a fundamental process in biology. For instance, the mating process in yeast is induced after haploid cells sense a gradient of mating pheromone, resulting in polarized cell growth (schmooing) toward the source. Later, we look at another important developmental transition in yeast—the switch to a filamentous growth form. We further see how yeast helps us understand mechanisms of prion formation and transmission as well as possible biological roles. Many of the articles presented also highlight the power of yeast in addressing genome-wide problems in biology. Along these lines, the final section introduces the reader to the rapidly expanding field of comparative genomics, which not only helps tease apart evolutionary kinship but also aids the discovery of functional elements throughout a genome.

It is hoped that this book will inform the reader of a variety of interesting studies undertaken in yeast, employing a broad range of methodology from mathematical modeling to next-generation sequencing.

In budding yeast, asymmetric cell division yields a larger mother and a smaller daughter cell, which transcribe different genes due to the daughter-specific transcription factors Ace2 and Ash1. Cell size control at the Start checkpoint has long been considered to be a main regulator of the length of the G1 phase of the cell cycle, resulting in longer G1 in the smaller daughter cells. Chapter 1, by Di Talia and colleagues, confirmed this concept using quantitative time-lapse microscopy. However, it has been proposed that daughter-specific, Ace2-dependent repression of expression of the G1 cyclin CLN3 had a dominant role in delaying daughters in G1. The authors wanted to reconcile these two divergent perspectives on the origin of long daughter G1 times. They quantified size control using single-cell time-lapse imaging of fluorescently labeled budding yeast, in the presence or absence of the daughter-specific transcriptional regulators Ace2 and Ash1. Ace2 and Ash1 are not required for efficient size control, but they shift the domain of efficient size control to larger cell size, thus increasing cell size requirement for Start in daughters. Microarray and chromatin immunoprecipitation experiments show that Ace2 and Ash1 are direct transcriptional regulators of the G1 cyclin gene *CLN3*. Quantification of cell size control in cells expressing titrated levels of Cln3 from ectopic promoters, and from cells with mutated Ace2 and Ash1 sites in the *CLN3* promoter, showed that regulation of *CLN3* expression by Ace2 and Ash1 can account for the differential regulation of Start in response to cell size in mothers and daughters. The study shows how daughter-specific transcriptional programs can interact with intrinsic cell size control to differentially regulate Start in mother and daughter cells. This work demonstrates mechanistically how asymmetric localization of cell fate determinants results in cell-type-specific regulation of the cell cycle.

Cells localize (polarize) internal components to specific locations in response to external signals such as spatial gradients. For example, yeast cells form a mating projection toward the source of mating pheromone. There are specific challenges associated with cell polarization including amplification of shallow external gradients of ligand to produce steep internal gradients of protein components (e.g. localized distribution), response over a broad range of ligand concentrations, and tracking of moving signal sources. In Chapter 2, Chou and colleagues investigated the tradeoffs among these performance objectives using a generic model that captures the basic spatial dynamics of polarization in yeast cells, which are small. The authors varied the positive feedback, cooperativity, and diffusion coefficients in the model to explore the nature of this tradeoff. Increasing the positive feedback gain resulted in better amplification, but also produced multiple steady-states and hysteresis that prevented the tracking of directional changes of the gradient. Feedforward/feedback coincidence detection in the positive feedback loop and multi-stage amplification both improved tracking with only a modest loss of amplification. Surprisingly, they found that introducing lateral surface diffusion increased the robustness of polarization and collapsed the multiple steady-states to a single steady-state at the cost of a reduction in polarization. Finally, in a more mechanistic model of yeast cell polarization, a surface diffusion coefficient between 0.01 and 0.001 $\mu m^2/s$ produced the best polarization performance, and this range is close to the measured value. The model also showed good gradient sensitivity and dynamic range. This research is significant because it provides an in-depth analysis of the performance tradeoffs that confront biological systems that sense and respond to chemical spatial gradients, proposes strategies for balancing this tradeoff, highlights the critical role of lateral diffusion of proteins in the membrane on the robustness of polarization, and furnishes a framework for future spatial models of yeast cell polarization.

Cell polarization is essential for processes such as cell migration and asymmetric cell division. A common regulator of cell polarization in most eukaryotic cells is the conserved Rho GTPase, Cdc42. In budding yeast, Cdc42 is activated by a single guanine nucleotide exchange factor, Cdc24. The mechanistic details of Cdc24 activation at the onset of yeast cell polarization are unclear. Previous studies have suggested an important role for phosphorylation of Cdc24, which may regulate activity or function of the protein, representing a key step in the symmetry breaking process. In Chapter 3, Wai and colleagues directly ask whether multisite phosphorylation of Cdc24 plays a role in its regulation. The authors identify through mass spectrometry analysis over thirty putative in vivo phosphorylation sites. They first focus on sites matching consensus sequences for cyclin-dependent and p21-activated kinases, two kinase families that have been previously shown to phosphorylate Cdc24. Through site-directed mutagenesis, yeast genetics, and light and fluorescence microscopy, we show that nonphosphorylatable mutations of these consensus sites do not lead to any detectable consequences on growth rate, morphology, kinetics of polarization, or localization of the mutant protein. The authors do, however, observe a change in the mobility shift of mutant Cdc24 proteins on SDS-PAGE, suggesting that they have indeed perturbed its phosphorylation. Finally, they show that mutation of all identified phosphorylation sites does not cause observable defects in growth rate or morphology. The study concludes that lack of phosphorylation on Cdc24 has no overt functional consequences in budding yeast. Yeast cell polarization may be more tightly regulated by inactivation of Cdc42 by GTPase activating proteins or by alternative methods of Cdc24 regulation, such as conformational changes or oligomerization.

Candida albicans is the most common human fungal pathogen, causing not only superficial infections, but also life-threatening systemic disease. *C. albicans* can grow in several morphological forms including unicellular yeast-form, elongated hyphae and pseudohyphae. In certain natural environments, *C. albicans* also exists as biofilms, which are structured and surface-attached microbial communities. Transcription factors play a critical role in morphogenesis and biofilm development. In Chapter 4, Du and colleagues identified four adhesion-promoting transcription factors (Tec1, Cph1, Ume6 and Gat2) by screening a transcription factor overexpression library. Sequence analysis indicates that Gat2 is a GATA-type zinc finger transcription factor. Here we showed that the *gat2/gat2* mutant failed to form biofilms on the plastic and silicone surfaces. Overexpression of *GAT2* gene promoted filamentous and invasive growth on agar containing Lee's medium, while deletion of this gene had an opposite effect. However, inactivation of Gat2 had no obvious effect on N-acetyl-glucosamine (GlcNAc) induced hyphal development. In a mouse model of systemic infection, the *gat2/gat2* mutant showed strongly attenuated virulence. Our results suggest that Gat2 plays a critical role in *C. albicans* biofilm formation, filamentous growth and virulence.

Heat shock transcription factor 1 (HSF1) plays an important role in the cellular response to proteotoxic stresses. Under normal growth conditions HSF1 is repressed as an inactive monomer in part through post-translation modifications that include protein acetylation, sumoylation and phosphorylation. Upon exposure to stress HSF1 homotrimerizes, accumulates in nucleus, binds DNA, becomes hyper-phosphorylated and activates the expression of stress response genes. While HSF1 and the mechanisms that regulate its activity have been studied for over two decades, our understanding of HSF1 regulation remains incomplete. As previous studies have shown that HSF1 and the heat shock response promoter element (HSE) are generally structurally conserved from yeast to metazoans, Chapter 4 has made use of the genetically tractable budding yeast as a facile assay system to further understand the mechanisms that regulate human HSF1 through phosphorylation of serine 303. Batista-Nascimento and colleagues show that when human HSF1 is expressed in yeast its phosphorylation at S303 is promoted by the MAP-kinase Slt2 independent of a priming event at S307 previously believed to be a prerequisite. Furthermore, the authors show that phosphorylation at S303 in yeast and mammalian cells occurs independent of GSK3, the kinase primarily thought to be responsible for S303 phosphorylation. Lastly, while previous studies have suggested that S303 phosphorylation represses HSF1-dependent transactivation, this study now shows that S303 phosphorylation also represses HSF1 multimerization in both yeast and mammalian cells. Taken together, these studies suggest that yeast cells will be a powerful experimental tool for deciphering aspects of human HSF1 regulation by post-translational modifications.

Chapter 6, by Li and Kowal, provides a short review of prions in yeast. They conclude that studying this topic can provide valuable information and that because yeast is so easy to study and manipulate, it is worth loking into further as an organism of prion research.

Prions are self-propagating conformations of proteins that can cause heritable phenotypic traits. Most yeast prions contain glutamine (Q)/asparagine (N)-rich domains that facilitate the accumulation of the protein into amyloid-like aggregates. Efficient transmission of these infectious aggregates to daughter cells requires that chaperones, including Hsp104 and Sis1, continually sever the aggregates into smaller "seeds." Yang and colleagues previously identified 11 proteins with Q/N-rich domains that, when overproduced, facilitate the *de novo* aggregation of the Sup35 protein into the [PSI +] prion state. Here, in Chapter 7, the authors show that overexpression of many of the same 11 Q/N-rich proteins can also destabilize pre-existing [PSI+] or [URE3] prions. They explore in detail the events leading to the loss (curing) of [PSI+] by the overexpression of one of these proteins, the Q/N-rich domain of Pin4, which causes Sup35 aggregates to increase in size and decrease in transmissibility to daughter cells. We show that the Pin4 Q/N-rich domain sequesters Hsp104 and Sis1 chaperones away from the diffuse cytoplasmic pool. Thus, a mechanism by which heterologous Q/N-rich proteins impair prion propagation appears to be the loss of cytoplasmic Hsp104 and Sis1 available to sever [PSI+].

Cells must adjust their gene expression in order to compete in a constantly changing environment. Two alternative strategies could in principle ensure optimal coordination of gene expression with physiological requirements. First, characters of the internal physiological state, such as growth rate, metabolite levels, or energy availability, could be feedback to tune gene expression. Second, internal needs could be inferred from the external environment, using evolutionary-tuned signaling pathways. Coordination of ribosomal biogenesis with the requirement for protein synthesis is of particular importance, since cells devote a large fraction of their biosynthetic capacity for ribosomal biogenesis. In Chapter 8, to define the relative contribution of internal vs. external sensing to the regulation of ribosomal biogenesis gene expression in yeast, Levy and colleagues subjected *S. cerevisiae* cells to conditions which decoupled the actual vs. environmentally-expected growth rate. Gene expression followed the environmental signal according to the expected, but not the actual, growth rate. Simultaneous monitoring of gene expression and growth rate in continuous cultures further confirmed that ribosome biogenesis genes responded rapidly to changes in the environments but were oblivious to longer-term changes in growth rate. Our results suggest that the capacity to anticipate and prepare for environmentally-mediated changes in cell growth presented a major selection force during yeast evolution.

Hundreds of RNA-binding proteins (RBPs) control diverse aspects of post-transcriptional gene regulation. To identify novel and unconventional RBPs, Scherrer and colleagues probed high-density protein microarrays with fluorescently labeled RNA and selected 200 proteins that reproducibly interacted with different types of RNA from budding yeast *Saccharomyces cerevisiae* in Chapter 9. Surprisingly, more than half of these proteins represent previously known enzymes, many of them acting in metabolism, providing opportunities to directly connect intermediary metabolism with posttranscriptional gene regulation. The authors mapped the RNA targets for 13 proteins identified in this screen and found that they were associated with distinct groups of mRNAs, some of them coding for functionally related proteins. They also found that overexpression of the enzyme Map1 negatively affects the expression of experimentally defined mRNA targets. The results suggest that many proteins may associate with mRNAs and possibly control their fates, providing dense connections between different layers of cellular regulation.

Chromatin regulation underlies a variety of DNA metabolism processes, including transcription, recombination, repair, and replication. To perform a quantitative genetic analysis of chromatin accessibility, the authors of Chapter 10, Lee and colleagues, obtained open chromatin profiles across 96 genetically different yeast strains by FAIRE (formaldehyde-assisted isolation of regulatory elements) assay followed by sequencing. While 5~10% of open chromatin region (OCRs) were significantly affected by variations in their underlying DNA sequences, subtelomeric areas as well as gene-rich and gene-poor regions displayed high levels of sequence-independent variation. They performed quantitative trait loci (QTL) mapping using the FAIRE signal for each OCR as a quantitative trait. While individual OCRs were associated with a handful of specific genetic markers, gene expression levels were associated with many regulatory loci. They found multi-target trans-loci responsible for a very large number of OCRs, which seemed to reflect the widespread influence of certain chromatin regulators. Such regulatory hotspots were enriched for known regulatory functions, such as recombinational DNA repair, telomere replication, and general transcription control. The OCRs associated with these multi-target trans-loci coincided with recombination hotspots, telomeres, and gene-rich regions according to the function of the associated regulators. The findings provide a global quantitative picture of the genetic architecture of chromatin regulation.

Comparative genomics is a formidable tool to identify functional elements throughout a genome. In the past ten years, studies in the budding yeast *Saccharomyces cerevisiae* and a set of closely related species have been instrumental in showing the benefit of analyzing patterns of sequence conservation. Increasing the number of closely related genome sequences makes the comparative genomics approach more powerful and accurate.

In Chapter 11, Liti and colleagues report the genome sequence and analysis of *Saccharomyces arboricolus*, a yeast species recently isolated in China, that is closely related to *S. cerevisiae*. The authors obtained high quality *de novo* sequence and assemblies using a combination of next generation sequencing technologies, established the phylogenetic position of this species and considered its phenotypic profile under multiple environmental conditions in the light of its gene content and phylogeny. They suggest that the genome of *S. arboricolus* will be useful in future comparative genomics analysis of the *Saccharomyces* sensu stricto yeasts.

Comparative genomics brings insight into sequence evolution, but even more may be learned by coupling sequence analyses with experimental tests of gene function and regulation. However, the reliability of such comparisons is often limited by biased sampling of expression conditions and incomplete knowledge of gene functions across species. To address these challenges, the authors previously systematically generated expression profiles in *Saccharomyces bayanus* to maximize functional coverage as compared to an existing *Saccharomyces cerevisiae* data repository. In Chapter 12, Guan and colleagues, take advantage of these two data repositories to compare patterns of ortholog expression in a wide variety of conditions. First, the authors developed a scalable metric for expression divergence that enabled them to detect a significant correlation between sequence and expression conservation on the global level, which previous smaller-scale expression studies failed to detect. Despite this global conservation trend, between-species gene expression neighborhoods were less well-conserved than within-species comparisons across different environmental perturbations, and approximately 4% of orthologs exhibited a significant change in co-expression partners. Furthermore, the analysis of matched perturbations collected in both species (such as diauxic shift and cell cycle synchrony) demonstrated that approximately a quarter of orthologs exhibit condition-specific expression pattern differences. Taken together, these analyses provide a global view of gene expression patterns between two species, both in terms of the conditions and timing of a gene's expression as well as co-expression partners. The results provide testable hypotheses that will direct future experiments to determine how these changes may be specified in the genome.

Human intervention has subjected the yeast *Saccharomyces cerevisiae* to multiple rounds of independent domestication and thousands of generations of artificial selection. As a result, this species comprises a genetically diverse collection of natural isolates as well as domesticated strains that are used in specific industrial applications. However the scope of genetic diversity that was captured during the domesticated evolution of the industrial representatives of this important organism remains to be determined. To begin to address this, Borneman and colleagues have produced whole-genome assemblies of six commercial strains of *S. cerevisiae* (four wine and two brewing strains). These represent the first genome assemblies produced from *S. cerevisiae* strains in their industrially-used forms and the first high-quality assemblies for *S. cerevisiae* strains used in brewing. By comparing these sequences to six existing high-coverage *S. cerevisiae* genome assemblies in Chapter 13, clear signatures were found that defined each industrial class of yeast. This genetic variation was comprised of both single nucleotide polymorphisms and large-scale insertions and deletions, with the latter often being associated with ORF heterogeneity between strains. This included the discovery of more than twenty probable genes that had not been identified previously in the *S. cerevisiae* genome. Comparison of this large number of *S. cerevisiae* strains also enabled the characterization of a cluster of five ORFs that have integrated into the genomes of the wine and bioethanol strains on multiple occasions and at diverse genomic locations via what appears to involve the resolution of a circular DNA intermediate. This work suggests that, despite the scrutiny that has been directed at the yeast genome, there remains a significant reservoir of ORFs and novel modes of genetic transmission that may have significant phenotypic impact in this important model and industrial species.

PART I

CELL POLARITY, GROWTH, AND CELL CYCLE

CHAPTER 1

DAUGHTER SPECIFIC TRANSCRIPTION FACTORS REGULATE CELL SIZE CONTROL IN BUDDING YEAST

STEFANO DI TALIA, HONGYIN WANG, JAN M. SKOTHEIM, ADAM P. ROSEBROCK, BRUCE FUTCHER, AND FREDERICK R. CROSS

1.1 INTRODUCTION

At the Start transition in G1, budding yeast cells integrate internal and external cues into an all-or-none commitment to a new round of cell division [1],[2]. Cell division is asymmetric, producing a smaller daughter cell and a larger mother cell [3]. Mother cells progress through Start more quickly than daughter cells [3],[4]. The regulation of G1 phase is composed of two independent modules separated by the nuclear exit of the transcriptional repressor Whi5 [5]: a cell size sensing module, which extends G1 in small cells to allow additional growth before Start [5], and a subsequent size-independent module [5],[6]. The fast and coherent transition between the two modules likely coincides with commitment to the cell cycle and is driven by transcriptional positive feedback [7]. The G1 cyclin Cln3 is the

This chapter was originally published under the Creative Commons Attribution License. Di Talia S, Wang H, Skotheim JM, Rosebrock AP, Futcher B, and Cross FR. Daughter-Specific Transcription Factors Regulate Cell Size Control in Budding Yeast. PLoS Biology 7,10 (2009), doi:10.1371/journal. pbio.1000221.

most upstream activator of the Start transition [8],[9],[10],[11],[12] and the main regulator of the size-sensing module. Cln3 initiates inactivation of Whi5 [13],[14] and expression of SBF/MBF dependent genes, including the G1 cyclins *CLN1* and *CLN2* [9],[11],[12],[15],[16]. Subsequent positive feedback of Cln1 and Cln2 on SBF/MBF dependent transcription ensures fast and coherent commitment to the cell cycle [7].

Cell size control is thought to regulate the length of the G1 phase of the cell cycle [4],[5],[17],[18]. In budding yeast, cell size control is readily detectable in daughter cells but much less obvious in mother cells. In part this is because mother cells are almost always born larger than daughters [3], but it has also been shown that daughters are slower to pass Start than mothers even when both are made equally large (greater than normal mother or daughter size) [19]. This finding suggested some asymmetry in Start control between mothers and daughters beyond that due to different cell size; differential gene expression in mothers and daughters could provide such asymmetry.

Regulation of gene expression is asymmetric in mother and daughter cells as a result of the daughter-specific localization of the transcription factors Ace2 and Ash1. Ace2 enters mother and daughter nuclei during mitotic exit [20],[21]. Asymmetric localization of Ace2 is due to the Mob2-Cbk1 complex [20],[21],[22], which promotes nuclear retention of Ace2 specifically in the newborn daughter nucleus, leading to daughter-specific expression of a number of genes [20],[21],[22],[23],[24]. Daughter-specific localization of Ash1 is achieved through active transport of *ASH1* mRNA to the bud tip and consequent preferential accumulation of Ash1 in the daughter nucleus [25]. Ash1 represses expression of the *HO* endonuclease gene responsible for mating type switching [26],[27], thus restricting *HO* expression to mother cells.

Recently, Ace2 was shown to cause a daughter-specific G1 delay, acting indirectly through "Daughter Delay Elements (DDE)" 5′ to the *CLN3* coding sequence to reduce *CLN3* expression in daughters [28]. In that work, it was proposed that this Ace2-dependent delay is the only reason that daughters have a longer G1 than mothers. Cell size was proposed to play no role in controlling the length of G1 [28]. This proposal is incompatible with our recent finding that small cells display very efficient size control, requiring a significantly longer period of growth to attain

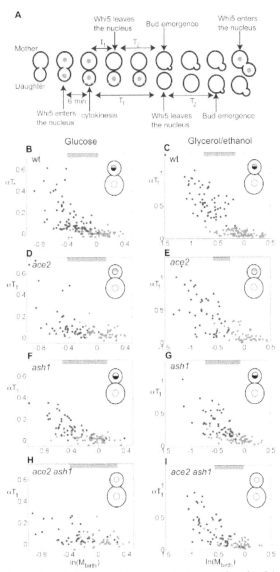

FIGURE 1: Differential regulation of Start is dependent on Ace2 and Ash1. (A) Illustration of the separation of G1 into two intervals, T_1 and T_2, by using Whi5-GFP. The total duration of G1 is T_1+T_2. (B–H) Correlation between αT_1 and $\ln(M_{birth})$ for cells grown in glucose or glycerol/ethanol. (B–C) wild-type, (D–E) *ace2*, (F–G) *ash1*, (H–I) *ace2 ash1*. Lighter dots, mothers; darker dots, daughters. Inset: cartoon illustrating presence of Ace2 or Ash1 in mother and daughter nuclei; black semicircle, Ace2; gray semicircle, Ash1. Gray bars indicate the region of size overlap used for the analysis presented in Table 1.

a sufficient size before exiting G1 [5]. Here, we resolve this conflict and further investigate the differences between mother and daughter cell cycle control by analyzing the interaction between daughter-specific transcriptional programs, cell size control, and irreversible commitment to the cell cycle at Start.

1.2 RESULTS

1.2.1 DIFFERENTIAL REGULATION OF START IN MOTHERS AND DAUGHTERS IS DEPENDENT ON ACE2 AND ASH1

G1 (defined operationally as the unbudded period of the cell cycle) can be decomposed into two independent steps, of duration T_1 and T_2, respectively, separated by exit from the nucleus of the transcriptional repressor Whi5 (Figure 1A) [5]. We previously used time-lapse fluorescence microscopy of yeast expressing *WHI5-GFP* and *ACT1pr-DsRed* [5] to simultaneously measure the duration of T_1, measured by the interval of Whi5 nuclear residence, and cell size, measured using total cell fluorescence expressed from the constitutive *ACT1pr-DsRed* [5]. T_2, the time between Whi5 nuclear exit and budding, is similar in mothers and daughters and is largely independent of cell size [5],[6]. T_1 is extremely short in mothers but of significant duration in daughters [5],[6]. G1 size control is readily detected in small daughter cells, and maps specifically to the T_1 interval [5].

Smaller cells have a longer T_1, allowing growth to a larger size before cell cycle entry. This links birth size to T_1 duration. Given exponential growth of single cells [5],[29], the size at Whi5 exit, M_1, is related to the size at birth, M_{birth}, through the period T_1 by the simple formula: $M_1 = M_{birth}e^{\alpha T1}$, where α is the growth rate for exponential growth. This expression yields: $\alpha T_1 = \ln(M_1) - \ln(M_{birth})$. The correlation between αT_1 and $\ln(M_{birth})$ characterizes the efficiency of size control. If there is efficient size control, then T_1 should become larger as $\ln(M_{birth})$ becomes smaller, because cells born smaller require a longer period of growth to promote Start. Specifically, the slope of the linear fit of the plot of αT_1

against $\ln(M_{birth})$ should be -1 in the case of perfect size control (that is, an exact size at which Start is invariably executed) and 0 in the absence of size control [5],[30].

The different duration of the period T_1 in mothers and daughters could in principle be solely a consequence of size control imposing a delay in the smaller daughter cells [3]. We analyzed the correlation between αT_1 and $\ln(M_{birth})$, comparing mothers and daughters binned for very similar size at birth (binning was necessary to ensure sufficient numbers of cells of a given size for statistical comparisons). This comparison demonstrates an increase in αT_1 in daughters compared to mothers of similar size (Figure 1B, 1C; regions marked with bars) (p values $<10^{-6}$; Table S4). This delay in Start is most readily detectable in glycerol-ethanol medium (Figure 1C) (p value $<10^{-70}$; Table S4), in which cell growth is much slower than in glucose medium. Slower growth means that the mother cell feeds less biomass into the daughter cell, resulting in smaller daughter size at the time of cell division [3]. The resulting population of very small daughters enhances detection of size control (Figure 1C) [5]. In glycerol-ethanol, across the domain of size overlap in mother-daughter size at birth, daughters exhibit clear size control (slope ~-0.8) while mothers exhibit essentially none (slope ~0). This increase in αT_1 in daughters with respect to mothers of equal size is consistent with previous findings of a daughter-specific delay, above and beyond the delay needed to achieve equivalent size [19],[28].

Laabs and collaborators had previously implicated the daughter-specific transcription factor Ace2 in delayed exit from G1 in daughters [28]. Ash1 is a second daughter-specific transcription factor [26],[27], and Ace2 contributes to the expression of *ASH1* in daughter cells [31]. Ash1 might therefore be the effector of the Ace2-induced daughter delay, or it could independently contribute to daughter delay. We analyzed the correlation between αT_1 and $\ln(M_{birth})$ in ace2 and ash1 single and double mutants (for a complete list of strains and plasmids used in this study, see Table S1 and Table S2).

ace2 ash1 mothers and daughters that were born at similar sizes exhibited similar αT_1 values, failing to display the daughter-specific delay seen in wild-type (Figure 1H, 1I, and Table 1). Furthermore, only very small *ace2 ash1* daughters from glycerol/ethanol cultures displayed efficient size control. It is important to note that the mutant still displayed efficient size control by our metric; the effect of the deletions was to shift the size domain where efficient size control could be detected, not to eliminate size control per se.

TABLE 1: Average daughter delay in newborn cells of the same size.

	Wild-Type	ace2	ash1	ace2 ash1
Daughter-mother delay in glucose	8 ± 1 min	2 ± 3 min (0.06)	6 ± 1 min (0.15)	3 ± 2 min (0.03)
Daughter-mother delay in gly/eth	87 ± 9 min	16 ± 13 min ($<10^{-5}$)	40 ± 8 min ($<10^{-4}$)	17 ± 9 min ($<10^{-7}$)
	Wild-Type	ACE2*	ASH1*	ACE2* ASH1*
Daughter-mother delay in glucose	8 ± 1 min	1.3 ± 0.9 min ($<10^{-5}$)	5 ± 1 min (0.03)	1.3 ± 0.9 min (10^{-5})
Daughter-mother delay in gly/eth	87 ± 9 min	37 ± 12 min ($<10^{-3}$)	19 ± 7 min (10^{-8})	5 ± 7 min ($<10^{-12}$)
	Wild-Type	Ace2/Swi5 Sites Mutated	Ash1 sites Mutated	Ace2/Swi5 and Ash1 Sites Mutated
Daughter-mother delay in glucose	8 ± 1 min	8 ± 2 min (0.82)	10 ± 2 min (0.37)	7 ± 1 min (0.48)
Daughter-mother delay in gly/eth	87 ± 9 min	46 ± 7 min ($<10^{-3}$)	47 ± 15 min (0.02)	54 ± 9 min (0.008)
	Wild-Type	cln3	ADH1pr-CLN3	nxCDC28pr-CLN3
Daughter-mother delay in glucose	8 ± 1 min	3 ± 1 min ($<10^{-3}$)	N/A	3 ± 1 min ($<10^{-3}$)
Daughter-mother delay in gly/eth	87 ± 9 min	9 ± 13 min ($<10^{-6}$)	22 ± 10 min ($<10^{-5}$)	33 ± 12 min ($<10^{-3}$)

The region of overlap in size at birth of mothers and daughters was evaluated for every genotype seperately (see gray bars in Figures 1, 2, 6, 7). Data for the duration of T_1 in this region were divided in small bins and the daughter delays (i.e. average excess in T_1 for daughters over mothers) were computed for every size bin with representation of both mothers and daughters. The results were averaged across all these size bins. This definition of daughter delay is largely independent of the uneven distribution of cell size at birth in the region of overalp. The p value, computed by t test, for the hypotheseis that the mutant daughter delay is the same as the wild-type daughter delay is indicated in parentheses. The statistical significance of difference in T_1 times between mothers and daughters in the region of size overlap is presented in Table S4. Asterisks indicate the dominant mutant forms. Data for the difference in T_1 in mother-daughter pairs are presented in Figures S4, S5, and S6.

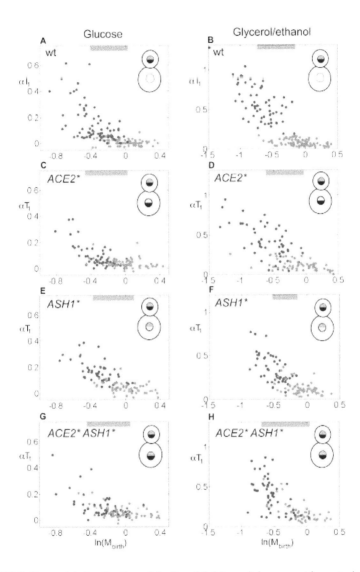

FIGURE 2: Symmetric localization of Ace2 and Ash1 result in symmetric control of Start in mothers and daughters. (A–H) Correlation between αT_1 and $\ln(M_{birth})$ for cells grown in glucose or glycerol/ethanol. (A–B) wild-type, (C–D) *ACE2**, (E–F) *ASH1**, (G–H) *ACE2* ASH1**. Lighter dots, mothers; darker dots, daughters. Black semicircle, Ace2; gray semicircle, Ash1. Asterisks indicate the dominant mutant forms. Gray bars indicate the region of size overlap used for the analysis presented in Table 1.

Single mutants (*ace2 ASH1* and *ACE2 ash1*) display a phenotype similar to but less extreme than *ace2 ash1* double mutants (Figure 1D–1G, Table 1). Ace2 contributes to transcriptional activation of *ASH1* [31], so some but not all of the effects of *ACE2* deletion may be a consequence of reduced *ASH1* expression. The characterized indirect effect of Ace2 on DDE sites 5' of *CLN3* coding sequence [28] likely accounts for at least some of the Ash1-independent effect of *ACE2* deletion; we argue below that there is likely an additional direct effect of Ace2 on *CLN3* transcription.

In strains with *ACE2* and/or *ASH1* deleted, little effect on mother cell size control is expected or observed, since mother cells naturally lack Ace2 and Ash1 due to differential segregation of the factors at cell division (see Introduction). *ace2 ash1* daughters exhibit efficient size control only when born at a size that mothers almost always exceed, due to the budding mode of growth (Figure 1B, 1C) [3].

To test whether Ace2 or Ash1 can affect size control when introduced into mothers, we used mutations resulting in symmetrical inheritance of the factors to mothers and daughters. For Ace2, we used *ACE2-G128E* (indicated as *ACE2** from here on). Ace2-G128E accumulates symmetrically and activates Ace2-dependent transcription in both mothers and daughters [20],[32] and was shown previously to reduce mother-daughter G1 asymmetry [28]. For Ash1, we used a mutant (*ASH1**) in which mutation of localization elements in *ASH1* mRNA results in accumulation of Ash1 in both mother and daughter nuclei [33].

As with *ace2 ash1* cells, *ACE2* ASH1** mothers and daughters that were born at similar sizes exhibited similar αT_1 values (Figure 2G, 2H; Tables 1, S4). Furthermore, *ACE2* ASH1** mothers, when born sufficiently small, exhibit size control, essentially as observed in similarly sized wild-type daughters. Such small mother cells are observed in significant numbers only in glycerol-ethanol culture (Figure 2H).

Thus, making the Ace2/Ash1 daughter-specific gene expression program symmetrical between mothers and daughters (either by deletion or by symmetrical introduction of the factors) results in effective size control (high negative slope in αT_1 versus $\ln(M_{birth})$ plots) over a similar cell size domain in mothers and daughters, eliminating the daughter-specific delay seen in wild-type. In wild-type daughters, size control is exerted at sizes where mothers do not experience size control.

Mother cell size control is in principle hard to detect in any case, because these cells "passed" size control in the previous cycle, and budding yeast cell division removes little or no material from the mother cell. For this reason, even in *ACE2* *ASH1* cells, which presumably all have daughter-type size control, mothers small enough to allow detection of size control are relatively rare.

Strains in which only Ash1 or Ace2 is symmetrically localized show intermediate phenotypes (Figure 2C–2F; Table 1), suggesting again that both transcription factors contribute to delay in T1 in partially independent ways. *ACE2* and *ASH1* had little effect on size control properties of daughter cells, as expected since these factors are already present in wild-type daughters.

Altogether, these results show that Ace2 and Ash1 define daughter-specific programs that shift size control responses to larger cell size. Ace2 and Ash1 appear to be necessary for this shift in size control in daughters compared to mothers; in addition they are sufficient for imposing daughter-like size control properties when introduced in mothers.

These results led to the idea that *ACE2* *ASH1* mothers should be "pseudo-daughters" with respect to size control, while *ace2 ash1* daughters should be "pseudo-mothers." To test this, we combined data for mothers and pseudo-mothers, and daughters and pseudo-daughters, in rich and poor medium (Figure 3A–3F). We define mothers and pseudo-mothers as "mother-like," and daughters and pseudo-daughters as "daughter-like." Remarkably, these combined data sets collapsed onto one plot for all mother-like cells and a different plot for all daughter-like cells (Figure 3E, 3F). The individual datasets fit well with the average behavior, as shown by plots separating out the various components (Figure S7, S8). The noise about the lines in these plots (size-independent variation) is of a magnitude consistent with previous results (Table S5) [5]. Further analysis showed that the daughter-like plot could be transformed to the mother-like plot simply by shifting the curve 0.2 units of $\ln(M_{birth})$ (Figure 3G, 3H). This implies that, with respect to Start, cells containing Ace2 and Ash1 interpret a given cell size as being ~20% smaller than cells lacking Ace2 and Ash1.

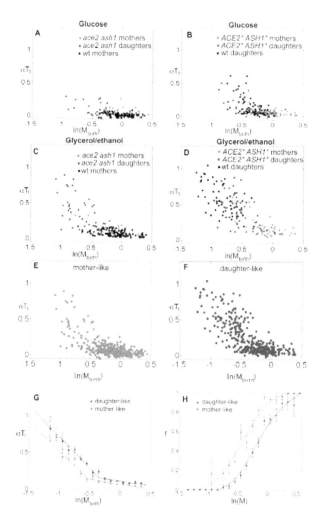

FIGURE 3: Daughter-specific localization of Ace2 and Ash1 results in asymmetric cell size control. Correlation between αT_1 and $\ln(M_{birth})$ for mothers and "pseudo-mothers" (in (A) cells grown in glucose, in (C) cells grown in glycerol/ethanol) and daughters and "pseudo-daughters" (in (B) cells grown in glucose, in (D) cells grown in glycerol/ethanol). (E) Correlation between αT_1 and $\ln(M_{birth})$ for mothers and "pseudo-mothers" grown in glucose and glycerol/ethanol (pulling together data from (A) and (C)). (B) Correlation between αT_1 and $\ln(M_{birth})$ for daughters and "pseudo-daughters" grown in glucose and glycerol/ethanol (pulling together data from (B) and (D)). (G) Correlation between αT_1 and $\ln(M_{birth})$ in mother-like and daughter-like cells. The graphs are obtained by binning all the data shown in (E) and (F). Error bars are standard errors of the mean. (H) Conditional probability of Whi5 nuclear exit as a function of $\ln(M)$ from data in (G). f is the probability that Whi5 will exit the nucleus at size $\ln(M)$ given that it had not exited at a smaller size.

These results can be interpreted in the classical framework of sizers and timers [18],[34] by defining the point at which cells switch from efficient size control to a timer control (the intersection between the two lines fitting the correlation between αT_1 and $\ln(M_{birth})$ in Figure 3C) as "critical size": a precise size that cells must attain to transit Start. This analogy is imperfect (the slopes are not -1 or 0, as required for perfect sizers and timers [5],[30], and the sharpness of the transition point cannot be rigorously determined) but provides a useful simplification using the terms of prior size control literature. Using this terminology, the effect of daughter-specific localization of Ace2 and Ash1 is to cause daughter cells to have a larger "critical size" than mother cells (increased by 0.2 units of $\ln(M_{birth})$, or ~20% larger). We emphasize that size control remains highly effective, independent of Ace2 and Ash1; essentially, Ace2/Ash1-containing cells read a given size as smaller than the same size read in the absence of Ace2 and Ash1.

Laabs and collaborators reported symmetrical G1 durations for ace2 mothers and daughters, and for *ACE2** mothers and daughters, independent of cell size [28]. In our experiments, the loss of asymmetrical "interpretation" of cell size caused by these mutations does indeed result in T_1 durations in mothers and daughters that are more similar than in wild-type (Figure S4, S5). However, our results differ in that in our experiments, size control remains present and effective despite deletion or mislocalization of Ace2 and/or Ash1. As a consequence, the average daughter T_1 is still significantly longer than the average mother T_1 (p values$<10^{-3}$ in glucose; p values$<10^{-14}$ in glycerol/ethanol) even in the mutants, since the budding mode of growth ensures that most daughters are born smaller than most mothers.

This discrepancy likely has a number of sources. First, our use of T_1, the time from cytokinesis to Whi5 nuclear exit as a landmark, rather than the differential time to budding for mothers versus daughters, as measured by Laabs and collaborators [28], greatly increases the sensitivity with which size control can be detected, since the interval from Whi5 exit to budding is quite variable, cell-size-independent, and very similar in mothers and daughters [5]. Inclusion of this noisy interval blurs the mother-daughter distinction, which is restricted to T_1. Second, the use of medium supporting slow cell growth (glycerol-ethanol) enhances the ability to detect size control, simply because daughters (of all genotypes) are born much smaller; the work of Laabs et al. [28] employed only rich glucose medium, making size control harder to detect. Our time resolution is also 3 min per frame rather than 10. Finally, our cell size estimates are based on the validated *ACT1-DsRed* marker [5], while Laabs et al. [28] employed volume estimations from geometry of cell images. We have found that the latter method gives on average similar results to *ACT1-DsRed* but increases noise in the detection of size control effects [5].

1.2.2 GENOME-WIDE ANALYSIS OF ACE2 AND ASH1 TARGETS

CLN3 was proposed as the relevant indirect transcriptional target of Ace2 to account for mother-daughter asymmetry [28]. Because Ace2 could affect other genes involved in cell size control or mother-daughter asymmetry, and because we had evidence for the involvement of an independent transcription factor, Ash1, we carried out an unbiased search for the transcriptional target(s) through which Ace2 and Ash1 modulate size control in daughters. We performed microarray analysis of synchronized cell populations, comparing cells lacking Ace2 and Ash1 to cells in which they localize symmetrically to both mother and daughter nuclei. Doing the comparisons in this way, rather than simply comparing wild-type to mutants, increases sensitivity of the analysis, since wild-type cultures always contain a mixture of mothers and daughters, reducing the detectable effects of manipulation of daughter-specific transcription factors. Our approach relies on three comparisons: *ace2 ash1* versus *ACE2* ASH1**, *ace2* versus *ACE2**, and *ash1* versus *ASH1** (see Dataset S2 for the microarrays raw data).

We also compared *swi5, ace2, swi5 ace2*, and wild-type in order to obtain insight into the set of genes regulated by one or both of these factors (see Dataset S1 for the microarrays raw data). Swi5 and Ace2 are closely related transcription factors that recognize the same DNA sequence and share many target genes [35],[36]. The best characterized Ash1 target, *HO*, is also a Swi5 target and its regulation by Swi5 and Ash1 is required for mother-daughter asymmetry in mating type switching [26],[27].

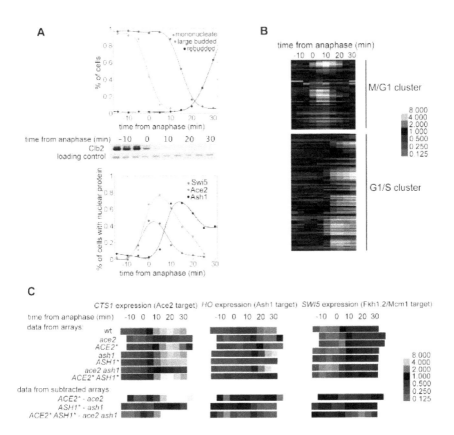

FIGURE 4: Genome-wide analysis of Ace2 and Ash1 targets. (A) Analysis of cell cycle synchronization and nuclear localization of Ace2, Swi5, and Ash1 in a *cdc20* block-release experiment. Top panel shows the percentage of mononucleate cells, large budded cells, and cells that have rebudded. The middle panel shows the levels of mitotic cyclin Clb2. The lower panel shows the dynamics of nuclear localization of fluorescently tagged Ace2, Swi5, and Ash1. (B) Expression data from the M/G1 and G1/S cell cycle regulated cluster of genes. (C) The regulation of *CTS1* (Ace2 target), *HO* (Ash1 target), and *SWI5* (Fkh1,2 Mcm1 target) expression from the microarray series, as well as data obtained by point-by-point subtraction of the arrays (*ACE2**−*ace2*, *ASH1**−*ash1*, *ACE2* ASH1**−*ace2 ash1*). In these graphs, the time of anaphase, which varies slightly between experiments, was used as the zero time to make the comparisons more accurate.

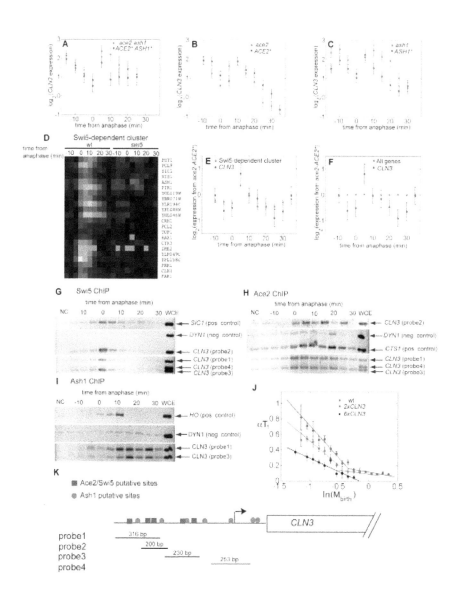

FIGURE 5: Ace2, Swi5, and Ash1 regulate the expression of the G1 cyclin *CLN3*. *CLN3* expression: (A) *ACE2* ASH1** versus *ace2 ash1*, (B) *ACE2** versus *ace2*, (C) *ASH1** versus *ash1*. The error bars were estimated from the variability in expression of the large number of genes that are not affected by Ace2 and Ash1. The expression levels of *CLN3*,

as well as any other gene in the genome, were estimated by at least four measurements from four distinct probes. These measurements showed a smaller variability than the presented error bars, suggesting that the reported error bars are a conservative estimate of measurement errors. (D) Expression of a cluster of Swi5-dependent genes. Arrays from *GALL-CDC20* block release time course experiments of wild-type and swi5 cells were hierarchically clustered. A Swi5-specific cluster is shown (see Text S1 for a complete list of genes regulated by Swi5). (E) *CLN3* expression compared with the average expression of the remaining genes belonging to the Swi5-specific cluster (error bars indicate s.e.m.) from the dataset obtained by subtracting the *ACE2** data from the *ace2* data. (F) *CLN3* expression compared with the average expression of the whole genome (error bars indicate s.e.m.) from the same dataset (i.e. *ace2−ACE2**). ChIP analysis of the interaction between Swi5 (G), Ace2 (H), Ash1 (I), and the *CLN3* promoter. Following cross-linking and immunoprecipitation, DNA was amplified by PCR. Amplification of a region of the ORF of DYN1 was used as negative controls, while regions of the *SIC1*, *CTS1*, and *HO* promoters were used as positive controls for Swi5, Ace2, and Ash1, respectively. All the strains were TAP-tagged (NC, negative control from an untagged strain; WCE, whole cell extract). The ChIP data were reproduced for Ace2 and Ash1. The Swi5 data are from a single experiment. (J) Correlation between αT_1 and $\ln(M_{birth})$ in daughter cells carrying different copy numbers of *CLN3*. (K) Representation of the Ace2/Swi5 and Ash1 putative binding sites on the *CLN3* promoter.

To synchronize cells during the critical M/G1 interval, we used strains expressing Cdc20 under the control of an inducible promoter (the truncated *GAL1* promoter, *GALL* [37]). Cells were arrested in metaphase by depletion of Cdc20 in glucose medium and released from the arrest by transfer to galactose medium to reinduce Cdc20. This synchronization procedure provides excellent synchrony in M/G1 (anaphase, cell division, and early G1) immediately following release, which is the time of nuclear localization of Ace2, Swi5, and Ash1 (Figure 4A) [36],[38].

About 15 min after release, cells of all genotypes complete anaphase and degrade the mitotic cyclin Clb2 (see Figure 4A). Subsequently, cells separate and rebud (Figure 4A). Both Swi5 and Ace2 enter the nucleus at about the time of anaphase (Figure 4A). On average, Swi5 nuclear entry precedes Ace2 nuclear entry by 2–3 min (see Text S1). A slightly longer

(10 min) Ace2 delay relative to Swi5 entry was recently reported [39]. Swi5 is rapidly degraded and disappears before cytokinesis and cell separation (Figure 4A and Text S1) [40]. Ace2 is quickly excluded from the mother nucleus but remains in the daughter nucleus for a significant period during G1 (Figure 4A and Text S1) [20]. Ash1 protein begins to accumulate a few minutes after Swi5 and Ace2 nuclear entry and localizes to the nucleus slightly before cytokinesis, remaining until about the time of budding (Figure 4A and Text S1) [26].

The microarrays for wild-type cells show well-defined M/G1 and G1/S clusters consistent with previous results (Figure 4B) [38]. Furthermore, well-characterized Ace2 and Ash1 targets, such as *CTS1* and *HO*, behave as expected upon transcription factor deletion or mislocalization (see Figure 4C). Cell-cycle-regulated genes that are unaffected by the two transcription factors behave very similarly in all arrays (Figure 4C). Note that the time of anaphase, which varies slightly between experiments, was used as the zero time to make the comparisons more accurate.

The high reproducibility of these microarray data allows us to do a time-point by time-point subtraction of the deletion mutant data from the mislocalization mutant data. This subtraction cancels out cell-cycle-regulated changes in gene expression that are independent of Ace2 and/or Ash1, allowing the hierarchical clustering algorithm [41] to efficiently detect changes that are specifically due to these transcription factors (see Figure 4C).

Clustering analysis of the subtracted data reveals a clear Ace2-dependent cluster composed of well-characterized Ace2-dependent genes, such as *CTS1*, *DSE1*, and *DSE2* (see Text S1 and Figure S1 for a complete list). Only two genes, *HO* and *PST1*, displayed strong changes in expression upon deletion versus mislocalization of Ash1 (see Text S1).

None of the genes whose expression was obviously and strongly Ace2- or Ash1-dependent appeared to be a good candidate to account for daughter-specific regulation of Start. We therefore performed a statistical analysis to obtain a list of genes specifically regulated by both Ace2 and Ash1. We imposed an "AND" logical condition that co-regulated genes should be detected as differential signals in the subtracted *ace2* versus *ACE2**, *ash1* versus *ASH1**, and *ace2 ash1* versus *ACE2* ASH1** comparisons. Additionally, we imposed a temporal requirement that the observed Ace2/

Ash1-dependent changes in expression be observed only at times when these factors have accumulated in wild-type nuclei (Figure 4A). This criterion excludes genes whose changes in expression are long-term, indirect consequences of mutation of Ace2 or Ash1. Using a p value cutoff sufficient for an expected false positive rate of less than one gene over the whole genome (see Text S1 and Table S3), we identified only five Ace2/Ash1 shared targets: *CLN3, HSP150, MET6, YRF1*-1, and *YRF1-5*.

Direct interactions between Ace2 or Ash1 and the promoters of three of these genes (Ace2 with *CLN3* and *HSP150*; Ash1 with *YRF1-1*) were previously observed in chromatin immuno-precipitation (ChIP)-chip experiments [42],[43], supporting the validity of our analysis (see Text S1).

Prominent in the list of genes affected by both Ace2 and Ash1 is the G1 cyclin, *CLN3*, a rate-limiting activator of the Start transition. Laabs and collaborators had likewise implicated *CLN3* as a gene repressed by Ace2, based on comparing *CLN3* RNA levels with and without Ace2, and examining mother versus daughter accumulation of GFP driven from a truncated *CLN3* promoter [28]. In that paper, it was also suggested that Ace2 might regulate *CLN3* indirectly through an unknown transcription factor that represses *CLN3* expression in daughters by binding to DDE sites on the *CLN3* promoter. Among all the identified Ace2 targets, Ash1 is the most likely candidate transcription factor for a repressive role on *CLN3* expression. We observe, however, that there is no obvious homology between the Ash1 consensus and the DDE. In the next sections we provide evidence that Ash1 binds to the *CLN3* promoter and that this binding is at least in part mediated by Ash1 consensus-binding sites that are different from the DDE.

Together, these findings suggested the hypothesis that differential regulation of Start in mothers and daughters due to Ace2 and Ash1 may be solely a consequence of differential regulation of *CLN3*.

CLN3 expression in M/G1 is from 1.5- to 2.5-fold higher in *ace2 ash1* cells (pseudo-mothers) than in *ACE2* ASH1** cells (pseudo-daughters) (Figure 5A), suggesting that *CLN3* is differentially regulated in wild-type mothers and daughters. Previously published data support this idea: in populations of cells containing both mothers and daughters, *CLN3* expression peaks at the M/G1 boundary [44], while in populations of size-selected daughters *CLN3* expression peaks later in G1 [45], or shows no peak

[12],[46], consistent with our conclusion that *CLN3* expression in M/G1 is higher in mothers than in daughters. M/G1 expression of *CLN3* is driven by Mcm1 through early cell-cycle box (ECB) elements [44]; our results and the results of Laabs and collaborators [28] suggest that in daughters, Ace2 and Ash1 antagonize this activation.

Hierarchical clustering of microarrays of wild-type and swi5 cells indicates that *CLN3* belongs to a cluster of genes whose expression is activated by Swi5 (Figure 5D). Analysis of *ace2* versus *ACE2** arrays (Figure 5B) shows that *CLN3* behaves similarly to the rest of this Swi5 dependent cluster upon manipulation of *ACE2* (see Text S1 and Figure S1 for a complete list of Swi5 and Ace2 targets). Expression of these genes in *ACE2** cells is lower than expression in ace2 at 5 min after anaphase, but higher from 15 min to 25 min (Figure 5E); that is, the genes appear to be repressed by Ace2 at early times, then activated by Ace2 at later times. This pattern is significantly different from a pattern assuming no regulation by Ace2 ($p < 10^{-11}$). *CLN3* expression depends on Ace2 similarly to these other Swi5 targets (probability that *CLN3* is regulated as the other Swi5/Ace2 targets: $p = 0.7$, Figure 5E; a model assuming that *CLN3* is not affected by Ace2 can be excluded with $p < 0.03$, Figure 5F).

Thus *CLN3* and a class of Swi5 dependent genes follow a pattern consistent with early repression and late activation by Ace2, and with early activation by Swi5, likely acting in concert with ECB regulation [44]. We do not know the mechanism underlying this complex pattern. We speculate that Ace2 may be an intrinsically poorer activator than Swi5, but it activates for a longer period due to its longer nuclear residence. Swi5 disappears from both mother and daughter nuclei a few minutes after anaphase, while Ace2 persists in daughter nuclei for about 20 min longer (Figure 4A). Competition between Ace2 and Swi5 for the same binding site [35] could then contribute to the differential expression observed in these arrays. Alternatively, Ace2 could directly repress expression of these genes; however, no previous evidence suggests a directly repressive role for Ace2.

Microarray analysis for *ash1* and *ASH1** shows that *CLN3* expression is repressed about 2-fold by Ash1 during the period from 10 min to 25 min after anaphase (Figure 5C). During this interval Ash1 is present in the nucleus (Figure 4A), suggesting that it could be a direct repressor of *CLN3* expression.

Many Swi5 and Ace2/Swi5 targets have moderately higher expression in the absence of Ash1 (Figure S2). The absolute repression of Swi5-dependent *HO* expression by Ash1 in daughter cells may thus be an enhancement of a common pattern of co-regulation.

Our data suggest that Ace2 and Ash1 cooperate to repress *CLN3* expression in daughters. Consistently, activation of the G1/S regulon controlled by Cln3 is delayed and/or happens at larger cell size in *cdc20*-synchronized cells containing these factors (Figure S3).

1.2.3 ACE2, SWI5, AND ASH1 MAY BE DIRECT TRANSCRIPTIONAL REGULATORS OF CLN3

We performed chromatin immuno-precipitation (ChIP) experiments in synchronized cell populations to ask if Ace2, Swi5, and Ash1 bind to the *CLN3* promoter. Genome-wide localization data in asynchronous cell populations suggested binding of Ace2 and Swi5 to the *CLN3* promoter but were statistically insufficient to definitively prove the association [42],. We used synchronized cell populations to provide dynamic information on the possible binding of Ace2, Swi5, and Ash1 to the *CLN3* promoter, providing a higher signal to noise ratio than can be obtained from asynchronous cells.

Swi5 and Ace2 bound to regions in the *CLN3* promoter around the time of anaphase, coincident with their nuclear entry (Figure 5G, 5H). Swi5 is on the *CLN3* promoter for only a few minutes (Figure 5G), while Ace2 is on the *CLN3* promoter for about 20 min (Figure 5H), also consistent with the time of Swi5 and Ace2 nuclear localization (Figure 4A and Text S1). Thus, Ace2 and Swi5 might directly regulate *CLN3* transcription by binding to multiple Ace2/Swi5 sites in the *CLN3* promoter.

A previous meta-analysis of multiple ChIP-chip experiments concluded that Swi5 and Ace2 both bound the *CLN3* promoter with high probability (data in Supp. Table S5 of Ref. [47]), consistent with our results.

Ash1 binds the *CLN3* promoter with kinetics similar to its nuclear localization (Figure 5I and Figure 4A). In contrast, Ash1 residence at the *HO* promoter is much briefer, consistent with previous results [20], despite persistence of Ash1 in the nucleus. We do not know the reason for this difference.

1.2.4 MUTATIONS OF ACE2/SWI5 AND ASH1 BINDING SITES ON THE CLN3 PROMOTER REDUCE THE ASYMMETRY OF START REGULATION

We noted three candidate Ace2/Swi5 sites (GCTGGS, consensus sequence: GCTGGT; [42]) in the *CLN3* promoter. The *CLN3* promoter also contains two possible variant sites (GCTGA); such sites are over-represented in Ace2 and Swi5 targets (B.F., unpublished data). There are eight candidate Ash1-binding sites (YTGAT) [48] in the *CLN3* promoter. We mutated these Ace2/Swi5 and/or Ash1 putative binding sites in the *CLN3* promoter by exact gene replacement (see Text S1 for details). To test if Ace2, Swi5, and Ash1 bind to these sites, we performed ChIP analysis in synchronized populations of heterozygous diploid strains containing a wild-type copy and a mutated copy of the *CLN3* promoter (Ace2/Swi5 and Ash1 putative binding sites mutated). Following immunoprecipitation, various regions of the *CLN3* promoter were amplified by PCR and analyzed by sequencing to obtain an estimate of the ratio of wild-type promoter sequences to mutated sequences (Figure 6). These experiments are internally controlled (as they do not require the comparison of two independent ChIP experiments). The measured ratio provides an indication of the preferential binding of Ace2, Swi5, and Ash1 to the identified putative binding sites. Ace2, Swi5, and Ash1 binding to the mutated *CLN3* promoter was reduced to about 60% relative to the wild-type promoter, assaying multiple sequences from −1,183 to −998 (ATG: +1) (Table 2 and Table S6). The binding of Ace2, Swi5, and Ash1 to sequences from −767 to −545 is not altered by mutation of the putative binding sites (Table 2 and Table S6). These results indicate that we have identified authentic Ace2, Swi5, and Ash1 binding sites in the 5′ region of the *CLN3* promoter. The residual binding signal from the mutant is consistent with either a low level of background precipitation, or to genuine residual binding of the factors to non-consensus sites in the promoter. (Due to uncertainties about such other sites, as well as variable shearing of the DNA in the ChIP procedure, we do not think we

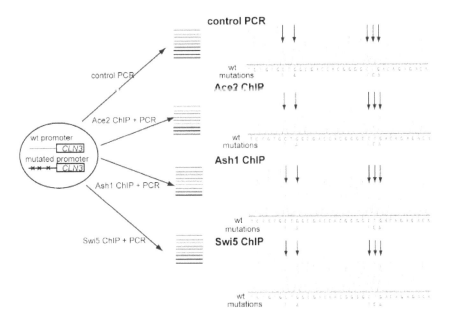

FIGURE 6: Binding of Ace2, Swi5, and Ash1 to the *CLN3* promoter is reduced by mutation of the Ace2/Swi5 and Ash1 consensus-binding sites. Experimental strategy to estimate the preferential binding of Ace2, Swi5, and Ash1 to their consensus-binding sites. Following ChIP, various regions of the *CLN3* promoter were amplified by PCR and analyzed by sequencing to obtain an estimate of the ratio of wild-type promoter sequences to mutated sequences. This ratio compared to the same ratio from PCR of genomic DNA provides an indication of the preferential binding of the factors to these sequences (Table 2).

can use these data to reliably map which candidate site(s) might be directly bound by Ace2, Swi5, or Ash1).

As a test to see if we might have missed significant binding sites in the *CLN3* promoter, we carried out a bioinformatics analysis (Text S1 and Figure S11) looking for regulatory motifs in the promoters of the identified Ace2 and Swi5 targets in *S. cerevisiae* and three closely related yeasts. Interestingly, we found only two strongly conserved sites: one was one of the candidate Ace2/Swi5 sites we mutated, at position −701, and the other was a similar but non-consensus site (GCTTGG) at position −569, which we did not mutate since it did not meet the consensus we used in designing the mutagenesis (see above). It is possible, although still untested, that this non-consensus site could account for residual binding of Ace2 to the mutant promoter.

The absence of a cluster of Ash1-dependent genes and the low information content of the known Ash1 consensus site (YTGAT) does not allow us to perform similar bioinformatics analysis; therefore, we cannot test the hypothesis that there are non-consensus Ash1 sites in the *CLN3* promoter that we did not mutagenize.

TABLE 2: Ace2, Ash1, and Swi5 Binding to the *CLN3* Promoter in Heterozygous Diploids.

	Ace2	Ash1	Swi5
Binding ratio (−1.183: −998)	0.66 ± 0.03 ($<10^{-20}$)	0.60 ± 0.05 ($<10^{-20}$)	0.74 ± 0.05 ($<10^{-5}$)
Binding ratio (−767: −545)	1.15 ± 0.16 (0.34)	1.01 ± 0.16 (0.94)	1.15 ± 0.14 (0.29)

The ratio of binding of Ace2, Ash1, and Swi5 to mutated and wild-type CLN3 *sequences in heterozygous diploids is reported (in parenthesis is the p value that the measured ratio is compatible with no change in binding). Binding to the* CLN3 *promoter region between −1183 and −998 (ATG: +1) is significantly reduced upoon mutation of the putative Ace2/ Swi5 and Ash1 binding sites (see Table S6 for details). Binding to the region from −767 to −545 is not affected.*

To test if the reduced binding of Ace2, Swi5, and Ash1 to the *CLN3* promoter also has an effect on the regulation of Start, we analyzed the correlation between αT1 and ln(Mbirth) in strains carrying mutations of the identified Ace2/Swi5 and/or Ash1 putative sites. These plots show that these mutations reduce the T1 delay in daughters compared to similarly-sized mothers (Figure 7, Table 1). The effect is easily detected and statistically significant in cells grown in glycerol-ethanol (Figure 7), a similar effect was observed in glucose, but this effect did not reach nominal statistical significance (Figure S9). In the Ace2/Swi5 sites mutant (Figure 7B, 7D) the duration of T1 in mothers is prolonged, consistent with the idea that Swi5 is an activator of *CLN3* (since mothers do not contain Ace2). Simultaneous mutation of Ace2 and Ash1 sites did not significantly enhance the phenotype of mutation only of one or the other (Figure 7, Table 1).

Although these promoter mutations have significant effects, they are less potent than deletion of *ACE2* and *ASH1* (compare Figure 1 with Figure 7, see Table 1). This may be in part due to the presence of additional

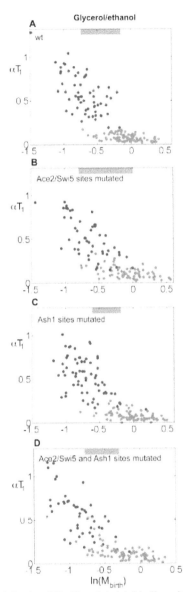

FIGURE 7: Mutation of the Ace2/Swi5 and Ash1 binding sites on the *CLN3* promoter reduces the asymmetrical regulation of Start. Correlation between αT_1 and $\ln(M_{birth})$ for cells grown in glycerol/ethanol in mutants lacking the Ace2/Swi5 and/or Ash1 sites on the *CLN3* promoter. (A) wild-type, (B) Ace2/Swi5 sites mutated, (C) Ash1 sites mutated, (D) Ace2/Swi5 and Ash1 sites mutated. Lighter dots, mothers; darker dots, daughters. Gray bars indicate the region of size overlap used for the analysis presented in Table 1.

non-consensus Ace2/Swi5 or Ash1 sites in the *CLN3* promoter (discussed above). Additionally, the comparison between mutating Ace2 sites and deleting *ACE2* is not exact because removing Ace2 sites perforce also removes Swi5 sites, and on the other hand, deletion of Ace2 alters *ASH1* expression.

The promoter mutants could also be less effective than deletion of *ACE2* and *ASH1* because Ace2 has indirect effects on *CLN3* expression. It was previously shown that "DDE" sites in the *CLN3* promoter play an important role in Ace2-dependent asymmetric control of Start, but these sites did not appear to be bound by Ace2, suggesting an indirect mechanism [28]. Interestingly, these sites are transcribed into the *CLN3* mRNA, and the Whi3 RNA binding protein binds to a repeated RNA sequence at the center of the DDE [49]. Whi3 is a regulator of cell size control thought to work by regulation of *CLN3* mRNA and protein [49],[50],[51],[52],[53]. At present, though, there is no information implicating Whi3 in mother-daughter asymmetry, nor is Ace2 known to regulate Whi3.

Finally, Ace2/Ash1 could regulate additional G1-regulatory genes at a level not detectable by our statistical analysis (see above).

The observation that reduced binding of Ace2, Swi5, and Ash1 to the *CLN3* promoter results in a significant reduction of asymmetric control of Start by cell size in mothers and daughters supports the idea that Ace2 and Ash1 directly repress *CLN3* expression in M/G1, accounting for a significant part of the regulation of G1 length by these transcription factors. We suggest that direct regulation of *CLN3* by Ace2 and Ash1 together with its indirect regulation by Ace2 through the DDE sites [28] can explain asymmetric control of Start by cell size in mothers and daughters.

1.2.5 CHANGES IN CLN3 EXPRESSION ARE SUFFICIENT TO ACCOUNT FOR MOTHER-DAUGHTER ASYMMETRY

CLN3 expression in M/G1 is ~2-fold higher in *ash1 ace2* cells (pseudo-mothers) than in *ASH1* ACE2** cells (pseudo-daughters) (Figure 5A). While this change is small, *CLN3* is a highly dosage-sensitive activator of Start. Previous measurements of cell sizes in cycling cell populations demonstrated effects on cell size upon 2-fold changes up or down in *CLN3* gene dosage [10],[44].

To increase the precision of this analysis, we analyzed the correlation between αT_1 and $\ln(M_{birth})$ in cells carrying either two or six copies of *CLN3*. This focuses the analysis on the critical interval, since Cln3 decreases cell size specifically by decreasing T_1 in wild-type mothers and daughters [5].

Daughter cells with two copies of *CLN3* exhibit efficient size control [high negative slope in αT_1 versus $\ln(M_{birth})$] over a size range that is shifted to smaller cell size by ~0.15 units of $\ln(M_{birth})$, compared to wild-type; this shift is similar to that in *ace2 ash1* daughter cells (mother-like cells) (compare Figure 5J to Figure 3C). Thus, the observed ~2-fold changes in *CLN3* expression upon deletion versus mislocalization of Ace2 and Ash1 could account for the observed changes in cell size control in these mutants.

Six copies of *CLN3* almost eliminate size control even in very small daughter cells (Figure 5J). Thus, size control is remarkably sensitive to *CLN3* gene dosage; it can only be modulated by altering *CLN3* expression in a narrow range before size control is lost.

1.2.6 ASYMMETRIC REGULATION OF CLN3 IS REQUIRED FOR ASYMMETRIC REGULATION OF START

We analyzed the correlation between αT_1 and $\ln(M_{birth})$ in cln3 cells and in cln3 cells expressing the *CLN3* ORF (without the upstream DDE sites) from constitutive promoters. It is important for this analysis that the constitutive promoters provide expression levels of Cln3 similar to those in wild-type cells and that the promoter-*CLN3* fusions complement the large-cell phenotype of cln3 mutants, without "overshoot" to a small-cell phenotype [8],[10],[54]. We screened a number of different constitutive promoters of different strengths [55] for these properties, examining both cell size and Cln3 protein levels using myc-tagged Cln3, compared to wild-type (including an approximate correction for cell cycle regulation of *CLN3* expression from the endogenous promoter) (Table 3; [44]).

The ACT1 and the *ADH1* promoters result in over-expression of Cln3 and in a small cell-size phenotype for cells grown in glucose-containing media (Table 3). Expression of Cln3 from the *CDC28* promoter is weaker than expression from the *CLN3* promoter and results in cell sizes bigger than wild-type and only slightly smaller than cln3 cells (Table 3). Integration into the yeast genome of six copies of the *CDC28pr-CLN3* construct results in a cell size distribution similar to that of wild-type cells. We also analyzed the effects of these constructs in glycerol-ethanol medium. Four tandemly integrated copies of *CDC28pr-CLN3* results in an overall cell size distribution similar to that of wild-type cells in glycerol-ethanol (Table 3). As a result of decreased *ADH1* expression in non-fermentable media [56], the *ADH1* promoter provides Cln3 levels similar to endogenous levels in glycerol-ethanol, resulting in a cell size distribution slightly (~10%) larger than wild-type (Table 3).

TABLE 3: Levels of Cln3 expression and average cell size for asynchronous cell populations expressing *CLN3* from various constitutive promoters

	Wild-type	*cln3*	*CDC28pr-CLN3*	*ACT1pr-CLN3*	*ADH1pr-CLN3*
Cln3 levels in D	1	0	0.4–0.6	5–7	8–10
Cln3 levels in g/e	1	0	0.2–0.5	8–10	1.5–2.0
Cell size in D (fl)	56	92	84	45	45
Cell size in g/e (fl)	47	88	60	41	51

The expression of CLN3 *is cell cyle regulated with a peak in expression at M/G1 characterized by a peak to trough ratio of order 3 (see Figure 5) [44]. Since the period of peak expression is brief, we consider a construct yielding ~3 times the average expression level of endogenous Cln3 to give approximately wild-type levels of expression during the critical interval.*

Measurements of Cln3 protein levels show that Cln3 overexpressors were smaller than wild-type, and underexpressors larger (Table 3); therefore, the measurements of Cln3 level were accurate over a physiologically relevant range. Based on results with a single copy of *CDC28pr-CLN3-myc*, four to six copies of *CDC28pr-CLN3* should produce approximately wild-type levels of Cln3 in M/G1, consistent with the observed cell size distributions (Table 3 and Figure 8).

We therefore used strains containing *6xCDC28pr-CLN3* in glucose medium, and strains containing *4xCDC28pr-CLN3* or *ADH1pr-CLN3* in glycerol-ethanol medium, to provide approximately endogenous levels of expression without mother-daughter asymmetry (and presumably without

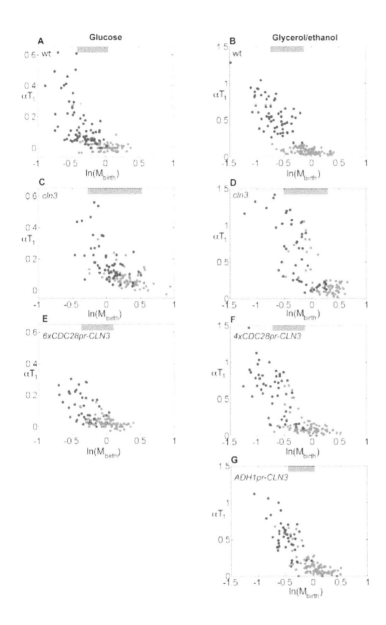

FIGURE 8: Symmetric regulation of *CLN3* expression result in symmetric control of Start in mothers and daughters. Correlation between αT_1 and $\ln(M_{birth})$ for cells grown in glucose or glycerol/ethanol. (A–B) wild-type, (C–D) cln3, (E) cln3 *6xCDC28pr-CLN3*, (F) cln3 *4xCDC28pr-CLN3*, (G) cln3 *ADH1pr-CLN3*. Lighter dots, mothers; darker dots, daughters. Gray bars indicate the region of size overlap used for the analysis presented in Table 1.

regulation by the cell cycle, Ace2, or Ash1; note that the 5′ DDE sites are not present in these constructs). In *6xCDC28pr-CLN3* cells the daughter-specific delay is almost entirely abolished (Figure 8C, 8E and Table 1). Similarly, in *4xCDC28pr-CLN3* and *ADH1pr-CLN3* cells grown in glycerol/ethanol, the daughter-specific delay is almost entirely abolished, and small mothers and daughters have similar size control properties (Figure 8D, 8F, and 8G and Table 1). Thus, similarly to the results obtained by placing Ace2 and Ash1 in mother nuclei, size control in small mother cells can be detected by eliminating differential mother-daughter control of *CLN3* expression.

Small *4xCDC28pr-CLN3* and *ADH1pr-CLN3* cells in glycerol/ethanol exhibit strong size control (slopes of ∼ −0.8, compared to a theoretical expectation for perfect size control of −1 [5],[30]) (Figure 8F, 8G), suggesting that while daughter-specific transcriptional regulation of *CLN3* specifies the cell size domain over which size control is effective, the intrinsic mechanism of size control is not dependent on mother-daughter regulation of *CLN3* transcription, or indeed on any transcriptional regulation acting through the *CLN3* promoter. We speculate that an M/G1 burst of *CLN3* expression from Mcm1 and/or Swi5 ([44]; Figure 5) may be sufficient to drive cells rapidly through T_1, as is observed in wild-type mothers of all sizes (Figure 1B, 1C; [5]); in daughters, this burst may be suppressed by Ace2 and Ash1.

The daughter-specific delay of wild-type cells depends on *CLN3*, since *cln3* mother and daughter cells of similar size have similar αT_1. Remarkably, cells deleted for cln3 still exhibit strong effects of cell size on G1 duration, although these effects are symmetrical between mothers and daughters of similar size (Figure 8C, 8D). Thus, while the cell size domain of effective size control in wild-type cells is set by *CLN3*, there may be an underlying Cln3-independent or parallel program of cell size control that acts or becomes detectable only upon deletion of *CLN3* [57]. In addition to loss of mother-daughter asymmetry, the response of *cln3* cells to cell size is shifted to about 1.5-fold larger cell sizes as measured using *ACT1pr-DsRed*; this finding confirms that cln3 cells are larger in terms of protein content than wild-type, in contrast to the proposal that the increase in cell size of cln3 cells [8],[10] is due primarily or entirely to increased vacuole size [58].

Our results are consistent with those of Laabs and collaborators, who reported that *cln3* cells and cells expressing *CLN3* from ectopic promoters lost mother-daughter asymmetry [28]. They also observed equal G1 durations for individual mother/daughter pairs [28]. In our analysis, in contrast, in almost all *cln3* mother-daughter pairs, with or without ectopic expression of *CLN3*, the daughters had a longer T_1 period (see Figure S6; p values$<10^{-5}$ in glucose; p values$<10^{-15}$ in glycerol/ethanol), although the daughter delay was reduced compared to wild-type, consistent with the results of Laabs and collaborators [28]. The symmetry that we observe in these mutants is restricted to mothers and daughters of similar size (more precisely, in the mother and daughter plots of αT_1 versus $\ln(M_{birth})$, in regions where the domains of mothers and daughters overlap). We assume this discrepancy arises from the same reasons discussed above.

1.3 DISCUSSION

1.3.1 CLN3, SIZE CONTROL, AND ASYMMETRIC TRANSCRIPTION

It was previously reported that asymmetric localization of Ace2 represses *CLN3* in daughter cells [28]. Our results extend this finding by showing that Ace2 regulation of *CLN3* is in part direct, mediated by Ace2 binding to the *CLN3* promoter. In addition, our results implicate Swi5 and Ash1 as well as Ace2 in *CLN3* regulation.

Neither asymmetric expression of *CLN3*, nor *CLN3* itself, is essential for size control [57] (Figure 8). However, *CLN3* sets the domain of cell sizes over which effective size control operates in wild-type cells. For this reason, negative control of *CLN3* by Ace2 and Ash1 allows differential Start regulation in mothers and daughters.

These findings provide empirical validation for one part of the theoretical cycle of transcriptional regulators proposed to account for a B-type cyclin-independent autonomous transcriptional oscillator [47].

1.3.2 DO MOTHERS DRIVE START WITH A BURST OF CLN3?

In the budding mode of growth, cell mass produced after budding goes to the daughter, but all pre-budding mass is retained by the mother [3]. As a result, a daughter that "passes" size control will retain this size through all subsequent (mother) budding cycles. This cell could thus be accelerated through Start by the M/G1 *CLN3* burst without a "need" for size checking. Given the high amount of noise in the mechanism of size control [5], this could prevent unnecessary delays in already full-sized mothers. The M/G1 *CLN3* burst, if experienced by daughters, would perturb the ability of daughters to effectively check their size (Figure 5J). This could result in the requirement for daughter-specific blockage to the burst. Thus, mother/daughter-specific *CLN3* regulation could simultaneously prevent unnecessary mother delays and prevent smaller daughters from passing Start prematurely.

In addition to repressing initial expression of *CLN3* in M/G1, Ace2 also induces *ASH1* expression; Ash1 represses later expression of *CLN3*. This is an example of "feed-forward" regulation, which may be a common regulatory structure for providing delayed response [59]—in this case, prolonged *CLN3* repression even after loss of Ace2 from the daughter nucleus. We speculate that this mechanism may allow daughter-specific delay over a broad range of timescales and growth rates.

The *CLN3* upstream region is unusually large (1.2 kb, compared to an average intergenic distance of 0.6 kb) and contains six ECB sites [45], multiple Ace2/Swi5 and Ash1 sites (this work), and four DDE sites [28]. How all of these sites and the factors that bind to them cooperate combinatorially to properly regulate *CLN3* is unknown. This is a large amount of regulatory machinery to provide a maximum peak-to-trough ratio only on the order of three [44]; however, since manipulation of *CLN3* gene copy number up or down by only a factor of two yields significant cell size phenotypes (Figure 5J) [10],[44], this level of control is likely to be physiologically significant, perhaps for the reasons cited above.

1.3.3 SIZERS, TIMERS, AND START

As noted above, our results can be interpreted in the classical framework of sizers and timers [18],[34] by defining the point at which cells switch from efficient size control to a timer control as "critical size" (Figure 3C): a precise size that cells must attain to transit Start. This point is marked by the intersection of the line of high negative slope with the line of low or zero negative slope in αT_1 versus $\ln(M_{birth})$ plots. Using this framework, we can summarize our results by stating that Ace2/Ash1-containing cells have a larger "critical size" than cells lacking these factors (normally, daughters and mothers, respectively). This formulation is inexact, primarily due to the evident non-zero slope in our data for the second component of the two-slope fit; remarkably, though, the effects of Ace2 and Ash1 shift not just the intersection point but the entire curve by 0.2 units of $\ln(M_{birth})$.

While increasing cell size and increasing Cln3 both decrease T_1 (i.e., accelerate Whi5 exit from the nucleus after cytokinesis) [5], Ace2 exit from the daughter nucleus occurs about 15 min (15±6 min) after cytokinesis, independent of cell size and *CLN3* (Text S1 and Figure S10). Thus, overall, Start control may consist of three distinct modules: Ace2 and Ash1-dependent but cell-size independent setting of the domain of cell size control; size control itself, leading to initiation of Whi5 nuclear exit; and a final size-independent step driven by CLN1,2-dependent transcriptional positive feedback, which rapidly completes Whi5 exit and drives the downstream events of Start [5],[7].

1.3.4 A NEW LINK BETWEEN DIFFERENTIATION AND CELL CYCLE IN BUDDING YEAST

In wild-type homothallic budding yeast, only mother cells express the *HO* endonuclease and switch mating type, due to Ash1 repression of *HO* expression in daughters [26],[27]. Phylogenetic analysis shows that in fungi, *ASH1* appeared before *HO*. This suggests that Ash1 may have functions predating *HO*, which may be important for asymmetrical cell division. It would be interesting to test whether Ash1 functions in cell cycle control in other fungi that can divide asymmetrically, such as *Candida albicans*, which lacks a *HO* homolog but expresses an Ash1 homolog that localizes specifically to the daughter cells [60],[61]. Ash1 also is found in *A. gossypii*, which undergoes asynchronous division in a multinucleate syncitium [62]; it would be interesting to evaluate the role of Ash1 in this asynchrony. Ace2 controls genes that confer diverse aspects of daughter cell biology [20],[23],[24]; here we elucidate how Ace2 also contributes to differential Start regulation in daughters [28].

There are interesting parallels and connections between *HO* control and *CLN3* control. Both are activated by Swi5 and inhibited by Ash1. Swi5 regulation of *HO* in mothers can be interpreted as feed-forward control, since Swi5 directly primes *HO* for expression [63] and also activates *CLN3* expression, which later yields efficient activation of the SBF factor that drives *HO* transcription [7],[63].

Cell cycle regulation and cell differentiation, often driven by asymmetric localization of cell fate determinants during cell division [64],[65],[66], are inter-regulated in many systems [67],[68],[69]. As the decision of cells to differentiate is often made in G1, cell differentiation and commitment to a stable G1 are often coregulated [67],[69],[70]. It would be interesting to examine cases in which stem cells produce one proliferating cell and one daughter that differentiates in G1 [65]; such cells might employ mechanisms similar to those we have uncovered in differential mother-daughter G1 control in budding yeast.

1.4 MATERIALS AND METHODS

1.4.1 STRAIN AND PLASMID CONSTRUCTION

Standard methods were used throughout. All strains are W303-congenic. All integrated constructs were characterized by qPCR. Mutations of the Ace2/Swi5 and Ash1 binding sites on the *CLN3* promoter were verified by sequencing.

1.4.2 TIME-LAPSE MICROSCOPY

Preparation of cells for microscopy and time-lapse microscopy were performed as previously described [5],[6]. Growth of microcolonies was observed with fluorescence time-lapse microscopy at 30°C using a Leica DMIRE2 inverted microscope with a Ludl motorized XY stage. Images were acquired every 3 min for cells grown in glucose and every 6 min for cells grown in glycerol/ethanol with a Hamamatsu Orca-ER camera. Custom Visual Basic software integrated with ImagePro Plus was used to automate image acquisition and microscope control.

1.4.3 IMAGE ANALYSIS

Automated image segmentation and fluorescence quantification of yeast grown under time-lapse conditions and semi-automated assignment of microcolony pedigrees were performed as previously described [6]. The nuclear residence of Whi5-GFP was scored by visual inspection of composite phase contrast-fluorescent movies. Cell size was measured as the total cell fluorescence from DsRed protein, expressed from the constitutively active ACT1pr, as previously described [5]. Cell size at every time point was extrapolated from a linear fit of the ln(M) as a function of time for cells grown in glucose and from a smoothing spline fit for cells grown in glycerol/ethanol. Individual cell growth in glycerol/ethanol appears to be intermediate between a linear and an exponential model (unpublished data); this deviation from exponentiality has very little effect on this analysis.

1.4.4 DATA ANALYSIS

Time-lapse fluorescence microscopy, microarray data, and sequencing data were analyzed with custom software written in MATLAB software (see Text S1 for details on the analysis of the microarray data) [5]. For cluster analysis, the log2 of the arrays data or of the subtracted arrays data were hierarchically clustered by the agglomerative algorithm [41].

Data were visually presented using JavaTreeView. For sequencing data, the area associated to every wild-type or mutated nucleotide was evaluated manually by using the MATLAB software.

1.4.5 CELL CYCLE SYNCHRONIZATION

YEP medium was used for all cell cycle synchronization experiments, supplemented with the appropriate carbon source as indicated below. Cell cycle synchronization by the *cdc20 GALL-CDC20* block release was achieved by growing cells to early log phase in YEP+galactose (3%), then filtering and growing them in YEP+glucose (2%) for 3 h to arrest cells in metaphase. Cells were released from the block by filtering back into YEP+galactose (3%). *GALL* is a truncated version of the *GAL1* promoter that shows inducible but significantly lower expression than the full-length *GAL1* promoter [37].

1.4.6 MICROARRAYS

Microarrays were performed as previously described [71] but using micro-arrays carrying PCR fragments from open reading frames of *S. cerevisiae*. Each array had each PCR fragment independently spotted four to eight times, leading to a high redundancy of data and small errors in expression ratios. RNA extraction, cDNA synthesis and labeling, and hybridization and scanning were carried out by the Stony Brook spotted microarray facility, as described previously [71].

1.4.7 CHIPS

Standard methods were used for ChIP experiments. Early log phase cells were fixed for 15 min in 1% formaldehyde at room temperature. Immuno-precipitations were performed with IgG Sepharose beads. Immunoprecipitated DNA was amplified by PCR.

REFERENCES

1. Hartwell L. H, Culotti J, Pringle J. R, Reid B. J (1974) Genetic control of the cell division cycle in yeast. Science 183: 46–51. doi: 10.1126/science.183.4120.46.

2. Nasmyth K (1993) Regulating the *HO* endonuclease in yeast. Curr Opin Genet Dev 3: 286–294. doi: 10.1073/pnas.95.25.14863.

3. Hartwell L. H, Unger M. W (1977) Unequal division in Saccharomyces cerevisiae and its implications for the control of cell division. J Cell Biol 75: 422–435. doi: 10.1083/jcb.75.2.422.

4. Johnston G. C, Pringle J. R, Hartwell L. H (1977) Coordination of growth with cell division in the yeast Saccharomyces cerevisiae. Exp Cell Res 105: 79–98. doi: 10.1016/0014-4827(77)90154-9.

5. Di Talia S, Skotheim J. M, Bean J. M, Siggia E. D, Cross F. R (2007) The effects of molecular noise and size control on variability in the budding yeast cell cycle. Nature 448: 947–951. doi: 10.1038/nature06072.

6. Bean J. M, Siggia E. D, Cross F. R (2006) Coherence and timing of cell cycle start examined at single-cell resolution. Mol Cell 21: 3–14. doi: 10.1016/j.molcel.2005.10.035.

7. Skotheim J. M, Di Talia S, Siggia E. D, Cross F. R (2008) Positive feedback of G1 cyclins ensures coherent cell cycle entry. Nature 454: 291–296. doi: 10.1038/nature07118.

8. Cross F. R (1988) DAF1, a mutant gene affecting size control, pheromone arrest, and cell cycle kinetics of Saccharomyces cerevisiae. Mol Cell Biol 8: 4675–4684.

9. Dirick L, Bohm T, Nasmyth K (1995) Roles and regulation of Cln-Cdc28 kinases at the start of the cell cycle of Saccharomyces cerevisiae. EMBO J 14: 4803–4813.

10. Nash R, Tokiwa G, Anand S, Erickson K, Futcher A. B (1988) The WHI1+ gene of Saccharomyces cerevisiae tethers cell division to cell size and is a cyclin homolog. EMBO J 7: 4335–4346.

11. Stuart D, Wittenberg C (1995) *CLN3*, not positive feedback, determines the timing of CLN2 transcription in cycling cells. Genes Dev 9: 2780–2794.

12. Tyers M, Tokiwa G, Futcher B (1993) Comparison of the Saccharomyces cerevisiae G1 cyclins: Cln3 may be an upstream activator of Cln1, Cln2 and other cyclins. EMBO J 12: 1955–1968.

13. Costanzo M, Nishikawa J. L, Tang X, Millman J. S, Schub O, et al. (2004) CDK activity antagonizes Whi5, an inhibitor of G1/S transcription in yeast. Cell 117: 899–913. doi: 10.1016/j.cell.2004.05.024.

14. de Bruin R. A, McDonald W. H, Kalashnikova T. I, Yates J 3rd, Wittenberg C (2004) Cln3 activates G1-specific transcription via phosphorylation of the SBF bound repressor Whi5. Cell 117: 887–898. doi: 10.1016/j.cell.2004.05.025.

15. Koch C, Schleiffer A, Ammerer G, Nasmyth K (1996) Switching transcription on and off during the yeast cell cycle: Cln/Cdc28 kinases activate bound transcription factor SBF (Swi4/Swi6) at start, whereas Clb/Cdc28 kinases displace it from the promoter in G2. Genes Dev 10: 129–141. doi: 10.1101/gad.10.2.129.

16. Wijnen H, Landman A, Futcher B (2002) The G(1) cyclin Cln3 promotes cell cycle entry via the transcription factor Swi6. Mol Cell Biol 22: 4402–4418. doi: 10.1128/MCB.22.12.4402-4418.2002.

17. Jorgensen P, Tyers M (2004) How cells coordinate growth and division. Curr Biol 14: R1014–R1027. doi: 10.1016/j.cub.2004.11.027.

18. Shields R, Brooks R. F, Riddle P. N, Capellaro D. F, Delia D (1978) Cell size, cell cycle and transition probability in mouse fibroblasts. Cell 15: 469–474. doi: 10.1016/j.cell.2008.02.007.

19. Lord P. G, Wheals A. E (1983) Rate of cell cycle initiation of yeast cells when cell size is not a rate-determining factor. J Cell Sci 59: 183–201.

20. Colman-Lerner A, Chin T. E, Brent R (2001) Yeast Cbk1 and Mob2 activate daughter-specific genetic programs to induce asymmetric cell fates. Cell 107: 739–750. doi: 10.1016/S0092-8674(01)00596-7.

21. Mazanka E, Alexander J, Yeh B. J, Charoenpong P, Lowery D. M, et al. (2008) The NDR/LATS family kinase Cbk1 directly controls transcriptional asymmetry. PLoS Biol 6: e203. doi:10.1371/journal.pbio.0060203.

22. Weiss E. L, Kurischko C, Zhang C, Shokat K, Drubin D. G, et al. (2002) The Saccharomyces cerevisiae Mob2p-Cbk1p kinase complex promotes polarized growth and acts with the mitotic exit network to facilitate daughter cell-specific localization of Ace2p transcription factor. J Cell Biol 158: 885–900. doi: 10.1083/jcb.200203094.

23. Knapp D, Bhoite L, Stillman D. J, Nasmyth K (1996) The transcription factor Swi5 regulates expression of the cyclin kinase inhibitor p40SIC1. Mol Cell Biol 16: 5701–5707.

24. Wang Y, Shirogane T, Liu D, Harper J. W, Elledge S. J (2003) Exit from exit: resetting the cell cycle through Amn1 inhibition of G protein signaling. Cell 112: 697–709. doi: 10.1016/0014-4827(77)90154-9.

25. Cosma M. P (2004) Daughter-specific repression of Saccharomyces cerevisiae *HO*: Ash1 is the commander. EMBO Rep 5: 953–957.

26. Bobola N, Jansen R. P, Shin T. H, Nasmyth K (1996) Asymmetric accumulation of Ash1p in postanaphase nuclei depends on a myosin and restricts yeast mating-type switching to mother cells. Cell 84: 699–709. doi: 10.1038/nrm758.

27. Sil A, Herskowitz I (1996) Identification of asymmetrically localized determinant, Ash1p, required for lineage-specific transcription of the yeast *HO* gene. Cell 84: 711–722. doi: 10.1016/S0092-8674(00)81049-1.

28. Laabs T. L, Markwardt D. D, Slattery M. G, Newcomb L. L, Stillman D. J, et al. (2003) *ACE2* is required for daughter cell-specific G1 delay in Saccharomyces cerevisiae. Proc Natl Acad Sci U S A 100: 10275–10280. doi: 10.1073/pnas.1833999100.

29. Elliott S. G, McLaughlin C. S (1978) Rate of macromolecular synthesis through the cell cycle of the yeast Saccharomyces cerevisiae. Proc Natl Acad Sci U S A 75: 4384–4388. doi: 10.1073/pnas.75.9.4384.

30. Sveiczer A, Novak B, Mitchison J. M (1996) The size control of fission yeast revisited. J Cell Sci 109(Pt 12): 2947–2957.

31. McBride H. J, Yu Y, Stillman D. J (1999) Distinct regions of the Swi5 and Ace2 transcription factors are required for specific gene activation. J Biol Chem 274: 21029–21036. doi: 10.1101/gad.9.22.2780.

32. Racki W. J, Becam A. M, Nasr F, Herbert C. J (2000) Cbk1p, a protein similar to the human myotonic dystrophy kinase, is essential for normal morphogenesis in Saccharomyces cerevisiae. EMBO J 19. 4524–4532. doi: 10.1038/nature06955.

33. Chartrand P, Meng X. H, Huttelmaier S, Donato D, Singer R. H (2002) Asymmetric sorting of ash1p in yeast results from inhibition of translation by localization elements in the mRNA. Mol Cell 10: 1319–1330. doi: 10.1016/j.cub.2004.11.027.

34. Wheals A. E (1982) Size control models of Saccharomyces cerevisiae cell proliferation. Mol Cell Biol 2: 361–368.

35. Dohrmann P. R, Butler G, Tamai K, Dorland S, Greene J. R, et al. (1992) Parallel pathways of gene regulation: homologous regulators SWI5 and *ACE2* differentially control transcription of *HO* and chitinase. Genes Dev 6: 93–104. doi: 10.1101/gad.6.1.93.

36. Voth W. P, Yu Y, Takahata S, Kretschmann K. L, Lieb J. D, et al. (2007) Forkhead proteins control the outcome of transcription factor binding by antiactivation. EMBO J 26: 4324–4334. doi: 10.1038/sj.emboj.7601859.

37. Mumberg D, Muller R, Funk M (1994) Regulatable promoters of Saccharomyces cerevisiae: comparison of transcriptional activity and their use for heterologous expression. Nucleic Acids Res 22: 5767–5768. doi: 10.1093/nar/22.25.5767.

38. Spellman P. T, Sherlock G, Zhang M. Q, Iyer V. R, Anders K, et al. (1998) Comprehensive identification of cell cycle-regulated genes of the yeast Saccharomyces cerevisiae by microarray hybridization. Mol Biol Cell 9: 3273–3297.

39. Sbia M, Parnell E. J, Yu Y, Olsen A. E, Kretschmann K. L, et al. (2008) Regulation of the yeast Ace2 transcription factor during the cell cycle. J Biol Chem 283: 11135–11145. doi: 10.1371/journal.pbio.0060203.

40. Tebb G, Moll T, Dowzer C, Nasmyth K (1993) SWI5 instability may be necessary but is not sufficient for asymmetric *HO* expression in yeast. Genes Dev 7: 517–528. doi: 10.1101/gad.7.3.517.

41. Eisen M. B, Spellman P. T, Brown P. O, Botstein D (1998) Cluster analysis and display of genome-wide expression patterns. Proc Natl Acad Sci U S A 95: 14863–14868. doi: 10.1073/pnas.95.25.14863.

42. Harbison C. T, Gordon D. B, Lee T. I, Rinaldi N. J, Macisaac K. D, et al. (2004) Transcriptional regulatory code of a eukaryotic genome. Nature 431: 99   104. doi: 10.1038/nature02800.

43. Simon I, Barnett J, Hannett N, Harbison C. T, Rinaldi N. J, et al. (2001) Serial regulation of transcriptional regulators in the yeast cell cycle. Cell 106: 697–708. doi: 10.1016/S0092-8674(01)00494-9.

44. McInerny C. J, Partridge J. F, Mikesell G. E, Creemer D. P, Breeden L. L (1997) A novel Mcm1-dependent element in the SWI4, *CLN3*, CDC6, and CDC47 promoters activates M/G1-specific transcription. Genes Dev 11: 1277–1288. doi: 10.1101/gad.11.10.1277.

45. MacKay V. L, Mai B, Waters L, Breeden L. L (2001) Early cell cycle box-mediated transcription of *CLN3* and SWI4 contributes to the proper timing of the G(1)-

to-S transition in budding yeast. Mol Cell Biol 21: 4140–4148. doi: 10.1128/MCB.21.13.4140-4148.2001.

46. Sillje H. H, ter Schure E. G, Rommens A. J, Huls P. G, Woldringh C. L, et al. (1997) Effects of different carbon fluxes on G1 phase duration, cyclin expression, and reserve carbohydrate metabolism in Saccharomyces cerevisiae. J Bacteriol 179: 6560–6565.

47. Orlando D. A, Lin C. Y, Bernard A, Wang J. Y, Socolar J. E, et al. (2008) Global control of cell-cycle transcription by coupled CDK and network oscillators. Nature 453: 944–947. doi: 10.1038/nature06955.

48. Maxon M. E, Herskowitz I (2001) Ash1p is a site-specific DNA-binding protein that actively represses transcription. Proc Natl Acad Sci U S A 98: 1495–1500. doi: 10.1073/pnas.98.4.1495.

49. Colomina N, Ferrezuelo F, Wang H, Aldea M, Gari E (2008) Whi3, a developmental regulator of budding yeast, binds a large set of mRNAs functionally related to the endoplasmic reticulum. J Biol Chem 283: 28670–28679. doi: 10.1038/nature06072.

50. Gari E, Volpe T, Wang H, Gallego C, Futcher B, et al. (2001) Whi3 binds the mRNA of the G1 cyclin *CLN3* to modulate cell fate in budding yeast. Genes Dev 15: 2803–2808.

51. Nash R. S, Volpe T, Futcher B (2001) Isolation and characterization of WHI3, a size-control gene of Saccharomyces cerevisiae. Genetics 157: 1469–1480.

52. Verges E, Colomina N, Gari E, Gallego C, Aldea M (2007) Cyclin Cln3 is retained at the ER and released by the J chaperone Ydj1 in late G1 to trigger cell cycle entry. Mol Cell 26: 649–662. doi: 10.1016/j.ymeth.2008.06.004.

53. Wang H, Gari E, Verges E, Gallego C, Aldea M (2004) Recruitment of Cdc28 by Whi3 restricts nuclear accumulation of the G1 cyclin-Cdk complex to late G1. EMBO J 23: 180–190. doi: 10.1038/sj.emboj.7600022.

54. Cross F. R (1989) Further characterization of a size control gene in Saccharomyces cerevisiae. J Cell Sci Suppl 12: 117–127.

55. Buchler N. E, Cross F. R (2009) Protein sequestration generates a flexible ultrasensitive response in a genetic network. Mol Syst Biol 5: 272. doi: 10.1038/msb.2009.30.

56. Denis C. L, Ferguson J, Young E. T (1983) mRNA levels for the fermentative alcohol dehydrogenase of Saccharomyces cerevisiae decrease upon growth on a nonfermentable carbon source. J Biol Chem 258: 1165–1171.

57. Schneider B. L, Zhang J, Markwardt J, Tokiwa G, Volpe T, et al. (2004) Growth rate and cell size modulate the synthesis of, and requirement for, G1-phase cyclins at start. Mol Cell Biol 24: 10802–10813. doi: 10.1128/MCB.24.24.10802-10813.2004.

58. Han B. K, Aramayo R, Polymenis M (2003) The G1 cyclin Cln3p controls vacuolar biogenesis in Saccharomyces cerevisiae. Genetics 165: 467–476.

59. Mangan S, Alon U (2003) Structure and function of the feed-forward loop network motif. Proc Natl Acad Sci U S A 100: 11980–11985. doi: 10.1038/sj.emboj.7600022.

60. Inglis D. O, Johnson A. D (2002) Ash1 protein, an asymmetrically localized transcriptional regulator, controls filamentous growth and virulence of Candida albicans. Mol Cell Biol 22: 8669–8680. doi: 10.1128/MCB.22.24.8669-8680.2002.

61. Munchow S, Ferring D, Kahlina K, Jansen R. P (2002) Characterization of Candida albicans *ASH1* in Saccharomyces cerevisiae. Curr Genet 41: 73–81. doi: 10.1007/s00294 002 0286-y.

62. Gladfelter A. S, Hungerbuehler A. K, Philippsen P (2006) Asynchronous nuclear division cycles in multinucleated cells. J Cell Biol 172: 347–362. doi: 10.1083/jcb.200507003.

63. Cosma M. P, Tanaka T, Nasmyth K (1999) Ordered recruitment of transcription and chromatin remodeling factors to a cell cycle- and developmentally regulated promoter. Cell 97: 299–311. doi: 10.1016/S0092-8674(00)80740-0.

64. Horvitz H. R, Herskowitz I (1992) Mechanisms of asymmetric cell division: two Bs or not two Bs, that is the question. Cell 68: 237–255. doi: 10.1016/0092-8674(92)90468-R.

65. Knoblich J. A (2008) Mechanisms of asymmetric stem cell division. Cell 132: 583–597. doi: 10.1016/j.cell.2008.02.007.

66. Roegiers F, Jan Y. N (2004) Asymmetric cell division. Curr Opin Cell Biol 16: 195–205. doi: 10.1016/s1097-2765(02)00709-8.

67. Buttitta L. A, Edgar B. A (2007) Mechanisms controlling cell cycle exit upon terminal differentiation. Curr Opin Cell Biol 19: 697–704. doi: 10.1016/j.ceb.2007.10.004.

68. Jensen R. B, Wang S. C, Shapiro L (2002) Dynamic localization of proteins and DNA during a bacterial cell cycle. Nat Rev Mol Cell Biol 3: 167–176. doi: 10.1038/nrm758.

69. Zhu L, Skoultchi A. I (2001) Coordinating cell proliferation and differentiation. Curr Opin Genet Dev 11: 91–97. doi: 10.1038/ncb945.

70. Lasorella A, Stegmuller J, Guardavaccaro D, Liu G, Carro M. S, et al. (2006) Degradation of Id2 by the anaphase-promoting complex couples cell cycle exit and axonal growth. Nature 442: 471–474. doi: 10.1038/nature04895.

71. Oliva A, Rosebrock A, Ferrezuelo F, Pyne S, Chen H, et al. (2005) The cell cycle-regulated genes of Schizosaccharomyces pombe. PLoS Biol 3: e225. doi: 10.1371/journal.pbio.0030225.

This article has supplemental information that is not featured in this version of the text. To view these files, please visit the original version of the article as cited in the beginning of this chapter.

CHAPTER 2

MODELING ROBUSTNESS TRADEOFFS IN YEAST CELL POLARIZATION INDUCED BY SPATIAL GRADIENTS

CHING-SHAN CHOU, QING NIE, AND TAU-MU YI

2.1 INTRODUCTION

Breaking symmetry is a fundamental process in biology [1]. Components that were previously uniformly distributed become asymmetrically localized. This anisotropy or polarization creates complexity of form and function. The challenge is polarizing in the right place at the right time to the proper extent under uncertain and changing conditions (i.e. robust polarization).

Cells localize components to specific locations leading to morphological changes in response to internal and external cues. For example, haploid cells of the yeast *Saccharomyces cerevisiae* typically form a new bud at the site of the previous bud (internal cue). In addition, haploid yeast cells can sense an external gradient of mating pheromone and form a mating projection (shmoo) toward the source. In both cases, a large number of signaling, structural, and transport proteins localize at the site of the morphological change [2], [3].

*Originally printed under the terms of the Creative Commons Attribution License. Ching-Shan Chou, Qing Nie, and Tau-Mu Yi. Modeling Robustness Tradeoffs in Yeast Cell Polarization Induced by Spatial Gradients. PLoS ONE **3**,9 (2008). doi:10.1371/journal.pone.0003103.*

There has been extensive mathematical modeling of cell polarization as a special case of pattern formation in living systems. Turing originally proposed that complex spatial patterns could arise from simple reaction-diffusion systems [4]. Meinhardt popularized the modeling of biological pattern formation using generic reaction-diffusion models. In particular, he introduced the idea that polar structures could arise from local autocatalysis (i.e. positive feedback) balanced by global inhibition [5]. Subsequently, researchers constructed more detailed models that incorporated information about specific molecular species and reactions in cells undergoing chemotaxis. One class of models used a local excitation, global inhibition (LEGI) mechanism [6], [7].

Sensing and responding to a chemical gradient present many challenges including sensitivity, dynamic range, tracking, and noise (Fig. 1A). The gradient may be shallow and the concentration difference between front and back small (sensitivity). The average concentration of the chemical ligand may be much higher or lower than the dissociation constant (Kd) of the ligand receptor (dynamic range). The source of the chemical signal may be moving (tracking). There may be noise in the gradient, and so forth. It is an open question how well these different performance objectives can be achieved simultaneously. In the literature, the focus has been on understanding how a shallow external gradient can be amplified to create a steep internal gradient of cellular components. High amplification can result in an all-or-none localization of the internal component to a narrow region. However, the tracking of a moving signal source has also been acknowledged to be important. Devreotes and colleagues [8] made the distinction between directional sensing (low amplification, good tracking) and polarization (high amplification, poor tracking). Meinhardt first highlighted the potential tradeoff between amplification and tracking [9].

This field possesses an extensive literature, and Dawes et al. [10] reviewed a number of previous models of eukaryotic gradient-sensing and cell polarization. Included were the models of Meinhardt [9], Narang [11], Levchenko-Iglesias [12], Postma-Van Haastert [13], Maly et al. [14], Haugh and colleagues [15], Gamba et al. [16], and Skupsky et al. [17]. Many of the models contained some type of positive feedback structure, as well as nonlinearities capable of generating ultrasensitivity to the input.

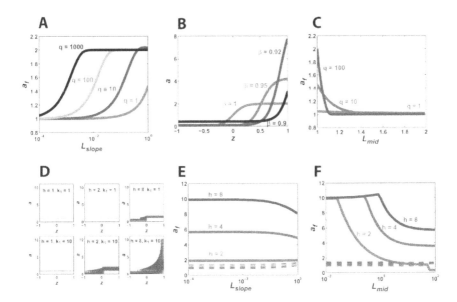

FIGURE 1: Schematic descriptions of performance objectives and model of polarization. (A) Performance objectives of sensing and responding to a gradient. The graphs depict the concentration of chemical ligand along the axial length of the cell. Below each graph is a picture of a cell in a chemical gradient (background shading) with the polarized component. The chemical gradient may be shallow (sensitivity), the average concentration may be low or high (dynamic range), and the direction of the gradient may be changing (tracking). In each case, the external gradient must be amplified to create a polarized distribution of some internal component. (B) In the model, the polarized species a becomes localized to the front of the cell through cooperative interactions (q is the Hill cooperativity parameter) in response to the input and through positive feedback (+). There is global negative feedback (integral control) mediated by the species b. (C) In feedforward/feedback coincidence detection, the positive feedback amplification of a depends on a feedforward component originating from the input u and a feedback component originating from a.

The models ranged from generic models (e.g. [9], [11], [13]) to more mechanistic models (e.g. [10], [14], [17]).

Dawes et al. categorized the models according to gradient-sensing, amplification, polarization, tracking of directional change, persistence when the stimulus is removed (i.e. multi-stability), etc. Among the 23 papers containing models mentioned in the article, only four [9], [10], [17], [18] simultaneously considered the issues of amplification, tracking, and multi-stability. Of these 4, the paper by Skupsky et al. [17] was most related to the work described here. Those authors defined 4 modes of gradient-sensing that depended on the strength of the positive feedback and the extent of translocation of signaling molecules from the cytoplasm to the membrane. These modes varied in the degree of amplification (polarization), presence of multiple steady-states, response to a rotating gradient, etc. However, a detailed characterization of the modes was hampered by the complexity of the mechanistic model. We have presented a more mathematical treatment using generic models motivated by yeast (small) cell gradient-sensing and polarization. These simple models motivated more complex mechanistic models later in our paper.

Here we investigated in a systematic fashion the tradeoff between amplification and tracking during gradient-sensing. We demonstrated the nature of these tradeoffs using a simple model and well-defined measures of performance. In particular, we focused on the roles of cooperativity and positive feedback on amplification and their effects on tracking. Although the tradeoff could not be eliminated, it could be fine-tuned through modifications to the model to ensure balanced performance in specific regimes of external conditions. In addition, we demonstrated that moderate lateral surface diffusion in the membrane increased the robustness of polarization. Finally, we used these findings to update our previous model of yeast spatial sensing of mating pheromone, and simulate polarization for a range of surface diffusion coefficients.

2.2 RESULTS

2.2.1 GENERIC MODEL AND MEASURES OF POLARIZATION AND AMPLIFICATION

Previously we constructed a model of yeast cell polarization that explicitly represented spatial dynamics [19]. In that model we explored the tradeoff between amplification of a shallow external gradient into a steeper internal gradient of intracellular components and tracking a gradient whose direction is changing. Both objectives were hard to achieve simultaneously.

The complexity of the model, however, prevented a thorough analysis of the tradeoff. Here, we constructed a simpler, generic model that captured the essence of the larger model.

$$\frac{\delta a}{dt} = D_s \nabla_s^2 a + \frac{k_0}{1 + (\beta u)^{-q}} + \frac{k_1}{1 + (\gamma a)^{-h}} - k_2 a - k_3 ba - k_5 \hat{a} \qquad \text{(M1.1)}$$

$$\frac{\delta b}{dt} = k_4 \hat{a} b \qquad \text{(M1.2)}$$

$$\hat{a} = \bar{a} - a_{ss}$$

$$\bar{a} = \frac{\int_s a \, ds}{\int_s ds}$$

$$u = L_{mid} + L_{slope}(z - z_0)$$

The default value for most of the parameters was 1: $k_0 = k_1 = k_2 = k_3 = k_4 = k_5 = 1$ s^{-1}; $\beta = \gamma = 1$; $a_{ss} = 1$. This default case assumes that all of the dynamics in the system are on the same time-scale. In the investigations below, we typically varied the values of k_0, k_1, q, h and D_s. We also explored varying the other parameters (data not shown) but found that they did not impact the steady-state behavior as significantly. The input u and the variables a and b were chosen to be unitless.

In this model (Model 1), the variable a represents the concentration of the species undergoing polarization and whose spatial dynamics are of interest (Fig. 1B). The second variable b represents the concentration of a negative regulator involved in a negative feedback loop that regulates a and behaves like a global inhibitor; it is uniformly distributed throughout the cell. The input u is a linear chemical gradient. The species represented by a is assumed to be bound to the membrane and the term $D_s \nabla^2 a$ describes its lateral surface diffusion in the membrane with diffusion coefficient D_s. The second term $(k_0/1+(\beta u)^{-q})$ in Eq. (1.1) represents the cooperative production of a which depends on the input u; the form of the term is a Hill expression possessing a Hill cooperativity parameter q and a Hill half-maximal constant $1/\beta$. The third term is a positive feedback term in which a stimulates its own production. This autocatalytic reaction is also a cooperative reaction possessing a Hill cooperativity parameter h and a Hill half-maximal constant $1/\gamma$.

Degradation is described by a first-order decay term (k_2a). Regulation is achieved through two negative feedback terms representing proportional feedback ($k_5\hat{a}$) and integral feedback (k_3ba) [20]. The variable b is involved in the integral feedback control loop with the second differential equation ensuring that the average steady-state levels of a ($\bar{a}$) will tend to the fixed value a_{ss}. The variable $\hat{a}$ represents the difference between $\bar{a}$ and a_{ss}. Because the integral feedback term k_3ba cannot be negative, the steady-state concentration of a will drop below a_{ss} for low input values. Note that we have assumed that there is fast mixing of the negative regulator represented by b in the cell interior; this assumption is likely to be valid for smaller cells. In addition, we point out that the production of the negative regulator b is autocatalytic, which prevents b from becoming negative. Finally, modifying the form of Eq. (1.2) by adding a constant basal synthesis rate (k_6) for b breaks the integral control, but did not significantly alter the steady-state behavior of the model.

Geometrically, we modeled the cell as a sphere with radius 1 µm. We applied a linear spatial gradient described by the concentration of ligand at the center of the cell, L_{mid}, and the gradient slope L_{slope} (which was relative to L_{mid}); z is the axial coordinate along the length of the cell in the direction of the gradient and z_0 is the position of the center of the cell. The response of the cell was measured by the spatial dynamics of a, and in particular, the polarized distribution of a. We represented these dynamics in one-dimension (1D) along the axial length because a sphere is rotationally symmetric around its axis (axisymmetric).

Biologically, we interpret this model as a signal transduction cascade in which the cooperative assembly of multi-protein signaling complexes can give rise to the cooperative input term. Positive feedback is found in many of these signaling systems. For example, in the yeast mating response the combined actions of the proteins Bem1p, Cdc42p, and Cdc24p create a positive feedback loop [21]. Negative feedback loops are also ubiquitous in signaling pathways and can act upstream at the level of receptor down-regulation to the more downstream transcriptional activation of negative regulators. Thus, we view the model as a simplification of more sophisticated models of cell polarity and chemotaxis from other authors such as the LEGI models previously mentioned [6]. It is important to note that for simplicity we chose a generic model formalism that does not obey mass-action. For example, the synthesis terms show no dependence on the "substrate" of a, implying that the level of substrate is constant. However, the fundamental spatial dynamics of the generic model are reproduced in mass-action models such as the model of yeast pheromone-induced polarization described later.

We investigated several measures of polarization. First was the value of a at the front of the cell, a_f, where the concentration of ligand is highest. The second was an approximation of the relative slope of a:

$$POL(a) \equiv \frac{a_f - \bar{a}}{\bar{a}R} \tag{1}$$

The average concentration of a is $\bar{a}$, and the radius of the cell is R. The third measure termed the polarization factor (PF) describes the "width" of the global distribution of the polarized component:

$$PF(a) \equiv 1 - 2\frac{S_p(a)}{SA} \tag{2}$$

$$S_p(a) = \min\left\{|C|, C \subset \Omega: \int_C a\, dS = \right.$$

$$\left. \frac{1}{2}\int_\Omega a\, dS, \Omega \text{ is the surface of the sphere}\right\}$$

$S_p(a)$ is the surface area at the front of the cell that encompasses 50% of the polarized component a and SA is the total surface area of the cell. An unpolarized cell would have a PF of 0 and an infinitely polarized cell would have a PF of 1. We concluded that in most cases, all three measures conveyed the same information (data not shown), and we have typically plotted a_f for convenience and consistency.

Amplification refers to the conversion of the external gradient signal into the polarization of the internal component. We defined the amplification factor (AF) as the ratio of polarization of a to the relative slope (i.e. polarization) of the external gradient of ligand L (POL(L)). A large AF indicates that the cell can amplify a shallow spatial gradient to produce significant internal polarization:

$$AF(a, L) \equiv \frac{POL(a)}{POL(L)} \qquad (3)$$

2.2.2 AMPLIFICATION IS PRODUCED BY COOPERATIVITY OR POSITIVE FEEDBACK

For the first half of this work, we explored the spatial dynamics when $D_s = 0$ (i.e. no surface diffusion). Initially we set $k_1 = 0$ in Eq (1.1) so that there would be no positive feedback. Amplification would arise from the cooperative production of a as a function of input ($k_0/(1+(\beta u)^{-q})$). With the parameter $\beta = 1$, and the average ligand concentration $L_{mid} = 1$, we varied the slope of the gradient (L_{slope}) for four different values of the Hill cooperativity parameter q. A maximum polarization of $a_f \sim 2$ (POL(a)~1 μm^{-1}) was achieved. Increasing q resulted in better polarization at smaller slopes (i.e. shallower gradients), and thus better amplification (Fig. 2A). We were able to increase polarization beyond $a_f = 2$ by fine-tuning β such that βL_f (L_f is the ligand concentration at the front of the cell) was closer to 1. For $\beta = 0.92$, $a_f = 8$ (Fig. 2B).

For a given value of L_{mid}, it was possible to obtain good polarization for a shallow slope using a high value of q and fine-tuning the value of β. What happens when we vary L_{mid} for a fixed β and q? Good polarization was observed only for a narrow range of concentrations. In Figure 2C, we varied L_{mid} (for a fixed L_{slope} relative to L_{mid}) over a 2-fold range from 1 to 2

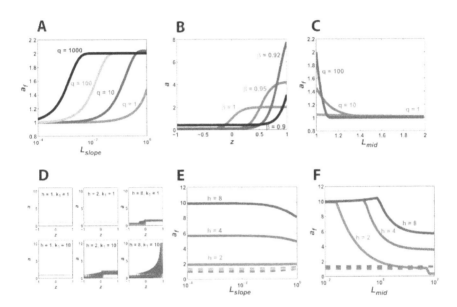

FIGURE 2: Cooperativity and positive feedback produced polarization and amplification. (A) Cooperativity alone ($k_1 = 0$) resulted in amplification of shallow gradients for larger values of q. Polarization was measured by a_f, the value of a at the front of the cell. We plotted polarization as the slope of the external gradient, L_{slope}, was varied ($L_{mid} = 1$). (B) With cooperativity alone, polarization increased by fine-tuning the Hill constant β. The variable a was plotted as a function of position along the axial length z. Lmid = 1, Lslope = 0.1 μm^{-1}, $k_0 = 10$ s^{-1}, q = 100. (C) Cooperativity alone showed limited dynamic range. We varied Lmid over a narrow range from 1 to 2 ($L_{slope} = (0.1 \times L_{mid})$ μm^{-1}) and determined af for different values of q. (D) Increasing the positive feedback gain ($k_1 > 0$, h≥1) enhanced polarization and produced multiple steady-states. We plotted the envelope of possible steady-state solutions. For cases with multiple solutions, the red trace represents the maximum polarization solution in the envelope. (E) Infinite amplification of shallow gradients with positive feedback. For $k_1 = 10$ s^{-1}, we plotted af for three values of h as we varied the gradient slope ($L_{mid} = 1$). The dashed lines represent the minimum polarization in the solution envelope for each value of h. (F) Positive feedback produced broad dynamic range. We plotted af as a function of Lmid ($L_{slope} = 0.01$ $\mu m{-}1$) for three values of h ($k_1 = 10$ s^{-1}). Dashed lines represent minimum polarization solutions.

for different values of the cooperativity parameter q. There was a tradeoff: higher values of q produced better polarization, but a reduced range of responsiveness. More importantly, the overall range was quite limited (less than 2-fold), thus indicating a very narrow dynamic range of the polarization response with $k_1 = 0$.

We added positive feedback by setting $k_1 > 0$; a acts autocatalytically to stimulate its own production. Within the positive feedback term, there is a Hill cooperativity parameter h. Both k_1 and h influenced the strength (gain) of the positive feedback. For $k_1 = 1$ s^{-1}, polarization improved for higher values of h (Fig. 2D). The increase in polarization was accompanied by the appearance of multiple steady-states (blue lines). We represented these steady-states by an envelope of possible solutions. We then explored different values of k_1 for fixed values of h. With h = 1, there was no enhanced amplification even for large values of k_1. Thus, substantial amplification required some degree of cooperativity in the positive feedback loop [22]. With h≥2, we saw increased maximum polarization for higher values of k_1. Thus, increasing k_1 or h resulted in dramatic polarization that was associated with multiple solutions.

When the positive feedback gain was sufficiently large, a decrease in the gradient slope did not cause a decrease in the maximum polarization solution. Indeed, the maximum polarization could be achieved as $L_{slope} \rightarrow 0$, indicating the presence of infinite amplification or what has more commonly been termed spontaneous polarization (i.e. polarization in response to an infinitesimal gradient) [11]: AF→∞ when POL(L)→0 and POL(a)→C>0 (Fig. 2E). Interestingly at higher gradient slopes there was actually a slight decline in maximum polarization. In Figure 2E, the envelope of possible solutions is indicated by the region between the solid lines (maximum polarization solution) and dashed lines (minimum polarization solutions).

Plotting af versus L_{mid} revealed a broad dynamic range for the maximum polarization solution spanning at least four orders of magnitude for higher values of h (Fig. 2F). At larger values of L_{mid}, polarization decreased but was still substantial for h = 4 and h = 8. The decrease was caused by the increased contribution of the input-dependent Hill term at all positions both front and back. In summary, one potential role of positive feedback in biological systems is to increase the amplification and dynamic range of gradient-induced polarization.

2.2.3 MULTIPLE STEADY-STATES ARISE FROM POSITIVE FEEDBACK

Given that there are multiple steady-states, how can one describe all such solutions? Simulations identify a subset of solutions one by one; analytic methods are needed to determine the range of possible solutions. At steady-state the time derivatives (left-hand side) of the differential Equations (1.1) and (1.2) in Model 1 go to 0, and then one can solve the resulting algebraic equations for a: $0 = f(a, b, z)$. The solution must also satisfy the integral constraint imposed by integral control: $0 = \bar{a}(b) - a_{ss}$. Because multiple values of b may satisfy the constraint, one scans for feasible b, b_s, and then solves for the roots of the polynomial equation $f(a, b_s, z) = 0$.

For didactic purposes, we explored a version of the model in which we let $\gamma = \gamma'(1/(1+(\beta u)^{-q}))$, $\gamma' = 1$ (see Section 2.5 for further description); the essential results did not depend on the particular model. For $k_1 = 0$, there was a single solution, and we obtained an expression in which a is a function of the input-dependent Hill cooperativity term. For $h = 1$ ($k_1 = 10$ s^{-1}), only one value of b satisfied the integral constraint, and the resulting quadratic equation in a possessed only one positive root. Thus, there was at most a single steady-state, which is shown in Fig. 3. For $h = 2$, there were multiple feasible values of b resulting in a family of root curves. The resulting polynomials were cubic, and depending on the parameter values, there could be one or three real roots, which could be stable or unstable. In Figure 3 ($h = 2$) for a given bs, we observed a lower stable root and an upper stable root and an overlapping region containing two stable roots and one unstable root. One forms a solution by connecting the stable points along the x-axis in a manner that satisfies the integral constraint, crossing between the lower and upper root curves in the overlapping region (blue lines). There were multiple solutions for each root curve given that one can cross between the lower and upper roots multiple times, but typically we were most interested in the solution with the highest polarization value, which is what is drawn in blue. The envelope of solutions represents the highest polarized solutions for each feasible b, and thus does not represent all possible solutions.

For values of h greater than 2, we solved for the roots numerically using MATLAB. As h increased, the plots became more curvy and "S"-

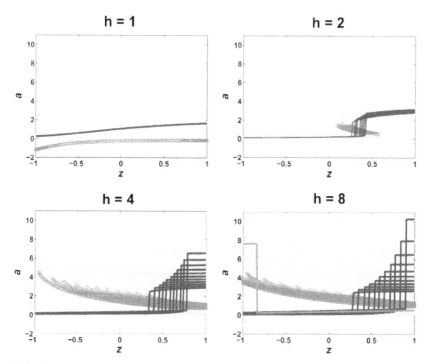

FIGURE 3: Root curves of steady-state equations define multiple steady-state solutions. The root curves displaying the steady-state solutions of one model for increasing values of h. Each curve represents the roots for a particular value of b that satisfies the integral constraint; both stable roots (green circles) and unstable roots (red circles) are present. The highest polarized solution for each root curve is traced in blue. For h = 8, a reversed polarization solution is shown in magenta, which arises from a "three-tier" root curve that is not contiguous within the dimensions of the cell.

shaped with a broader overlap region, and larger upper stable values. In addition, the range of feasible values of b increased resulting in more solutions and a broader envelope of polarized solutions. For h = 8, the overlap region of some root curves spanned the entire length of the cell. We termed such root curves "three-tier" because the root curve was no longer contiguous within the boundaries of the cell, resulting in three separate segments, the upper and lower stable solutions and the middle unstable solution. Such "three-tier" root curves allowed for reversed polarization solutions in which the intracellular component was concentrated at the wrong end of the cell where the ligand concentration was lower (Fig. 3, h = 8); such a situation may arise from flipping the gradient.

We examined these root plots as we varied other parameters. In general, increasing the contribution of the positive feedback to polarization (e.g. increasing k_1, decreasing L_{slope}, etc.) resulted in more "S" shaped root curves, a broader envelope of possible solutions, and greater maximum polarization. Thus, a more comprehensive picture of the spatial dynamics of the model emerges from the steady-state analysis, which highlights potential tradeoffs.

2.2.4 TRADEOFF BETWEEN TRACKING AND AMPLIFICATION

When the gain (strength) of the positive feedback was high, amplification was substantial when considering the most polarized solution. However, what happens when the gradient is flipped? Biologically, the source of a gradient (e.g. yeast mating partner) may be moving with respect to the sensing cell. We tested the ability of the model to track a 180° change in the gradient direction for different parameter values. In this section, we used simulations to select a single steady-state solution instead of using analytic methods to define all possible solutions. In the case of the pure cooperativity model with no positive feedback ($k_1 = 0$), tracking was perfect; the polarized distribution of a always aligned with the gradient (Fig. 4A, $k_1 = 0$). Adding positive feedback by increasing k_1 improved polarization, but at the cost of tracking. With $k_1 = 10$ s^{-1} and h = 4, flipping the gradient resulted in polarization that partially tracked the change, and when h = 8, the polarized species became stuck in the initial direction and did not track the 180° change in direction at all (Fig. 4A). Thus, there was a tradeoff between amplification and tracking.

A simple explanation is that tracking was impaired because of the presence of multi-stability (multiple steady-states) arising from the positive feedback, including steady-states in which the polarization was not correctly aligned in the same direction as the gradient. Cooperativity alone has a single-steady-state and hence can track perfectly, but without positive feedback, amplification is limited in terms of the magnitude and dynamic range. For moderate levels of positive feedback the polarization can be greater, but tracking was compromised because of the existence of partially polarized solutions that can be reached when the gradient

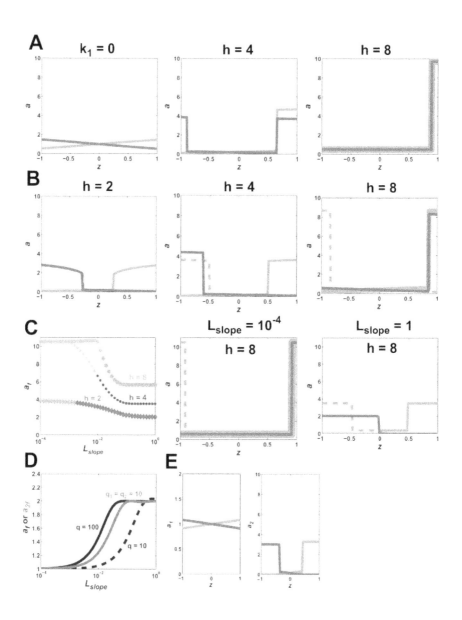

FIGURE 4: Tracking in standard positive feedback (SF), feedforward/feedback (FF), and multi-stage (MS) models. Simulations (not analytical solutions) determined the polarization in plots in which the gradient direction of the source was switched from the

right to the left (L_{mid} = 1, L_{slope} = 0.01 μm^{-1}). The forward gradient polarization solution is drawn in green and the reverse gradient solution in red. (A) In the standard feedback (SF) model, the no positive feedback case (k_1 = 0) is to the left. For (k_1 = 10 s^{-1}, h = 4 or 8), the SF model cannot track the directional change. (B) In the feedforward/feedback (FF) model (k_1 = 10 s^{-1}, q = 100), we observed better tracking but at the expense of the polarization. The dashed green line represents the mirror-image of the forward gradient polarization. (C) The maximum polarization solutions for the FF model as the gradient slope was varied for h = 2, 4, 8. For each h, at lower slope values, there was a transition denoted by the lighter shading to the presence of "three-tier" roots and higher polarization but reduced tracking. For h = 8, the polarization at the shallower slope (L_{slope} = 10^{-4} μm^{-1}) was greater than at the steeper slope (L_{slope} = 1 μm^{-1}), but some tracking was possible only at the steeper slope. (D) The polarization solution for the single-stage (Model 1) model with only cooperativity (k1 = 0, q = 10 or 100) is redrawn in black (solid and dashed, respectively). The multi-stage (MS) Model 2 with only cooperativity is drawn for q_1 = q_2 = 10 (red line). (E) Gradient directional switch in the MS model. There was amplification of the input gradient (L_{slope} = 10^{-3} μm^{-1}) to a steeper gradient of a_1 (h_1 = 1), and as a result, we observed tracking even after the more substantial amplification in the second stage (h_2 = 8).

direction was switched. For high levels of positive feedback, there was infinite amplification (spontaneous polarization), but also solutions in which the polarization was reversed with respect to the gradient.

Intuitively, the positive feedback overwhelms any dependence on the current input. As a result, hysteresis can arise in which the polarization depends on the past history of inputs to the cell, as well as the current input, so that tracking is impaired. Thus, positive feedback increases amplification, but also results in multi-stability and the loss of tracking.

2.2.5 FEEDFORWARD/FEEDBACK COINCIDENCE DETECTION IN POSITIVE FEEDBACK LOOP IMPROVES TRACKING

It would be desirable to obtain a compromise between the potent amplification obtained from high-gain positive feedback with the perfect tracking obtained from pure cooperativity in order to achieve good tracking and polarization under a range of environmental conditions. We developed a modified version of the model that could better balance amplification and tracking. We adjusted the positive feedback term to include a dependence on the input u: ,

$$\gamma = \gamma' \left(\frac{1}{1 + (\beta u)^{-q'}} \right) \tag{4}$$

$\gamma' = 1$ and $q' = q$. One can interpret this modification as a type of feed-forward/feedback coincidence detection [23] in the positive feedback loop. The result is that the positive feedback term has a dependence on both a and u (Fig. 1C). The input-dependence of the positive feedback is modulated by the cooperativity parameter q' in the Hill term. Thus, the feedback amplification of a has a feedforward component from u and a feedback component from a, and these must coincide to obtain the best amplification. Biologically, one can implement such a mechanism by the convergence of two signaling pathways, one of which is part of a positive feedback loop.

We tested the feedforward/feedback (FF) model by switching the gradient 180° using the default input values of $L_{mid} = 1$ and $L_{slope} = 0.01 \ \mu m^{-1}$. As before, increasing h resulted in better polarization, but decreased tracking. Compared to the standard positive feedback model (SF), however, the FF model displayed better tracking, but reduced polarization (Fig. 4B). For $h = 2$, the tracking was nearly perfect whereas in the standard model tracking was impaired for $h = 2$ (data not shown). For $h = 4$, again there was better tracking than the comparable SF model although the fact that the forward and reverse gradient solutions were not the same indicates multi-stability, which was associated with some loss of tracking. For $h = 8$, there was no tracking as was observed with the SF model.

Like the SF model, the FF model showed constant polarization and hence infinite amplification (spontaneous polarization) as the gradient slope approached 0. Interestingly, at steeper slopes the polarization actually decreased. We hypothesized that at the higher slopes there was stronger input-dependence of the polarization and hence reduced amplification, but better tracking. To check this possibility we examined the results from a 180° directional change for a small gradient slope and for a large gradient slope. As expected, when $L_{slope} = 0.0001 \ \mu m^{-1}$, there was no tracking but good polarization, whereas when L_{slope} = 1 µm^{-1}, there was some tracking but reduced polarization (Fig. 4C). In this figure, we also indicated the transition to the appearance of "three-tier" root curves described previously that can give rise to reversed polarization solutions. This transition occurred as Lslope decreased and the polarization jumped to a higher value. Thus, decreasing the input-dependence of the positive feedback by reducing the gradient slope, results in an increase in polarization but a loss in tracking.

2.2.6 MULTI-STAGE AMPLIFICATION CAN IMPROVE AMPLIFICATION OR TRACKING

In Model 1, the amplification resulting in polarization is achieved through the dynamics (i.e. positive feedback and cooperativity) of one species. We explored a model containing two polarized species in a cascade resulting in two amplification stages. The first polarized species a_1 serves as the input to the second stage which gives rise to the polarization of the second species a_2. We essentially duplicated Model 1 to form Model 2:

$$\frac{\delta a_1}{dt} = D_s \nabla_s^2 a_1 + \frac{k_0}{1 + (\beta u)^{-q_1}} + \frac{k_1}{1 + (\gamma a_1)^{-h_1}} - k_2 a_1 - k_3 b_1 a_1 - k_5 \widehat{a_1}$$

$$\text{(M2.1)}$$

$$\frac{db_1}{dt} = k_4 \hat{a}_1 b_1 \qquad\qquad \text{(M2.2)}$$

$$\frac{\delta a_2}{dt} = D_s \nabla_s^2 a_2 + \frac{k_0}{1 + (\beta a_1)^{-q_2}} + \frac{k_1}{1 + (\gamma a_2)^{-2}} - k_2 a_2 - k_3 b_2 a_2 - k_5 \widehat{a_2}$$

$$\text{(M2.3)}$$

$$\frac{db_2}{dt} = k_4 \widehat{a_2} b_2$$

$$\widehat{a_\iota} = \bar{a}_i - a_{iss}$$

$$\bar{a}_i = \frac{\int_s a_i ds}{\int_s ds}$$

$$u = L_{mid} + L_{slope}(z - z_0)$$

(M2.4)

One advantage of two stages is that amplification can be combined to achieve a larger net amplification. For example, a single reaction may produce a limited amount of cooperativity. Cascading two cooperative reactions together can result in a higher total cooperativity. In Model 2, we let $k_1 = 0$ so that there was no positive feedback. In Figure 4D we redrew the amplification of Model 1 using $q = 10$ or $q = 100$. It may be difficult for a single reaction to produce a Hill cooperativity of 100. However, when we cascaded two reactions with $q_1 = q_2 = 10$, then the final cooperativity approached that of 100. Indeed, it is common practice in engineering to link together amplifiers to attain greater amplification [24].

A second advantage for two stages is better tracking. The first stage can amplify the external gradient so that the input to the second stage is steeper than the original input. In Figure 4B, we observed that for $h = 8$ and $L_{slope} = 0.01$ μm^{-1}, there was no tracking (but excellent polarization) whereas at $L_{slope} = 0.1$ μm^{-1}, the polarization was reduced but the tracking was better (data not shown). We constructed a multi-stage model in which the first-stage amplification was approximately 100 (AF~100) so that the initial ligand slope $L_{slope} = 0.001$ μm^{-1} was transformed into a slope of a_1 that was approximately 0.1 μm^{-1}. In the second stage ($h = 8$), there was some tracking of the directional change by the polarized species a_2 because of the steeper input gradient of a_1 (Fig. 4E).

A third advantage, which is not investigated here, is that having two stages can produce a broader dynamic range. The input to the first stage produces a normalized input to the second stage. Thus, the negative feed-back in the first stage effectively shrinks the dynamic range being fed into the second stage. From a biological standpoint, one can propose that the cascaded arrangement of the heterotrimeric and the Cdc42p G-protein cycles results in multiple amplification stages, and we exploited this concept in our model of yeast cell polarization.

2.2.7 POLARIZATION AND TRACKING IN 2D SIMULATIONS

To this point, the simulations and analyses employed an axisymmetric 1D geometry so that it was only possible to change the direction of the gradient 180°. A greater challenge would be responding to more subtle directional changes. To this end, we constructed a two-dimensional (2D) model of the cell, which was represented as a circle, so that the gradient could be applied in any direction on the circle. Using this new model, cells were polarized with an initial gradient and then the direction of the gradient was changed. After the shift, we compared the direction of the polarization with the direction of the gradient as a measure of tracking.

Model 1 with only cooperativity ($k_1 = 0$) displayed perfect tracking for directional changes of 180°, 90°, and 45° as expected since there is a single solution (data not shown). For the feedforward/feedback model described above (h = 4, q = 100) we observed tracking of the 180° degree directional change (although a different polarization solution), but the polarization was not aligned with the gradient for the 90° and 45° changes (Fig. 5A). Thus, the 2D simulations offer a more stringent test of tracking, and more accurately reflects the conditions of a cell confronted with a shifting gradient.

2.2.8 SURFACE DIFFUSION IN MEMBRANE REDUCES NUMBER OF STEADY-STATES

Diffusion can exert a profound effect on spatial dynamics [4]. The small size of the yeast cell and the fast rate of diffusion in the cytoplasm for a freely diffusible protein caused us to focus on lateral surface diffusion in

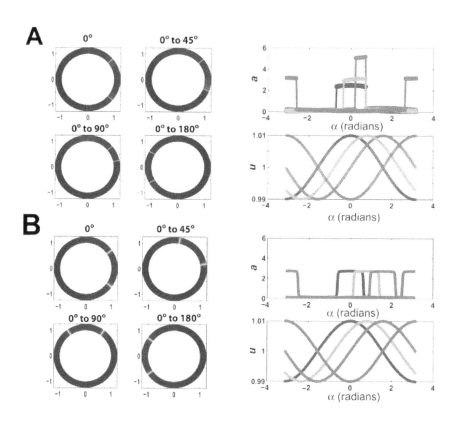

FIGURE 5: Tracking changes in gradient direction using simulations of two-dimensional polarization model. The gradient direction (L_{mid} = 1, L_{slope} = 0.01 μm⁻¹) was initially at 0° (3 o'clock position), and then shifted 45°, 90° or 180° in the counterclockwise direction. The response of the cell was depicted either on a disk (left figures) in which the value of a is color-coded or in a perimeter plot in which x-axis describes the radial position and the y-axis the value of a or the input u. In the perimeter plot, there is a curve for each new gradient direction. (A) Results of gradient directional change in FF model (h = 4) without diffusion. (B) Results of gradient directional change in FF model (h = 4) with diffusion (D_s = 0.001 μm²/s).

the membrane. Proteins in the membrane are able to diffuse laterally in the plane of the membrane [25]. One would expect surface diffusion would dampen cell polarization by allowing proteins to diffuse away from sharp concentration peaks. However, what would happen to the multiple steady-states in the presence of lateral diffusion? Would the envelope of solutions become less polarized or would the envelope be modified in some way?

Introducing diffusion caused the envelope of solutions to collapse to a single solution. In Figure 6A, we overlay the single solution with D_s = 0.001 $\mu m^2/s$ among the envelope of steady-state solutions with D_s = 0; the diffusion solution is positioned toward the rear of the envelope. It is important to note that the presence of diffusion prevents analytic solutions to the model. Instead, we employed exhaustive simulations from a wide variety of initial conditions to identify any stable steady-states, but simulations cannot guarantee that we have found all solutions.

However, the response of the FF model with diffusion to changes in gradient direction also argues for a single steady-state. When D_s = 0, a 180° change in direction resulted in a polarized solution different from the initial polarization. When D_s = 0.001 $\mu m^2/s$, the initial and final polarization were identical (Fig. 6B). A more stringent test was with the 2D models. In the presence of diffusion, the cell could accurately track directional changes in the gradient of 90° and 45° (Fig. 5B). Thus, in biological systems, lateral diffusion in the membrane may play an important role in preventing multi-stability during polarization.

Polarization in the presence of diffusion for the FF model increased when the gradient slope decreased (Fig. 6D) just as it did in the absence of diffusion (Fig. 4C). Again, we interpret this result in terms of the input-dependence of the polarization. Furthermore, polarization maintained a constant value even as $L_{slope} \rightarrow 0$ indicative of the infinite amplification (spontaneous polarization) that was observed in the model without diffusion. For the h = 8 case, decreasing the slope led to two additional solutions; one is the unpolarized solution with b = 0 and the other is a reversed polarization solution (Fig. 7B). Thus, at high levels of positive feedback gain, multiple steady-state solutions could arise with diffusion, but there was no envelope of solutions. Varying L_{mid} showed a peak at L_{mid} = 1 (Fig. 6E). When L_{mid}>1, there was a modestly polarized solution (af~2). For

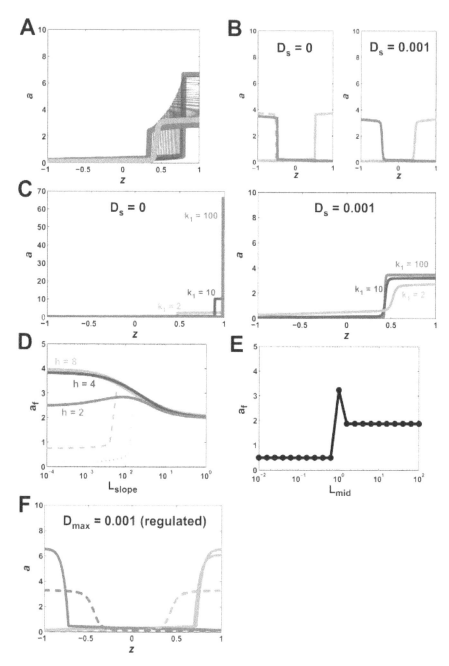

FIGURE 6: Lateral surface diffusion enhances the robustness of polarization. In the gradient, $L_{mid} = 1$ and $L_{slope} = 0.1 \ \mu m^{-1}$. (A) Surface diffusion selects a single solution

among multiple steady-states. The polarization envelope is shown for the FF model (k_1 = 10 s^{-1}, h = 8) with no diffusion. The single steady-state solution in the presence of diffusion (D_s = 0.001 µm^2/s). (B) Diffusion improves tracking. A 180° directional change is shown with and without diffusion. The overlap between the dashed lines in the presence of diffusion suggests a single steady-state solution and perfect tracking. (C) Presence of diffusion ensures that polarization is robust to variations in the positive feedback. When D_s = 0, increasing k_1 results in a dramatic increase in the maximum polarization solution. In the presence of diffusion there is almost no change in polarization for larger k_1 (note reduced scale of y-axis in right graph). (D) Polarization as a function of L_{slope} (L_{mid} = 1) in the FF model with diffusion (Ds = 0.001 µm^2/s). For h = 8, there were two additional solutions for smaller L_{slope} values: an unpolarized b = 0 solution and a reversed polarized solution. (E) Polarization as a function of L_{mid}, L_{slope} = (0.01×L_{mid}) µm^{-1}, in the FF model with diffusion (h = 8). (F) Regulating diffusion can produce stronger polarization. Using the regulated diffusion term described in the text, enhanced polarization seen compared to constant diffusion (dashed lines). There were two forward polarization solutions, and both gave the same reversed solution when the gradient was flipped.

L_{mid}<1, there was not sufficient activation to achieve polarization. The degree of polarization was much more modest when compared to the no diffusion case, but the dynamic range was still broad.

2.2.9 SURFACE DIFFUSION LIMITS EXTENT OF POLARIZATION IN A ROBUST FASHION

In the absence of surface diffusion, increasing the positive feedback gain increased the maximum possible polarization among the multiple steady-state solutions. Introducing membrane diffusion prevented the more extreme polarization states from being reached and reduced the number of steady-states. Indeed, one would expect diffusion to counteract the positive feedback concentrating components at the front.

Surprisingly, the presence of surface diffusion also caused the degree of polarization to become robust to changes in the gain of the positive feedback. In the D_s = 0 case, as we increased k_1, we dramatically increased the maximum polarized solution (Fig. 6C). For Ds = 0.001 µm^2/s, increasing k_1 had only a modest effect on the maximum polarized solution. Diffusion

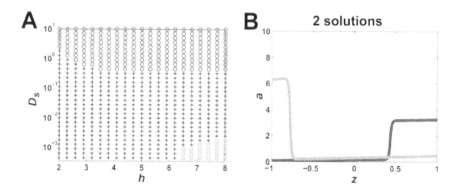

FIGURE 7: Bifurcation diagram of model behavior as a function of the parameters D_s (diffusion) and h (positive feedback). (A) This plot was constructed by exhaustive simulation of the FF model ($k_1 = 10$ s^{-1}; $L_{mid} = 1$, $L_{slope} = 0.01$ μm^{-1}) over a range of parameter pair values (Ds, h). Three behavior classes were observed. (1) One steady-state solution (+). (2) Two solutions (square). (3) Limit-cycle oscillations (circle). (B) An example of the two solutions class in which there is a forward (darker) and a reversed (lighter) polarization solution to a right-to-left gradient.

pushed the polarization back toward the least polarized solution in the envelope of steady-state solutions that exist in the absence of diffusion.

More generally, we found that when $D_s > 0$ the extent of polarization became relatively insensitive to changes in a wide range of internal and external parameters (e.g. L_{slope}, L_{mid}, k_1, h, etc.). Thus, surface diffusion adds robustness to polarization. From a biological standpoint, it may be beneficial to cells to have consistent polarization under different conditions. For example, yeast cells may not want the width of the mating projections to be too sensitive to variations in the concentration or gradient slope of mating pheromone.

2.2.10 REGULATING DIFFUSION ENHANCES POLARIZATION

The presence of lateral diffusion in the model, prevents the appearance of highly polarized states in Model 1. Yet, there are circumstances when a cell will want a particular protein localized to a narrow region at the front

[3]. One possibility is to regulate the diffusion coefficient in some manner. We postulated that the diffusion coefficient could depend on a and developed the following functional form:

$$D_S = \frac{D_{max}}{1 + (\gamma a)^m} \tag{5}$$

where

$$\gamma = \gamma' \left(\frac{1}{1 + (\beta u)^{-q'}} \right) \tag{6}$$

and we let $q = q' = 100$ and $m = 8$. This term effectively creates a diffusion barrier so that positions in the cell where a is high (front), $D_s \to D_{max}$, whereas at positions where a is small $D_s \to 0$.

For $D_{max} = 0.001$ $\mu m^2/s$, we examined polarization in the high positive feedback case ($k_1 = 10$ s^{-1}, $h = 8$). Compared to unregulated diffusion, we observed more pronounced polarization which peaked at af~9. There was also the appearance of more than one steady-state, but the number of steady-states was smaller than in the $D_s = 0$ case; only two steady-states were found from extensive simulations (Fig. 6F). Consistent with the fewer solutions, tracking of a 180° change in gradient direction was good. Thus, regulating diffusion balances polarization with tracking, and biologically this additional level of modulation could be important in optimizing the polarization response.

2.2.11 INCREASING DIFFUSION PROMOTES OSCILLATIONS

Increasing the positive feedback gain results in multi-stability and extremely polarized solutions. Increasing diffusion reduces multi-stability and polarization. What happens with high levels of both positive feedback and diffusion?

Using simulations, we constructed a bifurcation diagram summarizing the dynamical behaviors for different values of h and D_s in a version of the FF model possessing integral but not proportional negative feedback (i.e. $k_s = 0$). In Figure 7A, we explored values of D_s from 10^{-3} to $10 \ \mu m^2/s$ and values of h from 2 to 8. For lower values of h and D_s, we observed a single steady-state solution. Increasing h with small D_s resulted in two steady-state solutions; interestingly, the second solution was a reversed solution in which the polarization was at the rear of the cell relative to the gradient (Fig. 7B). Increasing D_s and to a lesser extent increasing h produced limit-cycle oscillations. Thus, there is a danger of instability for biological systems when diffusion and positive feedback are too high. It is important to emphasize that these results were derived from exhaustive simulations, and thus we cannot exclude the possibility that additional solutions exist.

2.2.12 CONSTRUCTING A NEW MODEL OF YEAST CELL POLARIZATION INDUCED BY MATING PHEROMONE GRADIENTS

This research was motivated by an interest in yeast cell polarization, and one of the primary goals was to apply the insights gained from the generic models to models more specific to yeast. Our past efforts modeling yeast cell polarization [19] were hampered by the complexity of the model. The work described above helped us to understand the model behavior and make improvements. This model was based on the spatial dynamics of the heterotrimeric and Cdc42p G-protein cycles. Receptor (R) binds ligand (L) and becomes activated (RL). Activated receptor converts heterotrimeric G-protein (G) into activated α-subunit (Ga) and free Gβγ (Gbg). All of these species are on the membrane. The connection between the two cycles is the fact that free Gβγ recruits cytoplasmic Cdc24p to the membrane. Membrane-bound Cdc24p (C24m) activates Cdc42p. Activated Cdc42p (C42a) recruits the scaffold protein Bem1p (B1) to the membrane. Finally, a positive feedback loop is created because membrane-bound Bem1p can bind and recruit Cdc24p to the membrane. All components residing on the membrane were subject to the same lateral diffusion. It is important to note that the model lacks an explicit consideration of ligand-stimulated endocytosis and polarized synthesis which are known to be crucial for many aspects of cell polarity [26], [27] and are the subjects of future research. For this work, we focused on the fast positive feedback loop mediated by Bem1p [28].

The connection between the yeast model and the generic model (Model 1) is best seen in equation describing the dynamics of membrane-bound, active Cdc24p (C24m, Eq. (3.5)). There, recruitment of Cdc24p to the membrane depends on a cooperative term that is a function of Gβγ, $(k_{24cm0}(Gbg_n*)[C24c])$, and a positive feedback term, $(k_{24cm1}(B1*)[C24c])$, that depends on Bem1p which in turn is a function of active Cdc42p and hence active Cdc24p. We made two important modifications to the previous model. First, we added a negative feedback loop for better regulation. The loop includes the protein kinase Cla4p which is activated by Cdc42p and which phosphorylates and inhibits Cdc24p resulting in negative feedback [29]. Second, there is a feedforward/feedback coincidence detection term in the positive feedback loop for better tracking that involves Gβγ. The input to the model was a gradient of the mating pheromone alpha-factor; the output was active Cdc42p ([C42a]).

The first four equations (M3.1 to M3.4) describe the spatial dynamics of the heterotrimeric G-protein cycle, and the next five equations (M3.5 to M3.9) describe the spatial dynamics of the Cdc42p G-protein cycle. The two-stage structure of the model was important for extending its dynamic range.

$$\frac{\delta[R]}{\delta t} = D_s \nabla_s^2 [R] - k_{RL}[L][R] + k_{RLm}[RL] - k_{Rd0}[R] + k_{Rs}$$
(M3.1)

$$\frac{\delta[RL]}{\delta t} = D_s \nabla_s^2 [RL] - k_{RL}[L][R] + k_{RLm}[RL] - k_{Rd1}[RL]$$
(M3.2)

$$\frac{\delta[G]}{\delta t} = D_s \nabla_s^2 [G] - k_{Ga}[RL][G] + k_{G1}[Gd][Gbg]$$
(M3.3)

$$\frac{\delta[Ga]}{\delta t} = D_s \nabla_s^2 [Ga] - k_{Ga}[RL][G] - k_{Gd}[Ga] \tag{M3.4}$$

$$\frac{\delta[C24m]}{\delta t} = D_s \nabla_s^2 [C24m] + k_{24cm0}(Gbg_n^*)[C24c] + k_{24cm1}(B1^*)[C24c]$$
$$- k_{24mc}[C24m] - k_{24d}[Cla4a][C24m] \tag{M3.5}$$

$$\frac{\delta[C42]}{\delta t} = D_s \nabla_s^2 [C42] - k_{42a}[C24m][C42] + k_{42d}[C42a] \tag{M3.6}$$

$$\frac{\delta[C42a]}{\delta t} = D_s \nabla_s^2 [C42a] + k_{42a}[C24m][C42] - k_{42d}[C42a] \tag{M3.7}$$

$$\frac{\delta[B1m]}{\delta t} = D_s \nabla_s^2 [B1m] + k_{B1cm}[C42a][B1c] - k_{B1mc}[B1m] \tag{M3.8}$$

$$\frac{\delta[Cla4a]}{\delta t} = k_{Cla4a}[C42a_t^*] - k_{Cla4d}[Cla4a] \tag{M3.9}$$

A more detailed description of the model is in the Appendix S1 (Supporting Information).

2.2.13 MODELING THE EFFECT OF SURFACE DIFFUSION ON THE ROBUSTNESS OF YEAST CELL POLARIZATION

Using the yeast model, we wished to estimate the range of surface diffusion coefficients that would permit robust polarization in yeast cells in response to mating pheromone. When $D_s = 0$, there was good polarization, but also multi-stability, which was manifested when we reversed the gradient and an alternative polarized solution appeared that was not identical to the initial polarization (Fig. 8). Adding lateral diffusion with $D_s = 0.001$ $\mu m^2/s$ resulted in a single steady-state with perfect tracking for the input conditions. Increasing D_s ten-fold ($D_s = 0.01$ $\mu m^2/s$) maintained a comparable level of polarization, although the shape of the distribution was altered. For $D_s = 0.1$ $\mu m^2/s$, polarization was abolished. Thus, in this model, the highest range of D_s that produced good polarization was between 0.01 and 0.001 $\mu m^2/s$. In a previous section, we demonstrated that larger values of diffusion were associated with better tracking and a reduced likelihood of multi-stability. In yeast, the measured value of D_s was 0.0025 $\mu m^2/s$ [30], and so our simulations suggest that membrane fluidity in yeast has been tuned for robust polarization.

We also examined the dynamic range and sensitivity to shallow gradients of the yeast model with Ds = 0.001 $\mu m^2/s$. Interestingly, the yeast model displayed similar qualitative behavior to the generic FF model with diffusion (compare Figs. 6 and 8). For h = 4 and h = 8, polarization in the yeast model was observed even at relative slopes less than 0.01 μm^{-1} (Fig. 8B). For h = 8, there was an additional steady-state solution in which the cell was unpolarized (Fig. 8B, dashed line). The dynamic range of gradient-sensing and polarization was excellent in the yeast model for h = 4 and h = 8, extending beyond L_{mid} = 1000 nM (Fig. 8C). Polarization was greater than in the generic model in part because of the two-stage (two-cycle) structure of the yeast model. For h = 2, polarization declined at higher values of L_{mid}, and for h = 8, there was an additional unpolarized solution; h = 4 represented a compromise. Finally, we ran 2D simulations of the yeast model (h = 4) which exhibited good tracking to changes in the gradient direction (data not shown) again resembling the generic FF model with diffusion. Thus, the yeast model could show robust performance with a balance of amplification, dynamic range, and tracking. It should be

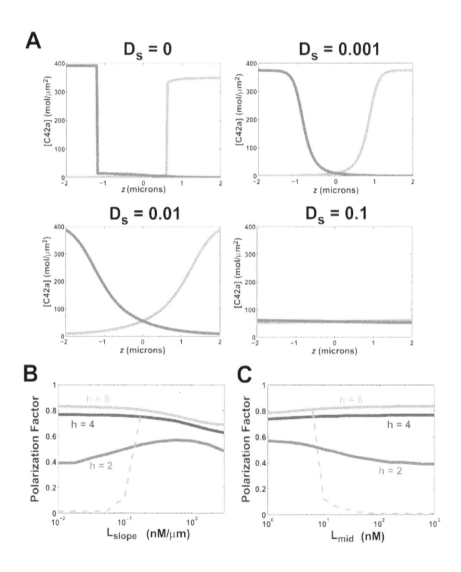

FIGURE 8: Simulations of spatial model of yeast cell polarization. (A) Effect of lateral surface diffusion on mating factor induced polarization in yeast cell model. The input was an alpha-factor gradient (L_{mid} = 10 nM and L_{slope} = 1 nM/μm) and the output was the steady-state concentration of active Cdc42p ([C42a]), which was plotted along the axial length

of the cell. The results were for different surface diffusion coefficients. (B) Polarization in yeast model as a function of L_{slope} (L_{mid} = 10 nM) in the yeast model with diffusion (D_s = 0.001 $\mu m^2/s$). Polarization is described in terms of the polarization factor (maximum polarization = 1; unpolarized = 0; PF = 0.8 corresponds to af~5; PF = 0.5 corresponds to af~2). Three values of h were examined: h = 2, h = 4, h = 8. For h = 8, there was an additional unpolarized solution (dashed line) for smaller L_{slope} values. (C) Polarization as a function of L_{mid}, L_{slope} = (0.01×L_{mid}) μm^{-1}, in the yeast model with diffusion for three values of h.

noted that recent experiments have demonstrated that yeast cells can sense pheromone gradients possessing relative slopes as shallow as 0.001 μm^{-1} and at concentrations as high as 1000 nM (T.I. Moore, C.S. Chou, Q. Nie, N. L. Jeon and T.-M. Yi, submitted), and so this modeling can help serve as a framework for future more realistic models that contain more detailed reaction mechanisms.

2.3 DISCUSSION

In this paper, we investigated the spatial dynamics of cell polarization induced by chemical gradients focusing on the tradeoff between amplification and tracking and on the impact of lateral surface diffusion on polarization. Previous work has noted this tradeoff, but we wished to explore its nature in greater depth by using a generic model and steady-state analysis. A highly cooperative response to the input resulted in good tracking of a moving signal source, but amplification to produce potent polarization was limited to a very narrow range of concentrations. Adding high-gain positive feedback resulted in strong amplification over a broad range of concentrations, but tracking was poor.

Intuitively, one can understand this tradeoff in terms of the input-dependence of the amplification. High input-dependence is necessary for tracking, but then weaker inputs (i.e. shallow gradients) will not be amplified as well. On the other hand, low input-dependence results in good amplification regardless of the input strength, but then tracking a directional change in the input becomes difficult (i.e. polarization becomes stuck).

An important technical tool was the application of steady-state analysis to the model. The positive feedback led to multiple steady-states, which we were able to describe by analytical solutions to the model equations. We could then see the connection between increased positive feedback, a larger envelope of steady-states, amplification that was not input-dependent, and the loss of tracking. With a single steady-state, tracking is perfect, whereas with multiple steady-states, there could exist solutions in which the polarization is not aligned in the same direction as the gradient.

Living systems evolve to find the appropriate balance for this tradeoff in a given environment. There must be sufficient amplification to induce the proper polarization for gradients typically encountered. Likewise, tracking is a significant consideration if the signal source is expected to move. In the context of the yeast mating response, there must be sufficient polarization to form a mating projection over a range of background pheromone concentrations, which may vary according to the number and proximity of mating partners, and at the same time, the ability to redirect the projection if the partner moves or mates with another cell. We constructed a modified model in which feedforward/feedback coincidence detection improved tracking with some loss in dynamic range. Tracking performance could be further improved using multi-stage amplification to split the amplification.

The presence of lateral surface diffusion significantly altered polarization behavior. First, at low diffusion coefficients, it collapsed the multiple solutions to fewer solutions, and in certain cases, to a single solution. As a result, tracking was improved, but the extent of polarization was reduced. When combined with the feedfoward/feedback coincidence detection, low levels of lateral diffusion produced perfect tracking over a range of input gradient conditions. A second effect of lateral diffusion was that the degree of polarization was quite robust to changes in the parameter values. It may be advantageous to cells that polarization is robust to variations in internal and external conditions. Third, high levels of diffusion coupled to high positive feedback resulted in oscillations. Together, these results argue that maintaining the proper level of diffusion in the membrane is critical for robust polarization. It is important to mention that there is a concern that some of these conclusions may depend on the particular model structure.

Although we attempted to formulate a "general" generic model structure, further research is needed to address this concern.

We took the lessons from the simple model and incorporated them into a more complex model of yeast polarization. In particular, we implemented feedforward/feedback coincidence detection via Gβγ influencing the Cdc24p-Cdc42p-Bem1p positive feedback loop, and also implemented negative feedback regulation of Cdc24p. The resulting model exhibited good polarization, gradient sensitivity, and dynamic range. In the future, we plan to improve the model by adding multi-stage amplification that takes advantage of polarized synthesis and endocytosis of the pathway components. In addition, we would like to add more mechanistic elements and evaluate the robustness of the models.

From this research, certain predictions and explanations arise. First, we expect the cellular polarization apparatus to contain elements that generate both cooperativity and positive feedback, and the amount of each depends on the appropriate amplification/tracking balance suitable for the cell in its natural environment. Second, we identify feedforward/feedback coincidence detection and multi-stage feedback as important strategies for improved tracking ability of cells. Third, we demonstrated that lateral surface diffusion contributes significantly to the robustness of polarization, and predict that this diffusion will be carefully regulated. Fourth, we used simulations of yeast cells to show that proper polarization was achieved using values of the diffusion coefficient between 0.01 $\mu m^2/s$ and 0.001 $\mu m^2/s$, and indeed Valdez-Taubas and Pelham [30] have measured a value of 0.0025 $\mu m^2/s$.

In the future, we will address additional robustness issues relating to cell polarization induced by spatial gradients. Foremost among these is handling the presence of noise. Stochastic noise arises from fluctuations in the gradient, Brownian motion of the cell, the random nature of the discrete binding events between ligand and receptor [31], etc. These stochastic variations must be distinguished from more meaningful changes in the gradient signal such as a directional change caused by the movement of the signal source. Separating signal from noise is a classic problem in engineering and requires some type of noise filtering [32]. In addition to external noise, there is internal noise arising from variations in the levels and functioning of system components. Regulatory systems must exist to ensure robust polarization in the presence of this internal uncertainty. Furthermore, it is important to investigate different control strategies for improving robustness. For example, an adaptive control strategy involving the self-tuning of key system parameters could make the system more robust to both internal and external variations. Finally, we would like to connect this research more closely to the biology of yeast cell gradient-sensing and polarization.

2.4 MATERIALS AND METHODS

2.4.1 SIMULATIONS

The surface diffusion of a quantity W on an axisymmetric surface in a three dimensional space has the following expression:

$$\nabla_s^2 W = \frac{\delta^2 W}{\delta s^2} + \frac{1}{r}\left(\frac{\delta W}{\delta s}\frac{\delta r}{\delta s}\right) \tag{7}$$

where

$$s = \sqrt{z^2(\alpha) + r^2(\alpha)}$$

is the arclength of the cell membrane. Consequently, the equations in Model 1 becomes one-dimensional in terms of the parameterization variable $\alpha\$$, even though the cell is a three dimensional axisymmetric ellipsoid.

For a system in the two-dimensional space, in which the cell surface is a curve, the expression of the surface diffusion of a quantity W becomes

$$\nabla_s^2 W = \frac{\delta^2 W}{\delta s^2} \tag{8}$$

where

$$s = \sqrt{z^2(\alpha) + r^2(\alpha)}$$

is the arclength of the cell membrane.

Numerical discretizations of each variable on the cell membrane were carried out in α for both cases. All spatial derivatives in the equations were approximated using a second-order finite difference discretization. The temporal discretization was carried out using a fourth order Adams-Moulton predicator-corrector method.

In a typical simulation, the number of grid points in space was 200 with a time-step of 5×10^{-4} s. We tested a range of grid and time-step sizes to assure convergence of the simulations. The simulations in this paper were well-resolved with the above discretization.

2.4.2 STEADY-STATE ANALYSIS

Without diffusion, the steady state equations 1.1 and 1.2 of Model 1 have a simple form,

$$k_0 1 + (\beta u)^{-q} + k_1 1 + (\gamma ap)^{-h} - (k_2 + k_3 b)a - k_5 \left(\int_s a \, ds \int_s 1 \, ds - a_{ss} \right) = 0 \tag{9}$$

$$k_4 \left(\int_s a \, ds \int_s 1 \, ds - a_{ss} \right) b = 0 \tag{10}$$

By rewriting Eq. (9) and eliminating the zero solution $b = 0$ in Eq. (10), the steady state system consists of a polynomial equation of a, with at most $h+1$ roots, and an equation for the integral control of a. The system was solved using the MATLAB polynomial solver 'ROOT'. We carried out linear stability analysis around each steady-state. We selected the stable steady-state solutions satisfying the integral control equation.

For the system with surface diffusion, a nonlinear Gauss-Seidel iteration procedure [33] was used for the simulations.

10.4.3 PERFORMING L_{SLOPE} AND L_{MID} SCANS IN MODELS CONTAINING DIFFUSION

We calculated the polarization as a function of L_{slope} and L_{mid} in the models containing diffusion by running a series of simulations. In the L_{slope} scan, we fixed L_{mid} and scanned through a series of L_{slope} values evenly distributed on a log scale. We first scanned from lower L_{slope} values to higher values. At each succeeding scan point, the initial values were taken as the steady-state computed at the previous scan point. The second scan started with the highest L_{slope} value and proceeded backwards. The L_{mid} scans were performed in an analogous manner.

REFERENCES

1. Thompson DW (1961) On Growth and Form. Cambridge, UK: Cambridge University Press.
2. Drubin DG, Nelson WJ (1996) Origins of cell polarity. Cell 84: 335–344. doi: 10.1016/S0092-8674(00)81278-7.
3. Pruyne D, Bretscher A (2000) Polarization of cell growth in yeast: I. Establishment and maintenance of polarity states. J Cell Sci 113: 365–375.
4. Turing AM (1952) The chemical basis of morphogenesis. Phil Trans Roy Soc Lond B237: 37–72.
5. Meinhardt H (1982) Models of biological pattern formation. London: Academic Press.
6. Iglesias PA, Levchenko A (2002) Modeling the cell's guidance system. Science's STKE 9: RE12. doi: 10.1126/stke.2002.148.re12.
7. Krishnan J, Iglesias PA (2004) Uncovering directional sensing: where are we headed? Syst Biol 1: 54–61. doi: 10.1049/sb:20045001.
8. Devreotes P, Janetopoulos C (2003) Eukaryotic chemotaxis: Distinctions between directional sensing and polarization. J Biol Chem 278: 20445–20448. doi: 10.1074/jbc.R300010200.
9. Meinhardt H (1999) Orientation of chemotactic cells and growth cones: Models and mechanisms. J Cell Sci 112: 2867–2874.
10. Dawes AT, Edelstein-Keshet L (2007) Phosphoinositides and Rho proteins spatially regulate actin polymerization to initiate and maintain directed movement in

a one-dimensional model of a motile cell. Biophys J 92: 744–768. doi: 10.1529/biophysj.106.090514.

11. Narang A (2006) Spontaneous polarization in eukaryotic gradient sensing: A mathematical model based on mutual inhibition of frontness and backness pathways. J Theor Biol 240: 538–553. doi: 10.1091/mbc.e04-12-1076.

12. Levchenko A, Iglesias PA (2002) Models of eukaryotic gradient sensing: Application to chemotaxis of amoebae and neutrophils. Biophys J 82: 50–63. doi: 10.1016/S0006-3495(02)75373-3.

13. Postma M, Haastert PJMV (2001) A diffusion-translocation model for gradient sensing by chemotactic cells. Biophys J 81: 1314–1323. doi: 10.1016/S0006-3495(01)75788-8.

14. Maly VI, Wiley HS, Lauffenburger DA (2004) Self-organization of polarized cell signaling via autocrine circuits: Computational model analysis. Biophys J 86: doi: 10.1016/S0006-3495(04)74079-5.

15. Haugh JM, Schneider IC (2004) Spatial analysis of 3′ phosphoinositide signaling in living fibroblasts: I. Uniform stimulation model and bounds on dimensionless groups. Biophys J 86: 589–598. doi: 10.1016/s0092-8674(00)81280-5.

16. Gamba A, Candia Ad, Talia SD, Coniglio A, Bussolino F, et al. (2005) Diffusion-limited phase separation in eukaryotic chemotaxis. Proc Natl Acad Sci USA 102: 16927–16932. doi: 10.1038/362167a0.

17. Skupsky R, Losert W, Nossal RJ (2005) Distinguishing modes of eukaryotic gradient sensing. Biophys J 89: 2806–2823. doi: 10.1529/biophysj.105.061564.

18. Maree AFM, Jilkine A, Dawes AT, Grieneisen VA, Edelstein-Keshet L (2006) Polarization and movement of keratocytes: A multiscale modeling approach. Bull Math Biol 68: 1169–1211. doi: 10.1007/s11538-006-9131-7.

19. Yi T-M, Chen S, Chou C-S, Nie Q (2007) Modeling yeast cell polarization induced by pheromone gradients. J Stat Phys 128: 193–207. doi: 10.1007/s10955-007-9285-1.

20. Yi T-M, Andrews BW, Iglesias PA (2007) Control analysis of bacterial chemotaxis signaling. Methods in Enzymology 422: 123–140.

21. Butty AC, Perrinjaquet N, Petit A, Jaquenoud M, Segall JE, et al. (2002) A positive feedback loop stabilizes the guanine-nucleotide exchange factor Cdc24 at sites of polarization. Embo J 21: 1565–1576.

22. Angeli D, Ferrell JE Jr, Sontag ED (2004) Detection of multistability, bifurcations, and hysteresis in a large class of biological positive-feedback systems. Proc Natl Acad Sci U S A 101: 1822–1827. doi: 10.1038/nrm1128.

23. Onsum M, Rao CV (2007) A mathematical model for neutrophil gradient sensing and polarization. PLoS Comp Biol 3: e36. doi: 10.1371/journal.pcbi.0030036.

24. Horowitz P, Hill W (1989) The Art of Electronics. Cambridge, UK: Cambridge University Press.

25. Jacobson K, Ishihara A, Inman R (1987) Lateral diffusion of proteins in membranes. Ann Rev Physiol 49: 163–175. doi: 10.1146/annurev.ph.49.030187.001115.

26. Marco E, Wedlich-Soldner R, Li R, Altschuler SJ, Wu LF (2007) Endocytosis optimizes the dynamic localization of membrane proteins that regulate cortical polarity. Cell 129: 411–422. doi: 10.1152/ajpcell.00457.2006.

27. Wedlich-Soldner R, Altschuler S, Wu L, Li R (2003) Spontaneous cell polarization through actomyosin-based delivery of the Cdc42 GTPase. Science 299: 1231–1235. doi: 10.1109/tsmc.1979.4310076.

28. Brandman O, Ferrell JE Jr, Li R, Meyer T (2005) Interlinked fast and slow positive feedback loops drive reliable cell decisions. Science 310: 496–498. doi: 10.1002/jemt.20069.

29. Gulli M-P, Jaquenoud M, Shimada Y, Niederhauser G, Wiget P, et al. (2000) Phosphorylation of the Cdc42 exchange factor Cdc24 by the PAK-like kinase Cla4 may regulate polarized growth in yeast. Mol Cell 6: 1155–1167. doi: 10.1016/S1097-2765(00)00113-1.

30. Valdez-Taubas J, Pelham HRB (2003) Slow diffusion of proteins in the yeast plasma membrane allows polarity to be maintained by endocytic cycling. Curr Biol 13: 1636–1640. doi: 10.1016/j.cub.2003.09.001.

31. Lauffenburger DA, Linderman JJ (1993) Receptors: Models for Binding, Trafficking, and Signaling. New York: Oxford University Press.

32. Andrews BW, Yi T-M, Iglesias PA (2006) Optimal noise filtering in the chemotactic response of E. coli. PLoS Comp Biol 2: doi: 10.1371/journal.pcbi.0020154.eor.

33. Briggs WL, Henson VE, McCormick SF (2000) A Multigrid Tutorial. Philadelphia: SIAM.

34. Leeuw T, Fourest-Lieuvin A, Wu C, Chenevert J, Clark K, et al. (1995) Pheromone response in yeast: Association of Bem1p with proteins of the MAP kinase cascade and actin. Science 270: 1210–1213. doi: 10.1126/science.270.5239.1210.

This article has supplemental information that is not featured in this version of the text. To view these files, please visit the original version of the article as cited in the beginning of this chapter.

CHAPTER 3

MULTISITE PHOSPHORYLATION OF THE GUANINE NUCLEOTIDE EXCHANGE FACTOR CDC24 DURING YEAST CELL POLARIZATION

STEPHANIE C. WAI, SCOTT A. GERBER, AND RONG LI

3.1 INTRODUCTION

Cell polarization is the process by which cells establish asymmetry along a single axis and is essential for processes such as cell migration and asymmetric cell division [1]. Budding yeast is an excellent model for the study of cell polarity, because it divides asymmetrically between the mother and bud and displays a characteristic cell and actin morphology at each cell cycle stage, facilitating the study of different polarity states [2]. Also, many proteins involved in cell polarity, such as the Rho GTPase Cdc42, are conserved from yeast to mammals [3], [4], [5]. Yeast cells are round and unpolarized in G1 phase, but after the G1-S transition, the actin cytoskeleton and localization of the Cdc42 GTPase are polarized to the presumptive bud site [2]. The cell is then set up for bud emergence, which occurs shortly afterwards. Bud formation through polarized growth is required for successful cell division.

This chapter was originally published under the Creative Commons Attribution License. Wai SC, Gerber SA, and Li R. Multisite Phosphorylation of the Guanine Nucleotide Exchange Factor Cdc24 during Yeast Cell Polarization. PLoS ONE 4,8 (2009). doi:10.1371/journal.pone.0006563.

The Cdc42 GTPase has been shown to regulate cell polarity in many organisms, including budding yeast [6]. Like most GTPases, Cdc42 is active when bound to GTP, and this is generally catalyzed by proteins called guanine nucleotide exchange factors (GEFs). In budding yeast, the only known GEF for Cdc42 is Cdc24. Both genes are essential, and the loss of either results in large unbudded multi-nucleate cells, indicative of an inability to undergo polarized growth [7], [8]. Activated Cdc42 signals through its downstream effectors to assemble and organize the actin cytoskeleton and the secretory machinery [2], [6]. Because Cdc42 is well conserved and important for cell polarity, much work has been done to characterize its function and its downstream effectors. Upstream regulation, however, particularly control of its GEF Cdc24, is not as well understood. Despite several studies, the mechanism of Cdc24 regulation remains unclear. Cdc24 belongs to the conserved family of Dbl-homology (DH) GEFs, which are characterized by adjacent DH and pleckstrin homology (PH) domains [9]. The DH domain is responsible for GEF activity, and the PH domain is thought to help localize or orient the GEF at the plasma membrane. Cdc24 has two other conserved functional domains: an N-terminal calponin-homology (CH) domain, the function of which is unclear, and a C-terminal PB1 domain, which mediates binding of Cdc24 to the adaptor protein Bem1 (see below).

Polarization of yeast cells must occur at the proper time in the cell cycle, suggesting a potential regulatory role for the cyclin-dependent kinase (CDK) Cdc28 on Cdc24, which leads to timely activation of Cdc42. Interestingly, Cdc24 is sequestered in the nucleus during G1 by the CDK-inhibitor Far1 [10], [11]. At bud emergence, Far1 is phosphorylated and degraded, releasing Cdc24 into the cytoplasm. Cdc24 mutants unable to bind to Far1 are constitutively located in the cytoplasm; however, Cdc28 activity is still required to activate these cytoplasmic mutants, indicating that nuclear export of Cdc24 is not sufficient to promote GEF activity [11], [12]. This suggests that Cdc28 directly or indirectly activates Cdc24. Cdc28 has been shown to phosphorylate Cdc24 in vitro [13], but it is unclear whether this occurs in vivo.

Yeast cell polarity is also regulated in a spatial manner. Haploid yeast cells generally bud adjacent to the previous bud scar, which is a remnant from the previous cell division site. The Ras family GTPase Bud1/Rsr1 recruits Cdc24 to the bud site in order to establish polarized growth [14]. Bud1 may also activate Cdc24 by inducing a conformational change [15]. Interestingly, however, *Δbud1* cells grow and divide at the same rate as wild type cells, although they bud in a random pattern [16]. This suggests that regulation of Cdc24 by Bud1 is not required for polarity per se. This may be due to redundant activation events mediated by the adaptor protein Bem1. In addition to possibly causing a conformational change by directly binding Cdc24, Bem1 also mediates the formation of a complex containing GTP-bound Cdc42, Cdc24, and the p21-activated kinase (PAK) Cla4, and enables Cla4 to phosphorylate Cdc24 [12], [17]. Our previous work showed that Bem1 is required for a mechanism of cell polarization that is independent of actin filaments [18]. Bem1 may induce cell polarization by mediating a positive feedback loop whereby Cdc24 is preferentially recruited/activated at the site of $Cdc42^{GTP}$ accumulation [19], [20], [21]. Cla4 phosphorylates Cdc24 in this complex likely at multiple sites; the function is not clear, but one study has suggested that this phosphorylation negatively regulates Cdc24 function [12], [17]. Because multisite phosphorylation has been shown in some cases to confer ultrasensitivity onto cell signaling processes [22], [23], one interesting possibility is that phosphorylation of Cdc24 by Cla4 may allow the Bem1 complex to function in a switch-like manner to induce cell polarization. However, despite all of these interesting ideas, there has been no direct study on the functional importance of this phosphorylation in vivo.

There is precedent for regulation of Dbl homology GEFs by phosphorylation [9]. The best characterized example is that of the GEF Vav1. It was shown that mouse Vav1 is regulated by an autoinhibitory fold and phosphorylation of a critical tyrosine residue releases this inhibition, activating Vav1 [24]. However, outside of the conserved DH and PH domains, Dbl homology GEFs are highly divergent, and so it is difficult to predict the exact mechanisms by which each GEF is regulated. In this study, we used mass spectrometry to map a large number of in vivo phosphorylation sites on Cdc24. Surprisingly, mutagenesis of these putative phosphorylation sites (including those matching consensus sequences for CDK or PAK family kinases and those not) has not resulted in any observable defects in cell polarization and growth by a variety of assays, suggesting that phosphorylation is largely dispensable for the regulation of Cdc24.

3.2 RESULTS

3.2.1 PURIFICATION OF CDC24-TAP

In order to better understand how Cdc24 may be regulated by multisite phosphorylation and, in particular, how this phosphorylation contributes to the Bem1-dependent and actin-independent cell polarization, we set out to purify Cdc24 from yeast cells in order to identify the in vivo phosphorylation sites by mass spectrometry analysis. This was challenging because of the following factors: a) Cdc24 is present in low quantities [25] and initial purifications contained many contaminants; b) the level of phosphorylation of Cdc24 changes during the cell cycle [12], [17]; and c) Cdc24 is only partially soluble in cell lysates [26], [27], [28], [29].

The cell lysis and purification protocols were systematically optimized in order to obtain a sufficient amount of phosphorylated protein for analysis. The purification protocol was modified from the tandem affinity purification (TAP) method [30], [31]. Because the mobility shift of Cdc24 on SDS-PAGE is maximal during bud emergence [12], [32] (Figure S1A and B), cell synchronization and release using *cdc28-13*, a temperature-sensitive allele of *CDC28*, was performed to enrich for small-budded cells, and thus phosphorylated Cdc24. Initially, cell lysis was performed using a French press, but Cdc24 in these lysates was mostly insoluble. Solubility of Cdc24 was increased by lysing frozen yeast cells with a coffee grinder and increasing the salt concentration of the lysis buffer (Figure S1C). Contaminants were decreased by precipitating Cdc24 from the cleared lysate with 40% ammonium sulfate precipitation. Most other proteins remained soluble and could thus be separated. More contaminants were removed by washing the IgG column with a buffer with high salt concentration. In order to shorten the time of the purification and decrease the loss of phosphorylated proteins, the calmodulin resin purification step of the TAP method was omitted. The bulk of the GST-TEV protease after cleavage was instead removed by passing the eluate over glutathione resin. The complete final purification scheme is described in Materials and Methods.

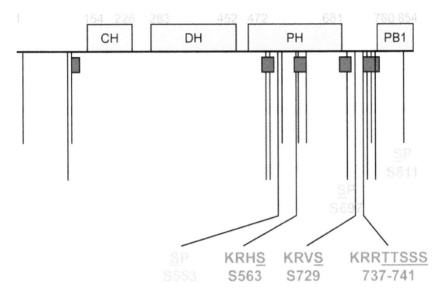

FIGURE 1: Cdc24 is phosphorylated on 35 residues. Schematic diagram of Cdc24 domain structure, regions of low complexity, and mapped phosphorylation sites. Regions of low complexity are denoted by the dark gray boxes under domains and correspond to residues 106–121, 523–545, 562–600, 680–710, 739–754, and 767–778. Adjacent residues were grouped together to facilitate analysis. Longer lines correspond to more highly ranked sites. Labeled are CDK consensus sites and PAK consensus sites. See Tables S1–S3 for details on sites and ranking information. CH, calponin homology; DH, Dbl homology; PH, pleckstrin homology; PB1, p67^{phox}-Bem1 homology.

3.2.2 CDC24 IS PHOSPHORYLATED ON A LARGE NUMBER OF RESIDUES

Cdc24-TAP was purified from 60 g of yeast cells (about 70% were small budded), resolved by SDS-PAGE (Figure S1D), and submitted for mass spectrometry analysis. Mass spectrometry analysis identified 35 residues that were putatively modified by phosphorylation (Figure 1 and Table S1). Most of these residues mapped to the PH domain or the linker region between the PH and PB1 domains. Interestingly, a sequence segmentation program identified sequences in both of these regions as being low-complexity or compositionally-biased [33], [34]. It is tempting to speculate that these intrinsically serine-rich sequences (39.5% serine by composition, on average) may have become favorable substrates for phosphorylation.

Adjacent residues were grouped together to facilitate analysis (e.g. TTSSS, residues 737–741). These single or clustered sites were then ranked based on their scores on NetPhos 2.0 [35] and Scansite [36] (Figure 1; see Tables S1, S2, S3 for more details). Cdc24 contains six CDK consensus sites (S/T P), three of which were mapped. Three of the most highly ranked phosphorylation sites were identified by Scansite to be PKA/PKC consensus sites; however, studies on various PAKs and their substrates have identified a similar consensus sequence [37], [38], and so it is possible that these are sites of PAK phosphorylation.

3.2.3 MUTATION OF CONSENSUS CDK OR PAK SITES ON CDC24 DOES NOT AFFECT CELL GROWTH OR MORPHOLOGY

We next used mutagenesis to study the in vivo function of the phosphorylation sites in Cdc24. Because of the large number of sites and because of previous work implicating CDK and PAK in Cdc24 regulation, we first examined the CDK and PAK sites. The different phosphorylation mutants are described in Table 1. We designed a construct that would allow homologous recombination of Cdc24 mutants directly into the endogenous locus, generating mutant strains expressing a phosphorylation mutant of Cdc24 as the sole source of Cdc24 in the cell (see Materials and Methods for details). When assessing whether the point mutations had been successfully introduced, we found that mutations of T134 and S245, the sites closest to the 5' end of *cdc24*, were lost in the CDK mutants, i.e. they had reverted back to the wild type sequence. S245A was retained in the CDK-A/PAK-A mutant. The other mutant residues were successfully mutated, and these are denoted in Table 1. The lost sites were included later in the 35A mutant (see Table 2).

TABLE 1: Mutation of CDK or PAK consensus sites.

Mutant alias	Mutated residues
CDK-3A	S553A, S697A, S811A
CDK-4A	S553A, S697A, T704A, S811A
PAK-A	S563A, S729A, TTSSS(737-741)AAAAA
CDK-A/PAK-A	S245A, S553A, S563A, S697A, T704A, S629A, TTSSS(737-741)AAAAA, S811A
CDK-DE	S553D, S697D, T704E, S811D
PAK-DE	S563D, S729D, T638E, S741D

TABLE 2: Mutation of remaining sites, including non-consensus sequences.

Mutant alias	Mutated residues
PH-A	All mapped residues within the PH domain
Linker-A	All mapped residues within the linker region and T704A
PH-A/Linker-A	All mapped residues with PH and linker regions and T704A
35A	All 35 mapped residues and T134A, S245A, T704A

The mutant strains were viable (see below), and we examined whether these mutations affected phosphorylation of Cdc24. Asynchronous cultures of the mutant strains were lysed and resolved by SDS-PAGE to assess the mobility shift of Cdc24, which was shown previously to be caused by phosphorylation [12], [17]. Mobility shift in each mutant strain was quantified and normalized to an appropriate wild type strain resolved on the same gel. Interestingly, a decrease in mobility shift roughly correlated with the increase in number of mutated residues (Figure 2A and B). Mutations of the CDK sites to alanine changed the mobility shift very little, but mutations of the PAK sites to alanine reduced the shift to about 45% of that observed in the wild type. Mutations of predicted phosphorylation sites to aspartate or glutamate did not significantly change the mobility shift. While it is possible that some of these mutations affect the phosphorylation of the Cdc24 protein without changing its mobility shift, it appears that the mutations to alanine, especially of the PAK sites, perturbed Cdc24 phosphorylation.

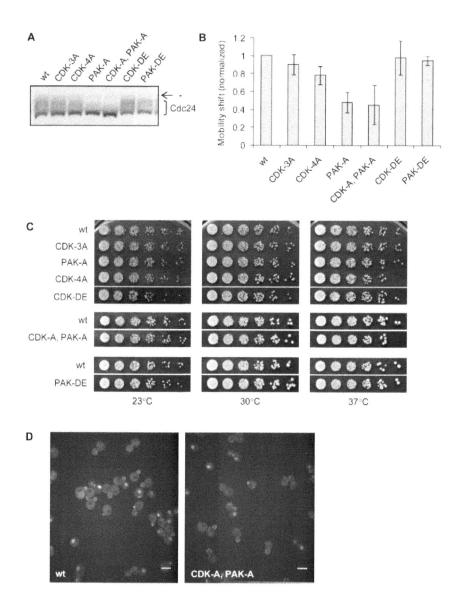

FIGURE 2: Phosphorylation site mutations decrease mobility shift of Cdc24, but do not affect cell growth or Cdc24 localization. A. Mobility shift of Cdc24 band. Whole cell extracts were prepared from asynchronous cultures of wild type control (wt, RLY2853), cdc24^{CDK-3A} (RLY2855), cdc24^{PAK-A} (RLY2858), cdc24^{CDK-4A} (RLY2934), cdc24^{CDK-DE} (RLY2938),

cdc24$^{CDK-A,PAK-A}$ (RLY3088), and cdc24^{PAK-DE} (RLY3089). Blots were probed with rabbit anti-Cdc24 antiserum. Faint band just above smear is a cross-reacting protein (*). B. Quantification and comparison of mobility shift. Mobility shift of band was calculated as (integrated intensity of smear)/(integrated intensity of entire band) and normalized against mobility shift of wild type control resolved on the same gel. Mean and standard deviation of at least three independent samples are shown. C. Serial dilutions of each strain were spotted onto YEPD agar plates and grown for 2 d at 23°C, 30°C, or 37°C, or 13 d at 16°C (not shown). Mutant strains are shown alongside a wild type control that was spotted on the same plate. Only strains relevant to this experiment are shown in this figure. See Materials and Methods for more information. D. Confocal images of strains containing Cdc24-GFP (RLY3096) and Cdc24$^{CDK-A,PAK-A}$-GFP (RLY3099). Scale bar represents 5 μm. Confocal images of other mutants are shown in Figure S2A.

Because *CDC24* is an essential gene and the only known GEF for Cdc42 in yeast, we expected to see a severe growth phenotype if these mutations affect Cdc24 function. However, these mutations, including the CDK and PAK sites mutated individually or in combination, did not confer cell lethality or temperature-sensitivity, as all strains grew like the wild type control at all temperatures tested (Figure 2C). Cell morphology and budding pattern of the mutants were also similar to the wild type (data not shown). GFP fusions of the mutant Cdc24 proteins localized like the wild type Cdc24, showing the expected localization to the nucleus, presumptive bud site, bud cortex, and bud neck (Figure 2D and S2A). We conclude that these mutations do not grossly affect Cdc24 function or protein localization.

Because these mutations did not affect cell growth on a gross level, we next asked if the mutations affected Cdc24 function on a subtler level. We focused on the strain containing the most mutations, cdc24$^{CDK-A/PAK-A}$ (RLY3088). Cell growth assayed by colony formation (as shown in Figure 2C) is at best a rough comparison of growth rates and might not be capable of detecting moderate defects in cell polarity. To this end, we directly assessed budding kinetics in these mutants. Wild type and cdc24$^{CDK-A/PAK-A}$ mutant cells were arrested in G1 by α-factor and synchronously released into the cell cycle, and the percentage of cells with a small bud was scored at each time point. Budding kinetics of the mutant were comparable to those of the wild type (Figure 3A).

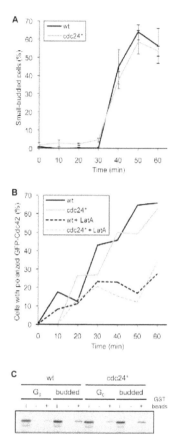

FIGURE 3: CDK-A/PAK-A mutation does not affect kinetics of bud formation of polarization, nor does it affect overall level of Cdc42 activation. The $cdc24^{CDK-A/PAK-A}$ strain is labeled here as cdc24* for simplicity. A. Comparison of kinetics of bud formation between wild type (wt, RLY2853) and cdc24* (RLY3088). Cells were arrested with α-factor, then washed and released. Samples of cultures were fixed every 10 min and scored for bud morphology. Percentage of small-budded cells was plotted over time. B. Comparison of kinetics of polarization between wild-type (wt, RLY2903) and cdc24* (RLY3072). Cells were arrested with α-factor, then treated briefly with LatA to depolarize GFP-myc-Cdc42. Cells were then washed and released +/− LatA and scored for polarized localization of GFP-myc-Cdc42. Percentage of cells with polarized GFP-myc-Cdc42 was plotted over time. C. Comparison of levels of active GTP-bound Cdc42 in wild type (wt, RLY2903) and cdc24* (RLY3072). Levels of active Cdc42 were determined at both stationary phase (G_0) and at the timepoint when most cells had small buds (budded) by measuring levels of GFP-myc-Cdc42 bound to GST-CRIB construct immobilized on glutathione sepharose beads. Small sample of the extract (i), pulldown with GST only beads (−), and pulldown with GST-CRIB beads (+) were resolved by SDS-PAGE. Blots were probed with mouse anti-myc antibody (9E10). Gel shown is representative of two independent experiments.

We previously showed that an actin-transport mediated feedback loop works in parallel with a Bem1-dependent polarization pathway [18]. In order to look at the function of the mutant Cdc24 specifically in the Bem1 feedback loop, we abrogated the actin feedback loop by treating the cells with latrunculin A (LatA). The strain was transformed with a plasmid containing GFP-myc-Cdc42 as a marker for polarization. Polarization kinetics of GFP-myc-Cdc42 in the $cdc24^{CDK-A/PAK-A}$ mutant is similar to that in the wild type, even when cells were treated with LatA (Figure 3B), indicating that Cdc24$^{CDK-A/PAK-A}$ functions normally in the Bem1-dependent cell polarization pathway.

Next, we asked if the CDK-A/PAK-A mutations had any effect on the GEF activity of Cdc24. We used a pull-down assay with the CRIB domain from Cla4, in order to measure the overall level of active GTP-bound Cdc42 in wild type and mutant strain. Both strains were grown to stationary phase, and half of each sample was harvested. The other half was resuspended in fresh media to reenter the cell cycle and harvested when most cells had formed small buds. Beads containing GST-CRIB construct were added to the lysate to pull down GTP-bound Cdc42. As expected, stationary cells contained a lower amount of GTP-bound Cdc42 than small-budded cells. The level of GTP-bound Cdc42 in the $cdc24^{CDK-A/PAK-A}$ mutant strain was comparable to that in the wild type at both time points (Figure 3C), suggesting that the phosphorylation site mutations do not have a gross impact on Cdc42 activation.

Finally, we examined whether the $cdc24^{CDK-A/PAK-A}$ mutant displayed any synthetic phenotypes with strains that had defects in polarization or in which the mobility shift of Cdc24 was affected. The $cdc42$-1 strain contains a temperature-sensitive allele of CDC42, resulting in minor morphological abnormalities at the 23–24°C and growth arrest as unbudded cells at 37°C [8]. The $\Delta bud1$ strain shows no overt polarization defects (apart from a random budding pattern), but we noticed that Cdc24 protein in the $\Delta bud1$ strain had a significantly reduced mobility shift, suggesting an abnormal phosphorylation state (Figure 4). The $\Delta bem1$ strain is viable at 23–30°C, but divides at a much slower rate than wild type cells and displays morphological heterogeneity, with morphological defects such as multiple buds and large unbudded cells [39]. $\Delta bem1$ is nonviable at 37°C. Cdc24 displays little to no mobility shift in the $\Delta bem1$ strain [12], [17].

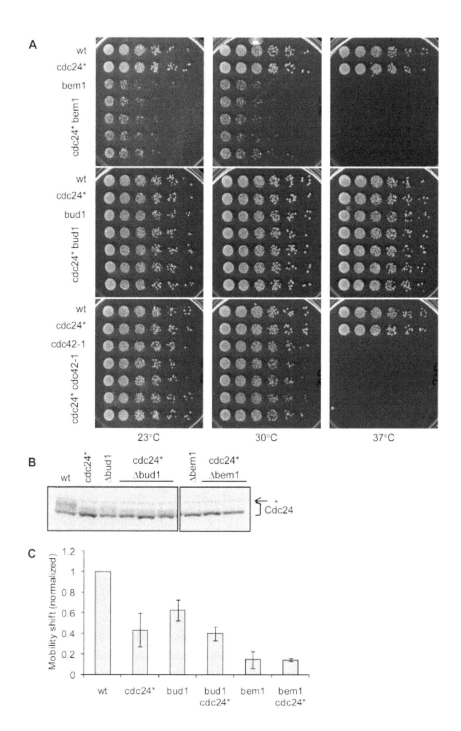

FIGURE 4: No synthetic effects on cell growth or mobility of Cdc24 are detected between CDK-A/PAK-A mutant and other polarity mutants. The *cdc24$^{CDK-A/PAK-A}$* strain is labeled here as cdc24* for simplicity. A. Serial dilutions of meiotic segregants from crosses between cdc24* (RLY2937) and polarity mutants *Δbem1* (RLY2771), *Δbud1* (RLY2773), and *cdc42-1* (RLY171). Double mutant strains were spotted onto YEPD agar plates and grown for 3 d at 23°C, 30°C, 35°C (not shown), and 37°C. B. Mobility shift of Cdc24 band. Whole cell extracts were prepared from asynchronous cultures of meiotic segregants from crosses between cdc24* (RLY2937) and Δbem1 (RLY2771) and Δbud1 (RLY2773). Blots were probed with rabbit anti-Cdc24 antiserum. Faint band just above smear is a cross-reacting protein (*). C. Quantification of mobility shift, calculated, normalized, and displayed as in Figure 2B.

cdc24$^{CDK-A/PAK-A}$ was crossed into the *cdc42-1, Δbud1*, and *Δbem1* strains, generating strains RLY3093, 3095, and 3102. These diploid strains were sporulated and dissected, and at least eight double mutant spores were analyzed for synthetic phenotypes (Figure 4). None of the double mutants showed temperature-sensitive phenotypes that were different from the single mutants, ruling out a synthetic effect on cell growth. Mobility shift of Cdc24$^{CDK-A/PAK-A}$ in the double mutants was identical to that in the most affected single mutant, ruling out a synthetic effect on phosphorylation (Figure 4). It is interesting to note that wild type Cdc24 in *Δbud1* cells have a less affected mobility shift than Cdc24$^{CDK-A/PAK-A}$ in *Δbud1* cells. This may suggest that phosphorylation sites affected in the CDK-A/PAK-A mutant encompass those that are perturbed in Δbud1.

Although we did not explicitly test mating efficiency of *cdc24$^{CDK-A/PAK-A}$*, the strain arrested in α-factor and mated with the *cdc42-1, Δbud1*, and *Δbem1* strains with efficiency indistinguishable from that of the wild type strain. Thus, *cdc24$^{CDK-A/PAK-A}$* did not show obvious defects in pheromone response. Taken together, the above results suggest that a lack of phosphorylation on the CDK and PAK consensus sites in Cdc24 has no overt functional consequences, not even when combined with mutations in proteins known to directly interact with Cdc24.

3.2.4 MUTATION OF REMAINING PHOSPHORYLATION SITES ON CDC24 ALSO DOES NOT AFFECT CELL GROWTH OR MORPHOLOGY

Because we were unable to identify a regulatory role for the consensus CDK or PAK sites, we mutated the remaining mapped sites in order to determine if any of these sites may be contributing to Cdc24 regulation. Recent studies have indicated the importance of either the location or the number of phosphorylated sites in the regulation of protein function [23], [32]. Because of the clustering of sites within the PH domain and within the linker region between the PH and PB1 domains, we generated mutants containing alanine mutations of all the mapped residues within either of those domains or in both. We also generated a mutant containing alanine mutations of all 35 mapped residues. In all of these mutants, we included any CDK consensus sites within the region, whether or not they were mapped. Table 2 summarizes these mutants. Because of our earlier difficulty in maintaining mutated sites near 5' end of *cdc24* by genomic integration, we introduced these mutants via a centromeric plasmid into a strain heterozygous for deletion of *CDC24*. In order to characterize these new *CDC24* mutants, the strains were sporulated and dissected, and spores were identified that carried a genomic deletion of *CDC24*, rescued by the mutant cdc24 on the plasmid.

Like earlier mutants, these mutants did not confer cell lethality or temperature-sensitivity (Figure 5A). Surprisingly, even the mutants containing the most mutations (PH-A/linker-A and 35A) grew like the wild type strain at all temperatures tested. Localization of GFP fusion proteins appeared to be normal and indistinguishable from wild type (Figure 5B and S2B). These results indicate that, like the earlier mutations, these mutations do not grossly affect Cdc24 function or localization in the cell.

We did, however, observe a decrease in the mobility shift of these mutants. Using the procedure described earlier, we quantified the mobility shift of the Cdc24 band from asynchronous cultures resolved by SDS-PAGE (Figure 5C and D). The PH-A/linker-A mutant showed the greatest decrease in mobility shift, retaining on average only 27.2% of the wild-type mobility shift. By comparison, *Δbem1* cells have on average 11.1% of a wild type mobility shift. It is unclear why the 35A mutant had a greater

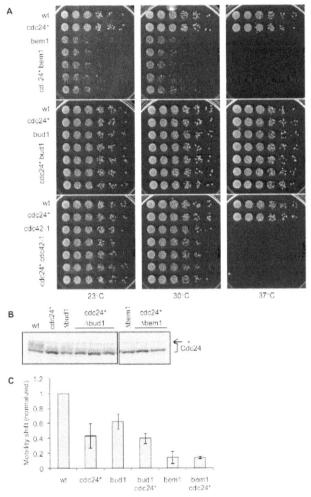

FIGURE 5: Mutations of non-consensus sites also do not affect cell growth or Cdc24 localization, but they further decrease mobility shift of Cdc24. A. Serial dilutions of wild type control (wt, RLY2530), *cdc24^PH-A* (RLY3391), *cdc24^linker-A* (RLY3393), *cdc24^PH-A.linker-A* (RLY3390), and *cdc24^35A* (RLY3461) were spotted onto YEPD agar plates and grown for 2 d at 23°C, 30°C, and 37°C, or 7 d at 16°C (not shown). Mutant strains are shown alongside a wild type control that was spotted on the same plate. Only strains relevant to this experiment are shown in this figure. See Materials and Methods for more information. B. Confocal images of strains containing Cdc24-GFP (RLY3437) and Cdc24^35A-GFP (RLY3468). Confocal images of other mutants are shown in Figure S2B. C. Mobility shift of Cdc24 band. Whole cell extracts were prepared from asynchronous cultures. Blots were probed with rabbit anti-Cdc24 antiserum. Faint band just above smear is a cross-reacting protein (*). D. Quantification of mobility shift, calculated, normalized, and displayed as in Figure 2B.

mobility shift (33.7% of wild-type) than the PH-A/linker-A mutant. Mobility shift may not be strictly caused by phosphorylation events, and it is possible that the additional mutations somehow affect the mobility of the protein. It is also possible that any compensatory phosphorylation on the 35A mutant may affect its mobility shift, more so than the compensatory phosphorylation on the PH-A/linker-A mutant. We conclude that we have perturbed most, if not all, of the native phosphorylation sites on Cdc24, but this perturbation did not affect the function of the protein in any obvious manner.

3.3 DISCUSSION

In the experiments described above, we investigated the potential role for Cdc24 phosphorylation in the establishment of cell polarity and polarized growth. These experiments were motivated by the suggestion that phosphorylation plays a key role in regulating the activity and/or localization of Cdc24 to achieve symmetry breaking in a cell cycle dependent manner. Although Cdc24 phosphorylation was abundantly observed in several studies, and two kinases, Cdc28 and Cla4, were shown to phosphorylate Cdc24 in vitro [13], [17], the in vivo significance of these phosphorylation events remained unclear. Here, we have mapped over thirty in vivo phosphorylation sites on Cdc24. However, to our surprise, mutation of sites matching the conserved consensus sequences for the CDK and PAK family kinases has not revealed any evidence for their functional importance. We have ruled out overt effects on temperature-sensitivity, cell growth, cell morphology, protein localization, and polarization kinetics in the presence and absence of actin. Our results on the putative CDK sites are consistent with a previous study reporting that mutation of these sites did not affect Cdc24 function in vivo [12]. In addition, mutation of the remaining mapped sites has not revealed any functional importance in regulation of Cdc24. Many of these mutations did affect the mobility shift of Cdc24 by SDS-PAGE, indicating that we have strongly perturbed phosphorylation.

Our results for multisite phosphorylation on Cdc24 differ from results of a study reporting multisite phosphorylation of Boi1, a protein that binds Bem1 and plays an essential but redundant role in cell polarization with

Boi2. In that study, the authors were able to identify phenotypes resulting from perturbation of phosphorylation. Mutation of the 12 consensus CDK sites or the 29 mapped phosphorylation sites on Boi1 resulted in temperature sensitive growth (in Δboi2 background) and impaired Boi1 localization [32].

Several studies on proteins that are phosphorylated on multiple sites highlight the importance of the number of phosphorylation sites, as opposed to the identity of the exact residues, on a regulation mechanism. A well known example is the CDK inhibitor Sic1, which must be phosphorylated on at least six consensus CDK sites before Cdc4 can bind and target it for degradation [22]. In a more recent study, it was demonstrated that the location of the phosphorylation sites, as well as resulting overall charge of the collective phosphate groups was important for a regulation mechanism [23], [40]. It was shown that Cln/Cdc28 phosphorylates Ste5 on eight CDK sites that flank a plasma membrane binding domain. The resulting negative charges are thought to prevent the interaction of Ste5 with phospholipids of the plasma membrane. However, our results showing the lack of even a subtle defect with the cdc24 mutant bearing 35 phosphorylation site mutations suggest that much of the phosphorylation on Cdc24 is dispensable for its function.

It is important to state that our data does not completely exclude a role for phosphorylation in Cdc24 regulation. A clear caveat is that low abundance phosphorylation sites may be overlooked in mass spectrometry analysis. It is also possible that, with further investigation, we may be able to detect a subtle polarization phenotype resulting from mutations of these sites. However, our data suggests that other modes of GEF or Cdc42 regulation may be equally or more important than Cdc24 phosphorylation, and these could mask the subtle effects of phosphorylation site mutations. For example, binding of Bud1 and Bem1 to Cdc24 has been proposed to allosterically activate Cdc24 by promoting an open conformation and/or be instrumental in bringing Cdc24 to the plasma membrane [14], [15], [19], [20]. Oligomerization of Cdc24 has also been proposed to regulate its activity [41]. Furthermore, two studies have characterized the phosphorylation of Cdc42 GTPase activating proteins (GAPs) and suggested that Cdc42 activation at the initiation of budding may be primarily regulated through downregulation of GAP activities through phosphorylation

[42], [43]. Even though our study has perhaps raised more questions that it has answered, it casts doubt on the simple model that phosphorylation of Cdc24 is the key step in the establishment of cell polarity and suggests that much of protein phosphorylation in the cell could be innocent by-products of kinase activation that occur at cellular transitions.

3.4 MATERIALS AND METHODS

3.4.1 PLASMID CONSTRUCTION

All plasmids used in this study are described in Table S4. Site-directed mutagenesis of phosphorylated residues was performed using the QuikChange II XL Site-Directed Mutagenesis Kit (Stratagene), and the final product was sequenced to ensure that there were no secondary mutations introduced into the CDC24 ORF.

3.4.2 YEAST STRAIN CONSTRUCTION

All yeast strains used in this study are described in Table S5. Techniques for yeast cell culture and genetics were essentially as described [44]. Transformation of plasmid DNA into yeast was performed based on the lithium acetate method [45]. Transformation of PCR fragments was performed by the same method, but after transformation, cells were resuspended in non-selective media and grown for the equivalent of at least two cell cycles before plating on selective media.

Yeast strains containing mutations of the CDK or PAK consensus sites (as described in Table 1) were constructed as follows. pSW44 or its derivatives were digested with XhoI and NotI and gel purified to obtain DNA fragments containing the *CDC24* and *LEU2* ORFs. A diploid strain heterozygous for deletion of *CDC24* (RLY2775) was transformed with these fragments. Transformants were replica-plated onto media lacking leucine or YPD media containing geneticin. Transformants that were Leu+ and

Gen^S (indicating that the fragment had replaced the $\Delta cdc24::KanMX$ locus) were sporulated and dissected. Integration of the *LEU2* marker into the proper locus was confirmed by PCR. Point mutations that were successfully integrated were determined by sequencing of genomic DNA purified from the resulting strain. PCR-based GFP tagging was performed as previously described [46], [47] in order to obtain strains expressing GFP-tagged Cdc24 mutants.

Yeast strains containing the mutations of non-consensus sites (as described in Table 2) were constructed as follows. Centromeric plasmids (pSW72-73 and their derivatives) containing these mutants were transformed into a strain heterozygous for deletion of *CDC24* (RLY2775). The resulting strains were sporulated and dissected. Spores were identified that were both Leu$^+$ and Gen^S, indicating that they carry the genomic deletion of *Cdc24* ($\Delta cdc24::KanMX$), rescued by the mutant Cdc24 on the plasmid.

3.4.3 WHOLE CELL EXTRACTS

Whole cell extracts for Western blot analysis were prepared by TCA precipitation. Cell pellets were harvested from 5–10 ml cultures grown to OD$_{600}$ ~0.5 and frozen in liquid nitrogen. Cell pellets were thawed, resuspended in 1 ml 20% TCA, and transferred to 1.5 ml eppendorf tubes. Cells were pelleted again by spinning for 1 min at 3000 rpm in a microcentrifuge. Glass beads (acid-washed, 425–600 μm, Sigma) and 100–200 μl of 20% TCA were added to the pellets. The mixture was then vigorously vortexed for 7 min to lyse the cells. 400 μl of 5% TCA was added to the mixture, and all of the liquid was transferred to new tubes. The protein pellet was harvested by spinning for 10 min at 3000 rpm in a microcentrifuge. After removal of liquid, the pellet was resuspended in SDS-PAGE sample buffer and titrated with Tris base if necessary.

3.4.4 WESTERN BLOTTING

Standard methods were used for SDS-PAGE and Western blotting. Protein samples were resolved on 7.5% polyacrylamide gels, and proteins smaller than 50 kD were run off the gel in order to allow maximum separation of Cdc24 and its phosphorylated species ("smear"). Blots were developed using ECL or ECL Plus Reagents (GE Healthcare Life Sciences). For quantification, blots were scanned using the Typhoon 9400 (GE Healthcare Life Sciences) and analyzed using ImageQuantTL.

Anti-Cdc24 antiserum was raised in rabbits (Cocalico Biologicals, Reamstown, PA) using an MBP-Cdc24$^{472-854}$ construct that was previously described [14] and generously provided by E. Bi (Univ. of Pennsylvania School of Medicine, Philadelphia, PA). Because the resulting antiserum was fairly clean (only a single cross-reacting protein was visibly detected), the antibody was not affinity purified. Anti-Arp3 antibody from goat (yG-18, to ensure equal loading of samples across lanes) and anti-myc antibody from mouse (9E10) were purchased from Santa Cruz Biotechnology.

3.4.5 PURIFICATION OF CDC24-TAP

Cells were synchronized by using a temperature-sensitive *cdc28-13* allele (RLY2194). The *cdc28-13* strain was arrested in G1 at the restrictive temperature (37°C) for no more than 3 h and released into the cell cycle by cooling down the culture to 25°C using an ice water bath. Cells were monitored for bud formation and harvested about 1 h after release, when most cells were small-budded. Cells were washed once with harvest buffer (50 mM HEPES pH 7.5, 0.1 M KCl, 3 mM MgCl$_2$ 6H$_2$O, 1 mM EGTA, 50 mM NaF, 120 mM β-glycerophosphate, 1 mM Na$_3$VO$_4$, and 1 mM DTT, supplemented with a protease inhibitor cocktail containing 0.5 mg/ml each of antipain, leupeptin, aprotinin, pepstatin, and chymostatin (Sigma)). Cell pellets were frozen in small strands in liquid nitrogen. Cell lysis was performed in a small coffee grinder (Krups), cooled with dry ice in order to keep yeast cells frozen. Lysed cells were resuspended in lysis buffer (harvest buffer brought up to high salt concentration (0.5 M KCl) and containing 1 mM PMSF). Extracts were immediately centrifuged at low speed (2430 g) in order to remove unlysed cells, then incubated on ice for 30 min. Extracts were then centrifuged at high speed (100,000 g) to remove nonsoluble components. Protein was precipitated using 40% ammonium sulfate, and the precipitate was stored at −80°C. Precipitate (which contained Cdc24-TAP) was resuspended in lysis buffer and centrifuged briefly to remove any undissolved precipitate. Extract was incubated with washed IgG Sepharose 6 Fast Flow beads (GE Healthcare Life Sciences) for 4 h at 4°C. Slurry was applied to Poly-Prep disposable columns (Bio-Rad), and beads were washed thoroughly with lysis buffer containing 0.1% NP-40. Beads were then washed with TEV buffer (harvest buffer without protease inhibitors), and incubated overnight at 4°C with purified GST-TEV (tobacco etch virus) protease to cleave off Cdc24. Beads were washed with small amount of lysis buffer, and the first two washes were pooled with the eluate. These were passed through Glutathione Sepharose 4B beads (GE Healthcare Life Sciences) to remove most of the GST-TEV protease. The flowthrough and first wash from the glutathione beads were pooled and precipitated by 20% TCA. The precipitate was resuspended in a small volume of SDS-PAGE sample buffer and resolved by SDS-PAGE on a 7.5% gel.

3.4.6 MASS SPECTROMETRY ANALYSIS

Coomassie-stained bands corresponding to Cdc24 were excised from SDS-PAGE gels, destained to clarity with 50 mM ammonium bicarbonate/50% acetonitrile solution, and digested in-gel using established procedures [48]. The digested samples were extracted, dried by vacuum centrifugation, resuspended in 60 μl of 250 mM acetic acid/30% acetonitrile and treated with immobilized metal affinity chromatography (IMAC) media (Phos-Select IMAC resin, Sigma) per manufacturer's directions [49]. The IMAC resin was eluted with 50 mM K_2HPO_4/20% isopropanol, pH 10.5 (with 4 M ammonia in ethanol), and the two combined extracts were acidified with 5% formic acid prior to drying by vacuum centrifugation. Finally, the dried extracts were desalted using STAGE tips as described [50].

   The dried, desalted samples were resuspended in 6 μl of a 1% formic acid/2.5% acetonitrile solution immediately prior to sequencing analysis. Sequencing was performed by nanoscale microcapillary LC-MS/MS essentially as described [51]. Ultimately, peptides were separated on a hand-pulled 125 μm×20 cm reverse-phase microcapillary column packed with 5 μm (200 Å) C18-AQ material as described [52]. Tandem mass spectrometry was performed on a hybrid linear (2D) ion trap – Fourier transform ion cyclotron resonance mass spectrometer over a 60 minute gradient of 5% to 32% of a 0.1% formic acid/95% acetonitrile solution as described [51]. Tandem mass spectra were de-isotoped using in-house written software and data searched using the SEQUEST algorithm with no enzyme specific-ity for the variable modifications+15.99491 on methionine and+79.96633 on serine, threonine and tyrosine [53], [54]. Results were initially filtered using only XCorr and mass accuracy, and candidate phosphopeptides were manually verified by inspection of the corresponding tandem mass spec-tra. In cases where adjacent acceptor residues precluded a definite assign-ment of phosphorylation locus, multiple candidate phosphorylation sites were reported.

3.4.7 SERIAL-DILUTION GROWTH ASSAYS ("SPOT TESTS")

Overnight cultures were diluted to an OD_{600} of 0.1 to make the start-ing cultures. Starting cultures were pipetted into the leftmost wells of a sterile 96-well plate, and serial dilutions of four-fold were made in the subsequent wells. After thorough mixing, a sterilized frogger was used to spot the array onto the desired number of agar plates. After the cells were allowed to grow for 2–3 d at the 23°C, 30°C, 35°C, or 37°C, or 7–13 d at 16°C, plates were imaged using a digital camera or scanner. Images of each plate were processed using Adobe Photoshop CS3. The contrast was increased using the Auto Levels command, in order to bet-ter display the yeast colonies against the dark background. The images were subsequently cropped to display only the strains that were relevant to the experiment.

3.4.8 FIXATION AND STAINING

Yeast cells were fixed by adding formaldehyde to a final concentration of 5% and incubating at least 2 h at room temperature with gentle shaking. Cells were then washed once in PBS, before storing at 4°C in PBS. Before scoring for morphology, fixed cultures were briefly sonicated in order to separate clumps. In order to assess bud scar patterns, Calcofluor staining was performed essentially as described [55], except that the stock solution was stored at −20°C and cells were stained with a 20- to 40-fold dilution of the Calcofluor stock solution.

3.4.9 MICROSCOPY

Widefield imaging was performed on a Nikon Eclipse E1000 fluorescent microscope with a 100X Plan-Apo TIRF NA 1.45 oil objective. Samples were illuminated with a mercury lamp through a 3-cube GFP filter, and images were acquired using a Hamamatsu ORCA-ER camera. Confocal imaging was performed on a Zeiss Axiovert 200 M inverted microscope with a 100X αPlan-Fluar NA 1.45 oil objective, attached to a Yokogawa spinning disk confocal and Hamamatsu EM-CCD C9100 camera. Images of GFP-tagged phosphorylation site mutants were collected as a series of 13 optical sections, with a step size of 0.3 μm. The final images shown are maximum z-projections. MetaMorph software (v. 6.3r5; Molecular Devices Corporation) was used to control both microscopes. Image J software (v. 1.37, http://rsb.info.nih.gov/ij/) was used to process the images.

3.4.10 RELEASE ASSAYS

Cells were arrested for 3 h using 5 μg/ml α-factor. In order to determine budding kinetics, cells were released into the cell cycle by washing three times in sterile water before resuspending in fresh media. Samples were taken every 10 min and fixed with formaldehyde as described above. At each time point, 100 cells were scored for budding morphology (small-budded or not). In order to determine polarization kinetics, all arrested cells were treated with 100 μM LatA (gift from P. Crews, University of California, Santa Cruz, CA) for 20 min to depolarize GFP-myc-Cdc42, and then released by washing three times in water before resuspending in fresh media with or without LatA. At each time point, >40 cells were scored for polarized localization of GFP-myc-Cdc42.

3.4.11 CRIB PULLDOWN FOR MEASUREMENTS OF CDC42^{GTP}

Bacteria expressing either a GST fusion construct containing the CRIB domain from Cla4 (RLB252) or just GST for control (RLB247) were lysed using a French Press (1700 psi), and the cleared lysate was loaded onto Glutathione Sepharose 4B beads (GE Healthcare). The beads were washed and stored at 4°C until use. Yeast strains expressing GFP-myc-CDC42 (pRL369) and either wild type Cdc24 (RLY2903) or mutant Cdc24 (RLY3072) were grown up overnight. 3×10^7 cells of each strain were spread into an even layer onto YPD plates. A total of three plates were spread for each strain and grown for 3 d at 30°C to reach stationary phase. For each strain, 1 L of YEPD media was inoculated with the cells from two plates. Half of the culture was immediately harvested. All centrifugation steps were performed at 4°C. Each pellet was washed in buffer consisting of 50 mM HEPES pH 7.5, 0.1 M KCl, 3 mM MgCl$_2$ · 6H$_2$O, 1 mM EGTA, 1 mM DTT, supplemented with a protease inhibitor cocktail. The cells were frozen in liquid nitrogen as small pellets and stored at −80°C. The remaining culture was incubated at 30°C and harvested as above when most of the cells had become small-budded. Yeast extracts were prepared by grinding the frozen cell pellets with a mortar and pestle that had been cooled with liquid nitrogen. Lysed yeast powder was stored at −80°C until all samples had been lysed. The powder from each sample was resuspended in a small volume of the same buffer described above, but also including 0.2% Triton X-100 and 1 mM PMSF. Extracts were spun for 30 min at 100,000 g to remove unlysed cells and insoluble material. The amount of protein was quantified by the Bradford Assay (Bio-Rad), and extracts were adjusted so that they had the same protein concentration. Each pull-down reaction contained 20 μl of GST-CRIB beads (or 20 μl of GST beads, diluted 1:20 with empty beads) and 400–500 μl of yeast extract. Pull-downs were incubated at 4°C for 1 h with gentle shaking. After washing the beads 3–4 times in the same buffer, the beads were resuspended in SDS-PAGE sample buffer. All samples were boiled and spun briefly before loading onto an acrylamide gel.

REFERENCES

1. Drubin DG, editor. (2000) Cell polarity. Oxford; New York: Oxford University Press.
2. Pruyne D, Bretscher A (2000) Polarization of cell growth in yeast. I. Establishment and maintenance of polarity states. J Cell Sci 113 (Pt3): 365–375.
3. Hall A (1998) Rho GTPases and the actin cytoskeleton. Science 279: 509–514.
4. Etienne-Manneville S (2004) Cdc42—the centre of polarity. J Cell Sci 117: 1291–1300.
5. Park HO, Bi E (2007) Central roles of small GTPases in the development of cell polarity in yeast and beyond. Microbiol Mol Biol Rev 71: 48–96.
6. Johnson DI (1999) Cdc42: An essential Rho-type GTPase controlling eukaryotic cell polarity. Microbiol Mol Biol Rev 63: 54–105.
7. Hartwell LH (1971) Genetic control of the cell division cycle in yeast. IV. Genes controlling bud emergence and cytokinesis. Exp Cell Res 69: 265–276.
8. Adams AE, Johnson DI, Longnecker RM, Sloat BF, Pringle JR (1990) CDC42 and CDC43, two additional genes involved in budding and the establishment of cell polarity in the yeast Saccharomyces cerevisiae. J Cell Biol 111: 131–142.
9. Rossman KL, Der CJ, Sondek J (2005) GEF means go: turning on RHO GTPases with guanine nucleotide-exchange factors. Nat Rev Mol Cell Biol 6: 167–180.
10. Nern A, Arkowitz RA (2000) Nucleocytoplasmic shuttling of the Cdc42p exchange factor Cdc24p. J Cell Biol 148: 1115–1122.
11. Shimada Y, Gulli MP, Peter M (2000) Nuclear sequestration of the exchange factor Cdc24 by Far1 regulates cell polarity during yeast mating. Nat Cell Biol 2: 117–124.
12. Gulli MP, Jaquenoud M, Shimada Y, Niederhauser G, Wiget P, et al. (2000) Phosphorylation of the Cdc42 exchange factor Cdc24 by the PAK-like kinase Cla4 may regulate polarized growth in yeast. Mol Cell 6: 1155–1167.
13. Moffat J, Andrews B (2004) Late-G1 cyclin-CDK activity is essential for control of cell morphogenesis in budding yeast. Nat Cell Biol 6: 59–66.
14. Park HO, Bi E, Pringle JR, Herskowitz I (1997) Two active states of the Ras-related Bud1/Rsr1 protein bind to different effectors to determine yeast cell polarity. Proc Natl Acad Sci U S A 94: 4463–4468.
15. Shimada Y, Wiget P, Gulli MP, Bi E, Peter M (2004) The nucleotide exchange factor Cdc24p may be regulated by auto-inhibition. Embo J 23: 1051–1062.
16. Bender A, Pringle JR (1989) Multicopy suppression of the cdc24 budding defect in yeast by CDC42 and three newly identified genes including the ras-related gene RSR1. Proc Natl Acad Sci U S A 86: 9976–9980.
17. Bose I, Irazoqui JE, Moskow JJ, Bardes ES, Zyla TR, et al. (2001) Assembly of scaffold-mediated complexes containing Cdc42p, the exchange factor Cdc24p, and the effector Cla4p required for cell cycle-regulated phosphorylation of Cdc24p. J Biol Chem 276: 7176–7186.
18. Wedlich-Soldner R, Wai SC, Schmidt T, Li R (2004) Robust cell polarity is a dynamic state established by coupling transport and GTPase signaling. J Cell Biol 166: 889–900.

19. Butty AC, Perrinjaquet N, Petit A, Jaquenoud M, Segall JE, et al. (2002) A positive feedback loop stabilizes the guanine-nucleotide exchange factor Cdc24 at sites of polarization. Embo J 21: 1565–1576.

20. Irazoqui JE, Gladfelter AS, Lew DJ (2003) Scaffold-mediated symmetry breaking by Cdc42p. Nat Cell Biol 5: 1062–1070.

21. Kozubowski L, Saito K, Johnson JM, Howell AS, Zyla TR, et al. (2008) Symmetry-breaking polarization driven by a Cdc42p GEF-PAK complex. Curr Biol 18: 1719–1726.

22. Nash P, Tang X, Orlicky S, Chen Q, Gertler FB, et al. (2001) Multisite phosphorylation of a CDK inhibitor sets a threshold for the onset of DNA replication. Nature 414: 514–521.

23. Strickfaden SC, Winters MJ, Ben-Ari G, Lamson RE, Tyers M, et al. (2007) A mechanism for cell-cycle regulation of MAP kinase signaling in a yeast differentiation pathway. Cell 128: 519–531.

24. Aghazadeh B, Lowry WE, Huang XY, Rosen MK (2000) Structural basis for relief of autoinhibition of the Dbl homology domain of proto-oncogene Vav by tyrosine phosphorylation. Cell 102: 625–633.

25. Ghaemmaghami S, Huh WK, Bower K, Howson RW, Belle A, et al. (2003) Global analysis of protein expression in yeast. Nature 425: 737–741.

26. Miyamoto S, Ohya Y, Sano Y, Sakaguchi S, Iida H, et al. (1991) A DBL-homologous region of the yeast CLS4/CDC24 gene product is important for Ca(2+)-modulated bud assembly. Biochem Biophys Res Commun 181: 604–610.

27. Michelitch M, Chant J (1996) A mechanism of Bud1p GTPase action suggested by mutational analysis and immunolocalization. Curr Biol 6: 446–454.

28. Toenjes KA, Sawyer MM, Johnson DI (1999) The guanine-nucleotide-exchange factor Cdc24p is targeted to the nucleus and polarized growth sites. Curr Biol 9: 1183–1186.

29. Toenjes KA, Simpson D, Johnson DI (2004) Separate membrane targeting and anchoring domains function in the localization of the S. cerevisiae Cdc24p guanine nucleotide exchange factor. Curr Genet 45: 257–264.

30. Rigaut G, Shevchenko A, Rutz B, Wilm M, Mann M, et al. (1999) A generic protein purification method for protein complex characterization and proteome exploration. Nat Biotechnol 17: 1030–1032.

31. Puig O, Caspary F, Rigaut G, Rutz B, Bouveret E, et al. (2001) The tandem affinity purification (TAP) method: a general procedure of protein complex purification. Methods 24: 218–229.

32. McCusker D, Denison C, Anderson S, Egelhofer TA, Yates JR 3rd, et al. (2007) Cdk1 coordinates cell-surface growth with the cell cycle. Nat Cell Biol 9: 506–515.

33. Wootton JC, Federhen S (1996) Analysis of compositionally biased regions in sequence databases. Methods Enzymol 266: 554–571.

34. Wootton JC (1994) Non-globular domains in protein sequences: automated segmentation using complexity measures. Comput Chem 18: 269–285.

35. Blom N, Gammeltoft S, Brunak S (1999) Sequence and structure-based prediction of eukaryotic protein phosphorylation sites. J Mol Biol 294: 1351–1362.

36. Obenauer JC, Cantley LC, Yaffe MB (2003) Scansite 2.0: Proteome-wide prediction of cell signaling interactions using short sequence motifs. Nucleic Acids Res 31: 3635–3641.

37. Tuazon PT, Spanos WC, Gump EL, Monnig CA, Traugh JA (1997) Determinants for substrate phosphorylation by p21-activated protein kinase (gamma-PAK). Biochemistry 36: 16059–16064.

38. Gururaj AE, Rayala SK, Kumar R (2005) p21-activated kinase signaling in breast cancer. Breast Cancer Res 7: 5–12.

39. Bender A, Pringle JR (1991) Use of a screen for synthetic lethal and multicopy suppressee mutants to identify two new genes involved in morphogenesis in Saccharomyces cerevisiae. Mol Cell Biol 11: 1295–1305.

40. Serber Z, Ferrell JE Jr (2007) Tuning bulk electrostatics to regulate protein function. Cell 128: 441–444.

41. Mionnet C, Bogliolo S, Arkowitz RA (2008) Oligomerization regulates the localization of Cdc24, the Cdc42 activator in Saccharomyces cerevisiae. J Biol Chem 283: 17515–17530.

42. Knaus M, Pelli-Gulli MP, van Drogen F, Springer S, Jaquenoud M, et al. (2007) Phosphorylation of Bem2p and Bem3p may contribute to local activation of Cdc42p at bud emergence. Embo J 26: 4501–4513.

43. Sopko R, Huang D, Smith JC, Figeys D, Andrews BJ (2007) Activation of the Cdc42p GTPase by cyclin-dependent protein kinases in budding yeast. Embo J 26: 4487–4500.

44. Burke D, Dawson D, Stearns T, Cold Spring Harbor Laboratory (2000) Methods in yeast genetics: a Cold Spring Harbor Laboratory course manual. Plainview, N.Y.: Cold Spring Harbor Laboratory Press.

45. Ito H, Fukuda Y, Murata K, Kimura A (1983) Transformation of intact yeast cells treated with alkali cations. J Bacteriol 153: 163–168.

46. Longtine MS, McKenzie A 3rd, Demarini DJ, Shah NG, Wach A, et al. (1998) Additional modules for versatile and economical PCR-based gene deletion and modification in Saccharomyces cerevisiae. Yeast 14: 953–961.

47. Huh WK, Falvo JV, Gerke LC, Carroll AS, Howson RW, et al. (2003) Global analysis of protein localization in budding yeast. Nature 425: 686–691.

48. Wilm M, Shevchenko A, Houthaeve T, Breit S, Schweigerer L, et al. (1996) Femtomole sequencing of proteins from polyacrylamide gels by nano-electrospray mass spectrometry. Nature 379: 466–469.

49. Gruhler A, Olsen JV, Mohammed S, Mortensen P, Faergeman NJ, et al. (2005) Quantitative phosphoproteomics applied to the yeast pheromone signaling pathway. Mol Cell Proteomics 4: 310–327.

50. Rappsilber J, Ishihama Y, Mann M (2003) Stop and go extraction tips for matrix-assisted laser desorption/ionization, nanoelectrospray, and LC/MS sample pretreatment in proteomics. Anal Chem 75: 663–670.

51. Haas W, Faherty BK, Gerber SA, Elias JE, Beausoleil SA, et al. (2006) Optimization and use of peptide mass measurement accuracy in shotgun proteomics. Mol Cell Proteomics.

52. Gatlin CL, Kleemann GR, Hays LG, Link AJ, Yates JR (1998) Protein identification at the low femtomole level from silver-stained gels using a new fritless electrospray interface for liquid chromatography microspray and nanospray mass spectrometry. Analytical Biochemistry 263: 93–101.
53. Elias JE, Gygi SP (2007) Target-decoy search strategy for increased confidence in large-scale protein identifications by mass spectrometry. Nat Methods 4: 207–214.
54. Eng JK, Mccormack AL, Yates JR (1994) An Approach to Correlate Tandem Mass-Spectral Data of Peptides with Amino-Acid-Sequences in a Protein Database. Journal of the American Society for Mass Spectrometry 5: 976–989.
55. Pringle JR (1991) Staining of bud scars and other cell wall chitin with calcofluor. Methods Enzymol 194: 732–735.

This article has supplemental information that is not featured in this version of the text. To view these files, please visit the original version of the article as cited in the beginning of this chapter.

PART II

YEAST REGULATORY CIRCUITS

CHAPTER 4

ROLES OF *CANDIDA ALBICANS* GAT2, A GATA-TYPE ZINC FINGER TRANSCRIPTION FACTOR, IN BIOFILM FORMATION, FILAMENTOUS GROWTH AND VIRULENCE

HAN DU, GUOBO GUAN, JING XIE, YUAN SUN, YAOJUN TONG, LIXIN ZHANG, AND GUANGHUA HUANG

4.1 INTRODUCTION

Candida albicans is the most common human fungal pathogen. With the increase in the number of immunocompromised patients, *Candida* infection is becoming more and more serious worldwide. *C. albicans* causes not only superficial infections, but also life-threatening systemic disease in immunocompromised hosts [1], [2]. Understanding the biology of this pathogen will definitely be helpful for developing new antifungal agents to combat this deadly pathogen.

C. albicans can grow in several morphological forms including unicellular yeast-form, elongated hyphae and pseudohyphae. The ability of switching between different growth modes is thought to be important for virulence [2], [3]. A variety of external and internal factors have been

This chapter was originally published under the Creative Commons Attribution License. Du H, Guan G, Xie J, Sun Y, Tong Y, Zhang L, and Huang G. Roles of Candida albicans Gat2, a GATA-Type Zinc Finger Transcription Factor, in Biofilm Formation, Filamentous Growth and Virulence. PLoS ONE 7,1 (2012). doi:10.1371/journal.pone.0029707.

shown to regulate morphogenetic transition in this yeast. For example, the addition of inducers such as serum and N-acetylglucosamine (GlcNAc) and increases in temperature and pH can promote filamentous growth [4]. *C. albicans* morphogenesis is also regulated by a number of signal transduction pathways as well as a number of key regulators such as kinases and zinc finger transcription factors. Two major pathways, the cAMP/PKA mediated pathway and the Cst20-Ste11-Hst7-Cph1 pathway, have been intensively investigated [2], [5], [6], [7], [8].

C. albicans can also exist as biofilms, which are surface-associated microbial communities [9]. Many infections are related to the formation of biofilms on implanted medical devices [9], [10]. The morphological transition between cell types plays a critical role in *C. albicans* biofilm development under different circumstances. Baillie et al. has shown that the hyphal growth defective mutant only produces the basal layer, while the yeast growth defective mutant develops a thick, hyphal biofilm [11]. Deletion of *UME6*, encoding a filament-induced transcription factor, leads to hyphal growth and biofilm development defects [12]. Interestingly, the quorum sensing molecule farnesol, which is a filamentous growth inhibitor, also inhibits biofilm formation [13]. These facts indicate that the morphogenesis regulated pathways are also involved in regulation of biofilms in *C. albicans* [10].

The filamentous development regulator Tec1 is involved in biofilm formation by controlling the expression of Bcr1, a C_2H_2 zinc finger transcription factor [14].

Recently, Sahni et al. found that Tec1 specifically regulates the pheromone induced biofilm development of the *C. albicans* white phenotype by screening a library for 107 transcription factors [15]. In order to get more insights into the regulation of biofilm formation and morphorgenesis in *C. albicans*, we did a more extensive screen by using the same library with modified methods. In the Sanhi's study, they did the screen at 25°C and only cultured the cells for 16 hours [15]. The major modification of this screen was the increase of culture temperature to 30°C and extension of incubation time to 48 hours. In this study, we identified three more adhesion-promoting transcription factors (Cph1, Ume6 and Gat2) in addition to Tec1 which has also been discovered in the previous study [15]. Since the roles of Cph1, Ume6 and Tec1 in morphogenesis and biofilm formation have been intensively investigated, in this study we focused on the biological roles of Gat2, a GATA-type zinc finger transcription factor.

4.2 RESULTS

4.2.1 SCREEN FOR THE ADHESION-PROMOTING TRANSCRIPTION FACTORS

C. albicans biofilm development includes a series of sequential steps: adherence→initiation→maturation→dispersal[10]. A lot of transcriptional regulators have been reported to control specific steps of the developmental process. By screening an overexpression library under the control of a doxycycline-inducible promoter [16], Sahni et al. have identified one adhesion-promoting transcription factor, that is, Tec1. Tec1 has been proved to be required for pheromone induced response in *C. albicans* white cells [15]. Given the complexity of biofilm development, we hypothesized that there would be more transcription factors or a transcription circuitry involved in the process. To prove this, we did another screen by using the same overexpression library constructed by the Soll lab [15]. We did the screen in 96-well plates at 30°C rather than at 25°C published in the previous study [15]. More importantly, we extended culture time to 48 hours since the culture time was critical for full biofilm development. These changes allowed mature biofilm development and lowered the threshold of screening the adhesion-promoting genes. Besides Tec1 identified in the early publication [15], we found 3 more adhesion-promoting transcription factors, including Cph1, Ume6 and Gat2 (Figure 1). Cph1 is a homolog of *S. cerevisiae* Ste12, which is required for mating and filamentous growth in the yeast. Deletion of CPH1 results in hyphal growth defect on solid Spider medium [8] and blocks mating in *C. albicans* [17], but does not affect biofilm formation [5], [13], [15]. Ume6 has been shown to be required for hyphal extension, adherence to plastic and virulence [12]. Gat2 is a GATA-type zinc finger transcription factor, which has been shown to regulate filamentous growth on Spider medium in a high-throughput

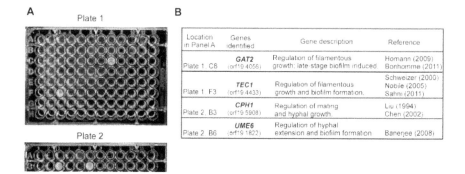

FIGURE 1: Biofilm development promoting transcription factors identified by a screen of an overexpression library. A. Screening of the library in 96-well plates. Overnight cultures were inoculated into a 96-well plate and then incubated for 48 hours with shaking at 30°C. The plates were washed with PBS and imaged. The parent strain transformed with the library was P37005, a natural MTLa/a isolate. B. Description of the four transcription factors identified.

screen [18]. However, the molecular mechanism of filamentous growth regulation of Gat2 and its roles in invasive growth, biofilm formation and virulence remain unclear.

4.2.2 OVEREXPRESSION OF ADHESION-PROMOTING GENES (GAT2, TEC1, CPH1 AND UME6) INDUCES FILAMENTOUS GROWTH

Filamentous growth ability directly relates to adhesion and biofilm formation in *C. albicans*. To further confirm the transcription factors we screened, we investigated the roles of the four transcription factors in promoting filamentous growth. Since the WT strain forms normal filamentous colonies at 37°C, the experiment was performed at 30°C, a temperature not favoring filamentous growth for *C. albicans*. As shown in Figure S1, overexpression of the transcription factors *GAT2, TEC1, CPH1* and *UME6* in a WT strain promoted filamentous growth dramatically. The strain WT+ vector served as a negative control.

4.2.3 ROLE OF GAT2 IN BIOFILM FORMATION

To validate the adhesion-promoting activity of Gat2, we first tested the ability of adherence to the plastic 96-well plate bottoms in the *GAT2*-over-expression strain (WT+TETp-GAT2), *gat2/gat2* mutant (*gat2/gat2*+v) and the GAT2-reconstituted strain (*gat2/gat2*+TETp-GAT2). The wild type strain (WT+ v) carrying an empty vector served as control. At 30°C, all the strains were unable to adhere to the plastic bottom in the absence of doxycyline, while WT+TETp-GAT2 and *gat2/gat2*+TETp-GAT2 showed enhanced adhesion in the presence of 100 µg/ml doxycycline. The WT+ v and *gat2/gat2*+v strains failed to adhere to the plastic bottom even in the presence of 100 µg/ml doxycycline. The cells adhered to the bottoms were released and quantified by counting (Figure 2A).

At 37°C, in contrast to the WT+ v and WT+TETp-GAT2 strains, the *gat2/gat2*+v mutant was unable to form biofilms on the plastic bottom either in the presence or in the absence of 100 µg/ml doxycycline. However, the reconstituted strain *gat2/gat2*+TETp-GAT2 adhered to the bottom almost as strongly as the WT+TETp-GAT2 strain did in the presence of 100 µg/ml doxycycline (Figure 2B).

We also tested for the ability of biofilm development on a silicone cob in the strains as indicated in Figure 2C and D. The *gat2/gat2*+v mutant failed to form biofilm on the silicone surface at both 30°C and 37°C either in the presence or in the absence of doxycycline. The *gat2/gat2*+TETp-GAT2 strain formed normal biofilm as the reference strain did in the medium containing 100 µg/ml doxycycline at both temperatures (Figure 2C, D). At 30°C, the WT+ v formed normal biofilms on the silicone surface, although the ratio of hyphal cells to yeast cells was much lower than that at 37°C (data not shown). The cells adhered to the surface were quantified (Figure 2C, D). Visualization of scanning electron microscopy (SEM) confirmed the inability of biofilm development of the *gat2/gat2* mutant on silicone material surface (Figure 3). The *gat2/gat2*+TETp-GAT2 and WT+ TETp-GAT2 strains underwent robust filamentous growth and formed thick biofilms under inducing condition (Figure 3), while their phenotypes were similar to the *gat2/gat2* mutant and WT strains, respectively, under non-inducing condition (data not shown). The biofilm ultrastructure indicated that the biofilm formed by the wild type reference strain was a mixture of

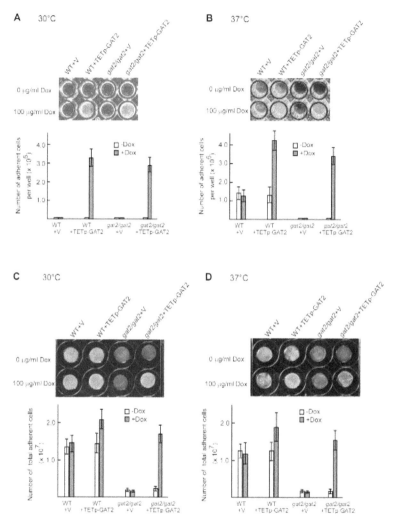

FIGURE 2: Role of Gat2 in biofilm formation. The parent strain for the gat2/gat2 mutant and WT was SN152 (MTLa/α) [18]. A. Biofilm formation on the plastic bottoms of the strains (WT+ vector, WT+TETp-GAT2, gat2/gat2 mutant+vector, gat2/gat2 mutant+TETp-GAT2) at 30°C. Overnight cultures were inoculated into a 96-well plate and then incubated for 48 hours with shaking. The plates were washed with water and imaged. The numbers of cells adhered to the bottom represented three independent experiments. B. Biofilm formation on the plastic bottoms at 37°C. The same assay was used as in Panel A. C. Biofilm formation on silicone rubbers at 30°C. The silicone was incubated in a 24-well plate. The number of cells adhered to the silicone represented three independent experiments. D. Biofilm formation on silicone rubbers at 37°C. The number of cells adhered to the silicone represented three independent experiments.

yeast cells and filamentous cells both at 30°C and at 37°C. However, the percentage of filamentous cells was much higher at 37°C than that at 30°C. The *gat2/gat2* mutant failed to undergo filamentous growth at 30°C and formed a few elongated cells at 37°C. Notably, the SEM images showed that the *gat2/gat2* mutant remained the basal level ability of adhering to the surface, although the number of adhered cells was much less than that of the WT.

4.2.4 DELETION OF GAT2/GAT2 RESULTS IN FILAMENTOUS GROWTH DEFECT ON LEE'S MEDIUM PLATES

Given the importance of hyphal development in biofilm formation, we hypothesized that Gat2 could play critical roles in filamentous growth. To test this, we first examined the hyphal growth ability of the strains WT+ v, *gat2/gat2*+v, WT+TETp-GAT2 and *gat2/gat2*+TETp-GAT2 on agar containing Lee's medium. In contrast to WT+ v, the mutant *gat2/gat2*+v was unable to form filamentous colonies at 37°C (Figure 4A). In the presence of 50 µg/ml doxycycline, the overexpression strain WT+TETp-GAT2 showed slightly stronger ability of filamentation than the WT+ v and *gat2/gat2*+TETp-GAT2 strains did (Figure 4A). At the cellular level, under inducing condition the WT+TETp-GAT2 was composed of over 98% of filamentous cells, while the WT+ v and *gat2/gat2*+TETp-GAT2 were composed of ~80% and 95% of filamentous cells, respectively. We also did similar experiments at 25°C, a temperature unfavorable for *C. albicans* filamentous growth. We observed that only the overexpression strain was able to form star-like filamentous colonies under inducing condition at this temperature (Figure 4B). The *gat2/gat2*+TETp-GAT2 strain showed weak filamentous growth. The cellular images were also presented in Figure 4B. The additive effect of the endogenous and the ectopic expression of GAT2 gene could result from the increased gene copies.

4.2.5 GAT2 IS NOT REQUIRED FOR SERUM AND GLCNAC INDUCED FILAMENTOUS GROWTH

Different environmental cues induce filamentous growth through distinct pathways. Serum is thought to be the most potent hyphal inducer.

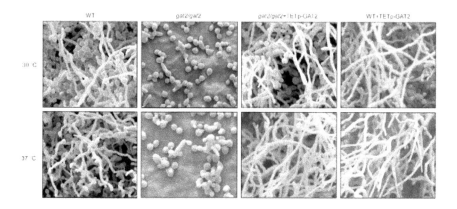

FIGURE 3: SEM of biofilms formed on silicone surfaces by the WT, *gat2/gat2* mutant, *gat2/gat2*+TETp-GAT2 and WT+TETp-GAT2 at 30 and 37°C. The images of *gat2/gat2*+TETp-GAT2 and WT+TETp-GAT2 under the inducing condition (100 μg/ml doxycycline) were shown. The data represented three independent experiments.

We therefore tested whether Gat2 was also required for serum induced hyphal development. As shown in Figure 5, deletion of *GAT2* obviously attenuated filamentous growth ability but did not block the effect of serum induction both on agar and in liquid medium at 37°C. Compared to the reference strain, the *gat2/gat2* mutant formed small and less branched hyphal colonies on agar+serum plates. Consistently, the *gat2/gat2* mutant formed shorter hyphal cells in liquid YPD+10% serum medium.

GlcNAc is a powerful filamentous growth inducer in *C. albicans* [19]. Recently, we have reported that GlcNAc also regulates white-to-opaque transition via Ras1-cAMP/PKA pathway in this organism [20]. To investigate whether Gat2 was essential for GlcNAc induced yeast-to-hyphal transition, we incubated the reference and the mutant strains on SD-GlcNAc plates at 37°C. We found that the *gat2/gat2* null mutant formed obviously wrinkled hyphal colonies, although they were not highly wrinkled as those formed by the reference strain (Figure 5). These results indicate that Gat2 is not required for serum or GlcNAc induced filamentous growth.

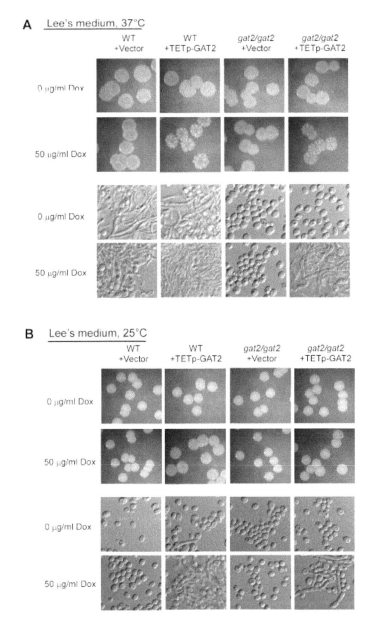

FIGURE 4: Gat2 is required for filamentous growth on Lee's medium. The strains indicated were incubated under non-inducing (0 μg/ml doxycycline) or inducing (50 μg/ml doxycycline) conditions. Colony and cellular pictures were shown. A. Filamentous growth at 37°C. B. Filamentous growth at 25°C.

4.2.6 GAT2 IS REQUIRED FOR INVASIVE GROWTH

The ability of *C. albicans* to undergo invasive growth is tightly linked to infection. To test the role of Gat2 in invasive growth, we performed the experiments at both 25 and 37°C. At 25°C, all the strains indicated in the figure were unable to undergo invasive growth under non-inducing condition, while the WT+TETp-GAT2 and *gat2/gat2*+TETp-GAT2 strains showed invasive growth under inducing condition (Figure 6). And the WT+TETp-GAT2 showed stronger invasive growth ability than the *gat2/gat2*+TETp-GAT2 strain did. At 37°C, under non-inducing condition the *gat2/gat2*+v and *gat2/gat2*+TETp-GAT2 failed to undergo invasive growth as the WT+ v and WT+TETp-GAT2 strains did. Under inducing condition, in contrast to the *gat2/gat2*+v strain, the *gat2/gat2*+TETp-GAT2 also underwent invasive growth as the WT+ v and WT+TETp-GAT2 strains did (Figure 6).

4.2.7 DELETION OF GAT2 ATTENUATES C. ALBICANS VIRULENCE IN A MOUSE MODEL OF SYSTEMIC INFECTION

Filamentous morphogenesis is important for *C. albicans* virulence. Therefore, we tested the virulence of *gat2/gat2* mutant in a mouse model of systemic candidiasis. All of the mice injected with the WT reference strain died within 6 days, whereas the mice injected with *gat2/gat2* mutant died after a longer time period. Three mice (37.5%) were still alive at 20 days post-infection (Figure 7). To confirm the decrease of virulence was due to the deletion of GAT2, we constructed a complemented strain by inserting a fragment containing GAT2 ORF and ~400 bp promoter on the original GAT2 locus. The filamentous growth ability of the complemented strains was restored (Figure 7A), although it was weaker than the WT. Notably, all the mice injected with the complemented strain were died within 8 days, suggesting that it was almost as virulent as the reference strain (Figure 7B). These results indicate that *gat2/gat2* mutant plays a role in virulence at least in a systemic infection model of mice.

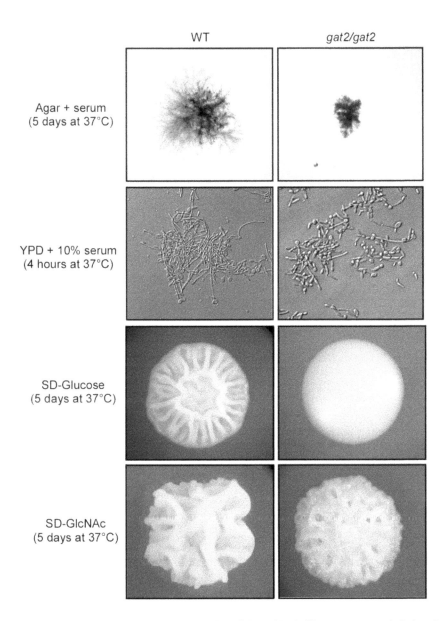

FIGURE 5: Deletion of GAT2 impaired, but did not block filamentous growth induced by serum or GlcNAc. On solid agar, the strains were cultured for 5 days at 37°C and then imaged. In liquid YPD+serum medium, the strains were cultured for 4 hours at 37°C with shaking.

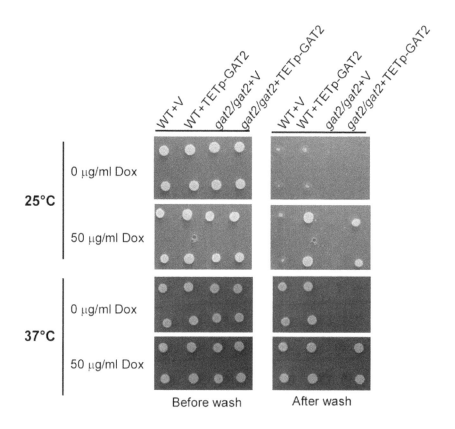

FIGURE 6: Role of Gat2 in invasive growth. The strains indicated were incubated under non-inducing (0 µg/ml doxycycline) or inducing (50 µg/ml doxycycline) conditions at 25°C or 37°C. The plates were imaged before and after washing.

4.3 DISCUSSION

Recently, Sahni et al. have identified *C. albicans* Tec1 as a key regulator of pheromone induced biofilm development through screening a transcription factor overexpression library [15]. To get more extensive insights into the molecular mechanism of biofilm and hyphal development, we did another screen using the same library with modified protocols in this study. We identified three more adhesion-promoting transcription factors (Cph1, Ume6 and Gat2) besides Tec1 that was also described in the previous report [15]. The roles of Tec1, Cph1 and Ume6 have been intensively

A

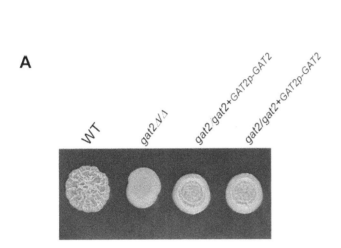

B

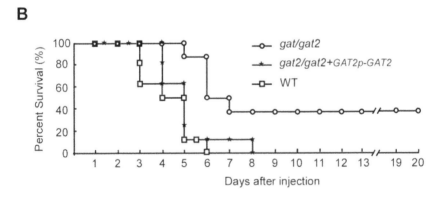

FIGURE 7: The *gat2/gat2* mutant is attenuated for virulence in a mouse systemic infection model. A. Filamentous growth of the reference strain, *gat2/gat2* mutant and the complemented strain *gat2/gat2*+GAT2p-GAT2. ~3000 cells of each strain in 3 μL of ddH2O were dropped onto solid Lee's medium plates and incubated at 37°C for 3 days. Two independently isolates of the complemented strains were shown. B. Survival curves for the reference strain, *gat2/gat2* mutant and the complemented strain *gat2/gat2*+GAT2p-GAT2. For each strain, 8 mice were used for infection.

investigated by us and others [8], [12], [15], [21]. Tec1 plays a critical role in both biofilm formation and filamentous growth [14], [15], [21]. The transcription of *TEC1* gene is regulated by Cph2, a Myc-bHLH family transcriptional activator of filamentous growth [22], while Tec1 controls the expression of biofilm regulator Bcr1 [14]. Cph1 regulates mating and filamentous growth in *C. albicans* [8], [17]. Ume6 has been proved to be required for hyphal extension, adhesion and virulence [12]. The GATA-type transcription factor Gat2 has been reported to be required for fila-mentous growth on Spider medium, a nutrient-poor medium for morpho-logical analysis [18]. Consistent with previous study, we found that all the four transcription factors (Tec1, Cph1, Ume6 and Gat2) promote filamen-tous growth on solid Lee's medium. Additional experiments indicate that Gat2 plays important roles in biofilm formation, filmentous and invasive growth, and also virulence.

Biofilm development is controlled by a number of transcription fac-tors including Tec1, Bcr1, Ume6, Efg1 and Zap1 [12], [14], [15], [23], [24]. Here, we added Gat2 to the list of biofilm regulators. Overexpression of GAT2 in a wild type strain promotes adhesion and biofilm formation, while deletion of GAT2 results in biofilm development defect.

Filamentous growth ability is thought to be important for biofilm de-velopment. Hyphae provide the structure integrity and multilayered archi-tecture feature of muture biofilms [9]. Although Gat2 is not essential for serum- and GlcNAc-induced hyphal growth, deletion of *GAT2* notably attenuated the ability of hyphal growth stimulated by these two inducers, especially by serum. Remarkably, deletion of *GAT2* completely blocked filamentous growth in Lee's medium. These data suggest that different environmental cues activate filamentous growth via different pathways. Gat2 plays critical roles in Lee's medium induced morphogenesis and is at least partially involved in regulation of serum- and GlcNAc-induced yeast-to-hyphal transition. Consistently, we found that Gat2 is also not required for GlcNAc induced white-to-opaque switching in an MTLa/a strain (data not shown). Given the importance of filamentous growth abil-ity in biofilm development, Gat2 possibly regulates biofilm development through filamentous growth control. Gat2 could be involved in regulation of the biofilm "initiation" and "maturation" steps, in which filamentous cells play critical roles [10].

In *S. cerevisiae*, ScTec1 binds to the promoter of *ScGAT2* [25]. By sequence analysis, we found two putative Tec1 binding sites on the promoter region of GAT2 gene (TCATTCT and ACATTCT) [26]. Interestingly, *tec1/tec1* mutant showed similar phenotypes on SD-glucose and SD-GlcNAc media as *gat2/gat2* mutant did. Both Tec1 and Gat2 were not required for filamentous growth induced by GlcNAc, but were essential for full hyphal development on SD-glucose medium (data not shown). Similar roles of Gat2 and Tec1 in adhesion and GlcNAc induced filamentous growth suggest that Gat2 possibly functions downstream of Tec1 in regulation of morphogenesis and biofilm development.

Our findings reveal that Gat2 is involved in regulation of biofilm development, morphogenesis and virulence. On Lee's medium plates which are characterized by neutral pH and poor nutrient, deletion of GAT2 gene completely blocked filamentous and invasive growth. However, Gat2 is not essential for filamentous growth induced by some environmental cues, such as serum and GlcNAc. We propose that Gat2 specifically regulates morphogenesis in some host niches.

4.4 MATERIALS AND METHODS

4.4.1 *STRAINS AND GROWTH CONDITIONS*

The transcription factor overexpression library was constructed by the Soll lab [15]. The *gat2/gat2* mutant and the reference strain were requested from the Johnson's lab [18]. While the strains used in Figure 1 and Supplemental figure S1 were homozygous at MTL locus (a/a), all the others were MTL heterozygous (a/α). Solid YPD medium (20 g/L Difco peptone, 10 g/L Yeast extract, 20 g/L glucose, 20 g/L Agar) and Lee's medium supplemented with 5 µg/ml phloxin B were used for routine growth. Lee's medium, SD-glucose, SD-GlcNAc and agar+serum plates were used for filamentous development [27]. In the SD-GlcNAc medium, 2% GlcNAc replaced glucose as carbon source. K_2HPO_4 (2.5 g/L) was added to the SD-glucose and SD-GlcNAc media for pH maintenance. The pH of Lee's and SD media was adjusted to 6.8 with 10% HCl.

To construct the complemented strain *gat2/gat2*+GAT2p-GAT2, we first generated a plasmid pGAT2res for transformation of the *gat2/gat2* mutant. A fragment of the 3-UTR of *GAT2* and a fragment containing the *GAT2* ORF and ~400 bp of the promoter region were subsequently inserted into the plasmid pNIM1, and yielded pGAT2res. The complemented strain was generated by transforming the *gat2/gat2* mutant with SalI digested pGAT2res fragments. The oligonucleotides used for PCR were listed below:

- GAT2-3F-xho: 5-aatcaaCTCGAGcgctgtaaattatatcctga-3;
- GAT2-3R-Bglsal: 5-aatcaaAGATCTatGTCGACatatctcagtgcagaaacagg-3;
- GAT2-5F-Sal: 5-aatcaaGTCGACaacccgtttaacatttgcagc-3;
- GAT2-5R-Bgl: 5-aatcaaAGATCTaattagatgtgtacattaatttctatg-3.

4.4.2 BIOFILM ASSAY

Biofilm experiments were performed as described previously with slight modifications [15]. For adhesion to the plastic bottoms, cells were cultured in Costar 96-well Cell Culture Plates at temperatures indicated in the main text. After 48 hours of incubation with shaking, the wells were gently washed with 1× PBS (phosphate-buffered saline). The bottoms were imaged. Biofilm growth on silicone material was performed as reported with slight modification [14]. Briefly, cells were incubated in a well of a 24-well cell culture plate containing a round silicone block with a diameter of 1 cm. Silicone blocks were cut from Cardiovascular Instrument silicone sheets. After 48 hours of culture, the silicone blocks were carefully washed and taken out for imaging. After gently washing, the cells adhered to the bottoms of 96-well plates or silicone material were treated with trypsin and collected for quantitation.

4.4.3 SCANNING ELECTRON MICROSCOPY (SEM)

For SEM, we developed *C. albicans* biofilms on silicone blocks. The SEM assay was performed as previously reported [11]. Briefly, the samples were gently washed with 1× PBS and fixed with 2.5% glutaraldehyde. Then, the samples were washed three times with 0.1 M Na_3PO_4 buffer (pH 7.2), dehydrated in increasing concentrations of ethanol (30% - 50% - 70% - 85% - 95% - 100%) and coated with gold. The surface of the biofilm was imaged with a scanning electron microscopy (FEI QUANTA 200).

4.4.4 INVASIVE AND FILAMENTOUS GROWTH ASSAYS

Lee's medium plates with or without doxycycline as indicated were used for invasive growth. 3 μl of liquid medium containing 2×10^4 cells was dropped onto the agar for 2 days (at 37°C) or 5 days (at 25°C) of incubation. The plates were imaged before and after washing with H_2O. Agar containing serum, Lee's medium, SD-glucose or SD-GlcNAc medium was used for filamentous growth analysis. The colonies were imaged after 5 days' culture at temperatures indicated.

4.4.5 VIRULENCE EXPERIMENTS

The virulence of *C. albicans* strains was performed as reported by Chen et al. [28]. ICR male mice (18–22 g) were used for the systemic infection experiments. 100 μl of 1× PBS containing 2×10^6 cells was injected into each mouse. All animal experiments were performed according to the guidelines approved by the Animal Care and Use Committee of the Institute of Microbiology, Chinese Academy of Sciences (permit number: IMCAS2011002). The present study was approved by the Committee.

REFERENCES

1. Pfaller MA, Diekema DJ (2007) Epidemiology of invasive candidiasis: a persistent public health problem. Clin Microbiol Rev 20: 133–163.
2. Whiteway M, Bachewich C (2007) Morphogenesis in *Candida* albicans. Annu Rev Microbiol 61: 529–553.
3. Lo HJ, Kohler JR, DiDomenico B, Loebenberg D, Cacciapuoti A, et al. (1997) Non-filamentous *C. albicans* mutants are avirulent. Cell 90: 939–949.

4. Cottier F, Muhlschlegel FA (2009) Sensing the environment: response of *Candida albicans* to the X factor. FEMS Microbiol Lett 295: 1–9.
5. Yi S, Sahni N, Daniels KJ, Lu KL, Srikantha T, et al. (2011) Alternative Mating Type Configurations (a/alpha versus a/a or alpha/alpha) of *Candida albicans* Result in Alternative Biofilms Regulated by Different Pathways. PLoS Biol 9: e1001117.
6. Csank C, Schroppel K, Leberer E, Harcus D, Mohamed O, et al. (1998) Roles of the *Candida albicans* mitogen-activated protein kinase homolog, Cek1p, in hyphal development and systemic candidiasis. Infect Immun 66: 2713–2721.
7. Leberer E, Harcus D, Broadbent ID, Clark KL, Dignard D, et al. (1996) Signal transduction through homologs of the Ste20p and Ste7p protein kinases can trigger hyphal formation in the pathogenic fungus *Candida* albicans. Proc Natl Acad Sci U S A 93: 13217–13222.
8. Liu H, Kohler J, Fink GR (1994) Suppression of hyphal formation in *Candida albicans* by mutation of a STE12 homolog. Science 266: 1723–1726.
9. Ramage G, Saville SP, Thomas DP, Lopez-Ribot JL (2005) *Candida* biofilms: an update. Eukaryot Cell 4: 633–638.
10. Finkel JS, Mitchell AP (2011) Genetic control of *Candida albicans* biofilm development. Nat Rev Microbiol 9: 109–118.
11. Baillie GS, Douglas LJ (1999) Role of dimorphism in the development of *Candida albicans* biofilms. J Med Microbiol 48: 671–679.
12. Banerjee M, Thompson DS, Lazzell A, Carlisle PL, Pierce C, et al. (2008) UME6, a novel filament-specific regulator of *Candida albicans* hyphal extension and virulence. Mol Biol Cell 19: 1354–1365.
13. Ramage G, Saville SP, Wickes BL, Lopez-Ribot JL (2002) Inhibition of *Candida albicans* biofilm formation by farnesol, a quorum-sensing molecule. Appl Environ Microbiol 68: 5459–5463.
14. Nobile CJ, Andes DR, Nett JE, Smith FJ, Yue F, et al. (2006) Critical role of Bcr1-dependent adhesins in *C. albicans* biofilm formation in vitro and in vivo. PLoS Pathog 2: e63.
15. Sahni N, Yi S, Daniels KJ, Huang G, Srikantha T, et al. (2010) Tec1 mediates the pheromone response of the white phenotype of *Candida* albicans: insights into the evolution of new signal transduction pathways. PLoS Biol 8: e1000363.
16. Park YN, Morschhauser J (2005) Tetracycline-inducible gene expression and gene deletion in *Candida* albicans. Eukaryot Cell 4: 1328–1342.
17. Chen J, Lane S, Liu H (2002) A conserved mitogen-activated protein kinase pathway is required for mating in *Candida* albicans. Mol Microbiol 46: 1335–1344.
18. Homann OR, Dea J, Noble SM, Johnson AD (2009) A phenotypic profile of the *Candida albicans* regulatory network. PLoS Genet 5: e1000783.
19. Simonetti N, Strippoli V, Cassone A (1974) Yeast-mycelial conversion induced by N-acetyl-D-glucosamine in *Candida* albicans. Nature 250: 344–346.
20. Huang G, Yi S, Sahni N, Daniels KJ, Srikantha T, et al. (2010) N-acetylglucosamine induces white to opaque switching, a mating prerequisite in *Candida* albicans. PLoS Pathog 6: e1000806.
21. Schweizer A, Rupp S, Taylor BN, Rollinghoff M, Schroppel K (2000) The TEA/ATTS transcription factor CaTec1p regulates hyphal development and virulence in *Candida* albicans. Mol Microbiol 38: 435–445.

22. Lane S, Zhou S, Pan T, Dai Q, Liu H (2001) The basic helix-loop-helix transcription factor Cph2 regulates hyphal development in *Candida albicans* partly via TEC1. Mol Cell Biol 21: 6418–6428.

23. Nobile CJ, Nett JE, Hernday AD, Homann OR, Deneault JS, et al. (2009) Biofilm matrix regulation by *Candida albicans* Zap1. PLoS Biol 7: e1000133.

24. Ramage G, VandeWalle K, Lopez-Ribot JL, Wickes BL (2002) The filamentation pathway controlled by the Efg1 regulator protein is required for normal biofilm formation and development in *Candida* albicans. FEMS Microbiol Lett 214: 95–100.

25. MacIsaac KD, Wang T, Gordon DB, Gifford DK, Stormo GD, et al. (2006) An improved map of conserved regulatory sites for Saccharomyces cerevisiae. BMC Bioinformatics 7: 113.

26. Lane S, Birse C, Zhou S, Matson R, Liu H (2001) DNA array studies demonstrate convergent regulation of virulence factors by Cph1, Cph2, and Efg1 in *Candida* albicans. J Biol Chem 276: 48988–48996.

27. Hnisz D, Tscherner M, Kuchler K (2011) Morphological and molecular genetic analysis of epigenetic switching of the human fungal pathogen *Candida* albicans. Methods Mol Biol 734: 303–315.

28. Chen J, Zhou S, Wang Q, Chen X, Pan T, et al. (2000) Crk1, a novel Cdc2-related protein kinase, is required for hyphal development and virulence in *Candida* albicans. Mol Cell Biol 20: 8696–8708.

This article has supplemental information that is not featured in this version of the text. To view these files, please visit the original version of the article as cited in the beginning of this chapter.

CHAPTER 5

DECIPHERING HUMAN HEAT SHOCK TRANSCRIPTION FACTOR 1 REGULATION VIA POST-TRANSLATIONAL MODIFICATION IN YEAST

LILIANA BATISTA-NASCIMENTO, DANIEL W. NEEF, PHILLIP C. C. LIU, CLAUDINA RODRIGUES-POUSADA, AND DENNIS J. THIELE

5.1 INTRODUCTION

All organisms are exposed to proteotoxic stresses that result in the accumulation of misfolded proteins. In response to these stresses cells have evolved adaptive responses to protect and stabilize cellular proteins until more favorable conditions for cell proliferation are encountered [1]. The heat shock transcription factor, HSF, is a homotrimeric transcription factor that activates gene expression in response to a variety of stresses including heat and oxidative stress, as well as inflammation and infection [2]. Recent evidence has shown that the *S. cerevisiae* HSF directly activates the expression of genes whose protein products are involved in protein folding and degradation, ion transport, signal transduction, energy generation, carbohydrate metabolism, vesicular transport, cytoskeleton formation and other cellular functions [3].

This chapter was originally published under the Creative Commons Attribution License. Batista-Nascimento L, Neef DW, Liu PCC, Rodrigues-Pousada C, and Thiele DJ. Deciphering Human Heat Shock Transcription Factor 1 Regulation via Post-Translational Modification in Yeast. PLoS ONE 6, (2011). doi:10.1371/journal.pone.0015976.

While mammalian cells express four distinct HSF proteins encoded by separate genes, HSF1 is the primary factor responsible for stress responsive gene transcription [2]. In the absence of stress, mammalian HSF1 is repressed through mechanisms that are not well understood. HSF1 is thought be maintained in an inactive monomeric state through intramolecular interactions between a hydrophobic coiled-coil domain in the carboxyl-terminus of the protein and three amino-terminal coiled-coils required for homotrimerization and transcriptional activation [4], [5], [6]. HSF1 is also thought to be bound and repressed by the protein chaperones Hsp90 and Hsp70, though it is not clear how these chaperones repress HSF1 activity [7], [8], [9], [10]. Studies suggest that during the initial phase of the stress response, the inactive HSF1 monomer dissociates from Hsp90, homotrimerizes, is transported to the nucleus and binds to heat shock elements (HSE) found in the promoters of HSF target genes [10], [11]. The DNA-bound homotrimer, remains relatively transcriptionally inert [12], potentially due to the continued interaction with Hsp70 and the HSF1-transactivation domain [9]. Stress-dependent hyperphosphorylation of HSF1 by potentially multiple protein kinases has been proposed to, in part, promote HSF1 dependent transactivation [13], [14], [15].

The activity of HSF1 is also thought to be negatively regulated through a number of post-translational modifications including phosphorylation, sumoylation and acetylation [16], [17], [18], [19]. Mass spectrometry analyses have shown HSF1 to be phosphorylated on at least 12 serine residues [13] and phosphorylation of S121, S303, S307 and S363 have been correlated with a repression in HSF1 activity [18], [20], [21]. The most comprehensively studied of these phosphorylation events are the phosphorylation of S303 and S307. However, much of what is known about S303 and S307 phosphorylation stems from in vitro phosphorylation experiments and in vivo studies using either lexA or Gal4-HSF1 fusion proteins lacking the native HSF1 DNA binding domain. As such, many of the earlier studies exploring S303 and S307-dependet regulation of HSF1 activity have resulted in conflicting results. For example, previous phosphorylation experiments suggested that S307 was phosphorylated by ERK which, in turn, acted as an essential priming step for GSK3-dependent phosphorylation of S303 [22]. However, subsequent in vitro studies suggested that S303 could also be phosphorylated by a variety of mitogen activated protein kinases (MAPK) including the stress responsive MAPK p38 [17], [18]. In addition, subsequent in vivo data suggested S303 phosphorylation could occur independently of S307 phosphorylation [16].

While the specific mechanism by which S303 and S307 phosphorylation repress HSF1 activity remains unclear, evidence has suggested that S303 and S307 phosphorylation represses the transactivation potential of HSF1 [18], [22], [23]. S303 and S307 are constitutively phosphorylated in the absence of stress and S303 phosphorylation levels increase after exposure to stress, suggesting that this phosphorylation event might also contribute to HSF1 inactivation during the recovery phase [16], [17]. Interestingly, phosphorylation of S303, but not S307, promotes sumoylation of K298 [16] which, like S303 phosphorylation, also increases in response to stress exposure and represses HSF1-dependent transactivation [24]. However, it remains unclear if the repressive effects of S303 phosphorylation on HSF1 activity are exclusively mediated through K298 sumoylation or occur through additional mechanisms.

While HSF1 and the cognate HSEs are quite well conserved from yeast to humans, our previous results demonstrated that human HSF1 expressed in S. cerevisiae is unable to complement for the loss of the essential yeast HSF protein [25]. Further analysis showed that human HSF1 expressed in yeast was unable to form a homotrimer and consequently unable to activate HSE-dependent gene expression to support cell viability. Human HSF1 homotrimerized, became active and complemented for the loss of yeast HSF when three derepressing mutations, collectively known as LZ4m, were introduced into the repressive carboxyl-terminal coiled-coil domain [6], [25]. Further studies in yeast identified an amino-terminal linker-domain as well as a loop in the DNA binding domain as repressive elements that contributed to HSF1 repression in both yeast and mammalian cells [26], [27]. We have also used to the yeast assay system to screen for and indentify novel pharmacological activators of human HSF1 [28]. Together, these results suggest that human HSF1 expressed in yeast is maintained in a constitutively repressed state through mechanisms similar to those of mammalian cells and that the yeast system can serve as a simplified assay system to decipher the complex mechanisms regulating human HSF1 activity.

Here we report the use of the yeast assay system to further understand the mechanisms that regulate human HSF1 through phosphorylation of serine 303. Our results suggest that S303 phosphorylation blocks human HSF1 homotrimerization thereby preventing human HSF1 activation and complementation of the loss of yeast HSF. Furthermore, we demonstrate that S303 phosphorylation also blocks HSF1 homotrimerization in mammalian cells. We show that phosphorylation of HSF1 S303 in yeast occurs via the action of the MAPK Slt2 and not via the action of GSK3 and we extend these findings to show that S303 phosphorylation also occurs independent of GSK3 in mammalian cells.

5.2 RESULTS

5.2.1 PHOSPHORYLATION OF S303 CONTRIBUTES TO REPRESSION OF HUMAN HSF1 IN YEAST

When human HSF1 is expressed in yeast it is unable to homotrimerize, promote gene expression and complement for the loss of the essential yeast HSF protein [25]. Because our previous work suggested that when HSF1 is expressed in yeast it exists in a constitutively repressed monomeric state, we sought to use the yeast assay system to better understand the complex mechanisms regulating HSF1 activity in mammalian cells. An important component of HSF1 repression occurs through the phosphorylation of serine 303 and serine 307 [17], [18]. Because S303 and S307 are constitutively phosphorylated in mammalian cells and alanine substitution of S303 or S307 promotes constitutive activation of HSF1 in mammalian cells in reporter assays [17], [18] we tested whether S303 and/or S307 contribute to HSF1 repression in yeast. Wild-type HSF1 or the individual S303A, S307A or S303/307A double mutants were expressed in yeast strain PS145 which lacks a chromosomal copy of the essential yeast HSF gene and constitutively expresses yeast HSF episomally from a galactose inducible and dextrose repressible promoter [29]. When PS145 is grown in the presence of dextrose as the sole carbon source, yeast HSF expression

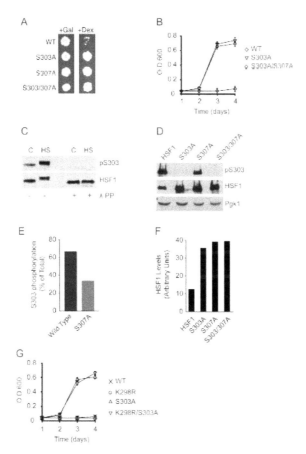

FIGURE 1. S303 phosphorylation represses HSF1 activity in yeast. (A) PS145 yeast strains expressing wild-type HSF1 (WT) or the S303A, S307A or S303/307A mutants were plated on either galactose or dextrose supplemented medium. (B) PS145 expressing either wild type HSF1 or the S303A or S303/307A mutants were grown in dextrose containing medium for 4 d. Growth was monitored by measuring O.D.$_{600}$. (C) HeLa cells were grown at 37°C (C) or heat shocked for 2 h at 42°C (HS). Total protein extracts were treated with lambda protein phosphatase and analyzed for phospho-S303 (pS303) and total HSF1 levels by immunoblotting. (D) PS145 was transformed with a plasmid expressing wild-type HSF1 (WT) or mutant alleles of HSF1 and grown on galactose containing medium. Total protein extracts were analyzed for pS303, HSF1 and Pgk1 by immunoblotting. (E) Levels of HSF1 phosphorylated at S303 were quantified and are shown as a percent of total HSF1, from panel D. (F) Protein levels of HSF1 were normalized to Pgk1, from panel D. (G) PS145 expressing either wild-type HSF1 or mutant HSF1 alleles were assayed for HSF1-dependent growth as in B.

is extinguished and growth becomes solely dependent on HSF1 which is episomally expressed [28]. While wild-type human HSF1 was unable to complement for the loss of yeast HSF, expression of the S303A, S307A or S303/307A HSF1 mutants allowed for human HSF1-dependent yeast growth (Figure 1A, B). Interestingly, the S303/307A double HSF1 mutant did not display enhanced activity over the S303A mutant (Figure 1B) suggesting that phosphorylation of both S303 and S307 modulate HSF1 repression through similar mechanisms.

To ascertain whether HSF1 is being phosphorylated in yeast, we employed a commercially available antibody specific for phospho-S303 (pS303). Because this antibody has not previously been characterized in the literature, we tested its specificity in human HeLa cells where S303 is known to be constitutively phosphorylated [17]. As shown in Figure 1C, using this antibody we detected that endogenous HSF1 was constitutively phosphorylated in HeLa cells in the absence of stress. We also observed an increase in S303 phosphorylation in response to a heat shock, which correlated with previous reports [16]. Importantly, the pS303-specific antibody did not detect HSF1 when HeLa extracts were treated with lambda protein phosphatase prior to immunoblot analysis (Figure 1C), nor does it detect HSF1 when S303 is mutated to alanine (Figure 1D). Together, these data suggest that the antibody is specific for HSF1 that is phosphorylated on S303. The detection of HSF1 using a polyclonal anti-HSF1 antibody demonstrates that there are no significant differences in the steady state levels of HSF1 either treated or untreated with lambda phosphatase (Figure 1C).

Consistent with a contribution to HSF1 repression (Figure 1A, B) S303 is robustly phosphorylated when HSF1 is expressed in yeast (Figure 1D, E). Interestingly, phosphorylation of S303 was also observed when the S307A mutant was expressed in yeast though it was reduced by approximately 50% when compared to wild-type HSF1 (Figure 1D, E). While this observation supports a previous report indicating that S303 phosphorylation could occur independently of S307 phosphorylation in mammalian cells [16], these data also suggests that under certain circumstances S303 phosphorylation may be enhanced by S307 phosphorylation. In addition, although a correlation between S303 and S307 phosphorylation and HSF1 protein stability has not been previously reported, we repeatedly observed two to three-fold higher steady state levels of HSF1 when the S303A,

S307A and S303/307A mutants were expressed in yeast (Figure 1D, F). While an antibody specific for phospho-S307 is commercially available, we have been unable to detect S307 phosphorylation of human HSF1. As such we focused our investigation on S303-phosphorylation dependent repression of human HSF1.

Human HSF1 S303 phosphorylation is known to promote sumoylation of lysine 298, which also contributes to the repression of HSF1 activity [16]. Therefore, to further investigate the idea that human HSF1 is being actively repressed in yeast, we explored the possibility that K298, like S303, contributes to HSF1 repression in yeast. However, unlike the S303A HSF1 mutant, the K298R mutant did not promote HSF1-dependent growth (Figure 1E), suggesting that at least in yeast, K298 does not significantly contribute to HSF1 repression. We also did not observe a reduction in human HSF1-dependent yeast growth for the S303A/K298R double mutant, indicating that K298 is also not required for HSF1 activity in yeast (Figure 1E).

5.2.2 S303 REPRESSES TRIMER FORMATION OF HSF1 IN YEAST AND MAMMALIAN CELLS

Previous reports have suggested that phosphorylation of S303 represses the ability of HSF1 to transactivate gene expression [18], [22], [23]. Here we show that the HSF1 S303A mutant functionally complements for the lost of yeast HSF (Figure 1A, B). Based on our previous work this indicates that S303 phosphorylation might also regulate the ability of human HSF1 to homotrimerize [25]. To test this hypothesis we carried out EGS cross-linking experiments in conjunction with immunoblot analysis to ascertain if S303 phosphorylation regulates the homotrimerization of human HSF1 in yeast. When the S303A HSF1 mutant was expressed in yeast we detected approximately 2-fold higher levels of trimerized HSF1 at the intermediate EGS concentration than when wild-type HSF1 was expressed in yeast (Figure 2A). However, trimerization of the S303A HSF1 mutant was lower than trimerization of the LZ4 HSF1 mutant, previously demonstrated to be constitutively trimerized in yeast and mammalian cells and able to complement for the loss of yeast HSF [6], [25]. We observed similar

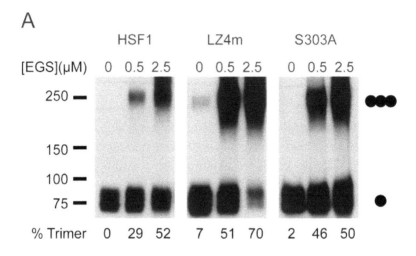

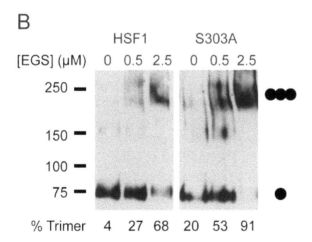

FIGURE 2: S303 represses trimer formation of HSF1 in yeast and mammalian cells. (A) PS145 was transformed with wild-type HSF1, the LZ4m mutant or the S303A mutant and grown on galactose containing medium. Total protein extracts were evaluated for HSF1 multimerization by EGS crosslinking, SDS-PAGE, and immunoblotting using an HSF1 specific antibody. The positions of molecular weight markers are indicated on the left, and circles indicating the expected migration of HSF1 monomers and trimers are on the right. Levels of HSF1 trimer as percent of total HSF1 are shown below. (B) hsf1[−/−] MEFs were transfected with a plasmid expressing wild-type HSF1 or the S303A mutant and analyzed for HSF1 multimerization by EGS cross-linking as in A.

results for the S307A and S303/307A HSF1 mutants (data not shown) further supporting the notion that S303 and S307 phosphorylation repress HSF1 activity through similar mechanisms. We next evaluated whether HSF1 S303 phosphorylation could also function to repress homotrimer formation in mammalian cells. To test this hypothesis we expressed wild-type HSF1 or the S303A, S307A or S303/307A mutants in hsf1$^{-/-}$ mouse embryonic fibroblasts (MEF) [30] and assayed for HSF1 trimerization in the absence of thermal stress by EGS crosslinking and immunoblotting. While wild type HSF1 could be detected as a multimer in these extracts, we observed approximately 2-fold higher levels of the HSF1 trimer for the HSF1 S303A mutant (Figure 2B) as well as the S307A and S303/307A mutants (data not shown).

5.2.3 S303 PHOSPHORYLATION AND COILED-COIL INTERACTIONS SYNERGIZE IN HSF1 REPRESSION

In addition to post-translational modifications, HSF1 activity is also thought to be repressed through intramolecular interactions between carboxyl- and amino-terminal coiled-coil domains and mutations in these domains render HSF1 constitutively trimerized, nuclear localized and bound to DNA in mammalian cells [6]. Because our results suggest that S303 phosphorylation might also regulate homotrimer formation, we tested the combined affects of both the S303A as well as the LZ4m mutations on human HSF1 activity in yeast. A human HSF1 mutant containing both the S303A and LZ4m mutations was created and its ability to promote human HSF1-dependent yeast growth was compared to the individual HSF1 mutants as well as wild-type HSF1 in quantitative cell growth assays. The individual S303A and LZ4m HSF1 mutants promoted human HSF1-dependent yeast growth to a similar extent, though neither the LZ4m nor the S303 mutant were fully derepressed, as the S303A/LZ4m double mutant displayed enhanced human HSF1-dependent yeast growth (Figure 3A). While we currently do not know if the S303A/LZ4m double HSF1 mutant has an increased propencity to trimerize, previous studies have shown that the LZ4m mutant, when expressed in yeast is not maximally trimerized and trimerization can be further enhanced via the addition of pharmacological HSF1 activators [28]. While we observed higher steady state protein levels for both the S303A and LZ4m mutants in comparison to wild-type HSF1 when expressed in yeast, no further increases in protein levels were observed for the double mutant (Figure 3B, C). These results suggest that while both HSF1 S303 phosphorylation and coiled-coil interactions regulate human HSF1 multimerization in yeast, they do so via distinct mechanisms. We also did not observe changes in HSF1 S303 phosphorylation when the LZ4m mutant was expressed in yeast, consistent with the notion that HSF1 trimerization does not affect HSF1 S303 phosphorylation.

5.2.4 GSK3 REGULATES HUMAN HSF1 ACTIVITY IN YEAST INDEPENDENT OF S303 PHOSPHORYLATION

Previous reports using in vitro phosphorylation experiments have suggested that HSF1 is phosphorylated at S303 by glycogen synthase kinase 3 (GSK3) [20], [22], [31]. However, it remains unclear if GSK3 phosphorylates and represses HSF1 via S303 phosphorylation in vivo. To test if GSK3 contributes to HSF1 repression, we assayed human HSF1-dependent yeast growth in a strain also lacking the yeast GSK3 homolog *Rim11*. Supporting the notion that yeast GSK3 can repress human HSF1 activity in yeast we observed human HSF1-dependent yeast growth as well as HSF1 multimerization in the *rim11Δ* strain (Figure 4A, B). However, HSF1-dependent yeast growth in the *rim11Δ* strain was less robust than growth of a wild-type strain expressing the S303A HSF1 mutant (Figure 4A). Furthermore, when we expressed the S303A HSF1 mutant in the *rim11Δ* strain we observed HSF1-dependent growth at a rate similar to the growth of the S303A mutant in wild-type cells. This suggested the possibility that HSF1 might not be fully derepressed in the *rim11Δ* strain. Consistent with this idea, we did not detect a reduction in S303 phosphorylation in the rim11Δ strain (Figure 4C). *S. cerevisiae* encodes four separate yet partially functionally redundant GSK3 homologues [32], suggesting the possibility that S303 remains phosphorylated in the *rim11Δ* strain due to phosphorylation through other GSK3 proteins. To test this hypothesis we assayed the phosphorylation state of HSF1 at S303 in a yeast strain lacking all four

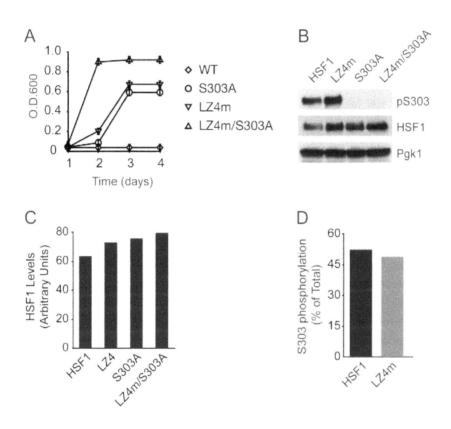

FIGURE 3: Phosphorylation of S303 and coiled-coil domains synergize in the repression of HSF1 in yeast. (A) PS145 expressing either wild-type HSF1 or mutant alleles of HSF1 were grown in dextrose supplemented medium for 4 d. Growth was monitored by measuring O.D.$_{600}$. (B) PS145 was transformed with a plasmid expressing wild-type HSF1 (WT) or mutant alleles of HSF1 and grown on galactose containing medium. Total protein extracts were analyzed for pS303, total HSF1 and Pgk1 by immunoblotting. (C) Protein levels of HSF1 were normalized to Pgk1, from panel B. (D) Levels of HSF1 phosphorylated at S303 were quantified and are shown as a percent of total HSF1, from panel B.

isoforms of yeast GSK3. As shown in Figure 4D, no reduction in S303 phosphorylation was observed in the *4xgsk3Δ* strain suggesting that while yeast GSK3 does contribute to HSF1 repression, it does so independently of S303 phosphorylation.

Results shown here for the S303A HSF1 mutant and previously published for the LZ4m HSF1 mutant suggest that mechanisms that regulate HSF1 in mammalian cells are at least partially conserved with regulation of human HSF1 expressed in yeast cells. Therefore, we carried out experiments to ascertain if GSK3 might also repress HSF1 independently of S303 phosphorylation in mammalian cells. To explore this possibility HeLa cells were treated with the GSK3 inhibitor SB-216763 [33] and assayed for HSF1 S303 phosphorylation as ascertained by immunoblotting with the anti-pS303 antibody. While SB-216763 treatment strongly inhibited GSK3 activity as shown by increased β-catenin levels [34], no reduction in S303 phosphorylation was observed (Figure 5A). However, similar to the results obtained from our yeast experiments, SB-216763 did promote activation of HSF1 under normal growth conditions, as determined by immunoblot analysis of Hsp70 expression (Figure 5A). This result is consistent with a previous report showing increased Hsp70 expression in response to lithium treatment, which also inhibits GSK3 function [35]. siRNA mediated knock-down of the two GSK3 isoforms in mammals, GSK3α and GSK3β, either singly or in combination, further confirmed that, while β-catenin expression was elevated, HSF1 S303 was not appreciably phosphorylated by GSK3 in unstressed mammalian cells (Figure 5B). Together, data from experiments in both yeast and mammalian cells support a model in which GSK3 inhibits HSF1 activity through a mechanism that is independent of S303 phosphorylation.

5.2.5 SLT2 REPRESSES HUMAN HSF1 ACTIVITY VIA S303 PHOSPHORYLATION IN YEAST

To begin to identify which protein kinase(s) in yeast phosphorylate human HSF1 at S303 to promote HSF1 repression, we assayed S303 phosphorylation in several previously generated protein kinase deletion strains obtained from the yeast gene deletion collection [36]. One strain in which

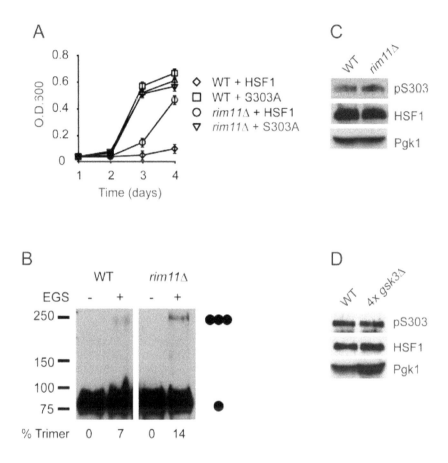

FIGURE 4: GSK3 represses HSF1 activity in yeast independent of S303. (A) PS145 (WT) expressing wild-type HSF1 or the S303A HSF1 mutant and LNY1 (rim11Δ) expressing wild-type HSF1 were grown in dextrose supplemented medium for 4 d. Growth was monitored by measuring O.D.600. (B) PS145 (WT) and LNY1 (rim11Δ) expressing wild-type HSF1 were grown on galactose containing medium and were evaluated for HSF1 multimerization by EGS crosslinking, SDS-PAGE, and immunoblotting using an HSF1 specific antibody. The positions of molecular weight markers are indicated on the left, and circles indicating the expected migration of HSF1 monomers and trimers are on the right. Levels of HSF1 trimer as percent of total HSF1 are shown below. (C) PS145 (WT) and LNY1 (rim11Δ) were transformed with a plasmid expressing wild-type HSF1 and were grown on galactose containing medium. Total protein extracts were analyzed for pS303, total HSF1 and Pgk1 by immunoblotting. (D) YPH499 (WT) and LNY3 (4xgsk3Δ) were transformed with a plasmid expressing wild-type HSF1 and were grown in dextrose containing medium. Total protein extracts were analyzed for pS303, total HSF1 and Pgk1 by immunoblotting.

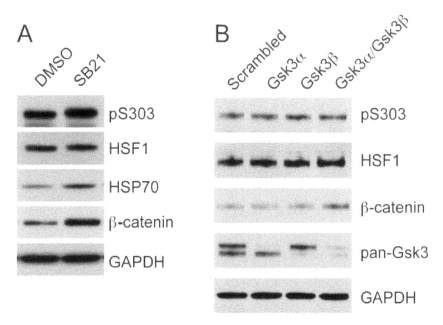

FIGURE 5: GSK3 represses HSF1 activity in HeLa cells independent of S303 phosphorylation. (A) HeLa cells were treated with DMSO solvent or the GSK3 inhibitor SB-216763 (25 μM) for 15 h. Total protein was analyzed for pS303, HSF1, and β-catenin by immunoblotting. GAPDH serves as a loading control. (B) HeLa cells were treated with siRNA specific for GSK3α and GSK3β either individually or together or a scrambled siRNA for 72 h. Total protein was analyzed for pS303, total HSF1, β-catenin, GSK3α/β and GAPDH by immunoblotting.

we detected severely reduced levels of human HSF1 S303 phosphorylation was a strain deleted for the *SLT2* gene, encoding a stress-responsive MAPK [37], [38], consistent with S303 lying within a consensus site for MAPK-dependent phosphorylation (Figure 6A) [39]. This suggests that Slt2 either directly or indirectly promotes the phosphorylation of human HSF1 expressed in yeast. This hypothesis was further supported by the observation that an *slt2Δ* strain allowed wild type human HSF1-dependent yeast growth at a rate similar to the HSF1 S303A mutant, while no growth was observed in the *SLT2* wild-type strain (Figure 6B). Homotrimerization of wild-type human HSF1 was observed in the *slt2Δ* strain at levels similar to the S303A and LZ4m HSF1 mutants, further supporting the notion that the Slt2 MAPK represses human HSF1 multimerization in yeast

(Figure 6C). In mammalian cells the most closely related homolog of Slt2 is the MAPK ERK5 [40]. However, using siRNA-mediated knock-down of ERK5 we were unable detect an effect of ERK5 on HSF1 S303 phosphorylation in mammalian cells (data not shown). This may suggest that in mammalian cells S303 can be phosphorylated by multiple MAPKs. This hypothesis is supported by previous data showing that ERK1/2 as well as the stress-responsive MAPK p38 could phosphorylate HSF1 at S303 in vitro [18]. In addition, our data showing reduced, but not eliminated phosphorylation of S303 in the *slt2Δ* strain (Figure 6A) also support a model where S303 may be phosphorylated by multiple MAPKs.

5.2.6 EXPRESSION OF S303A AND S307A MUTANTS IN HSF1□/□ CELLS RESULTS IN CONSTITUTIVE ACTIVATION OF HSP70 EXPRESSION

Previous studies have assayed the function of S303 and S307 phosphorylation in HSF1 regulation via in vitro phosphorylation experiments [22], in vivo using lexA/Gal4-HSF1 fusion proteins lacking the native HSF1 DNA binding domain [17], [18] or via overexpression of a S303A HSF1 mutants in mammalian cells expressing endogenous wild-type HSF1 [16]. We tested the consequences of loss of S303 and S307 phosphorylation on HSF1 activity in the context of the entire protein using hsf1$^{-/-}$ MEFs which lack endogenous HSF1. When we expressed S303A, S307A or S303/307A HSF1 mutants in hsf1$^{-/-}$ MEFs we observed a modest elevation of Hsp70 expression under normal growth conditions (Figure 7A, B) consistent with the hypothesis that S303 phosphorylation modulates both homotrimerization as well as transactivation by HSF1. However, HSF1 was not fully activated through the S303A and S307A mutations, as expression of Hsp70 was further enhanced when the transfected cells were exposed to low levels of the proteasome inhibitor MG132 (Figure 7A, B). This is consistent with our data generated in yeast demonstrating that while the S303A mutation did activate human HSF1-dependent yeast growth, this was further enhanced when the S303A HSF1 mutant was combined with the LZ4m mutation (Figure 3A). Interestingly, in hsf1$^{-/-}$ cells we observed a faster electrophoretic mobility on SDS-PAGE gels for

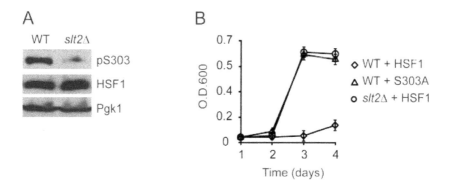

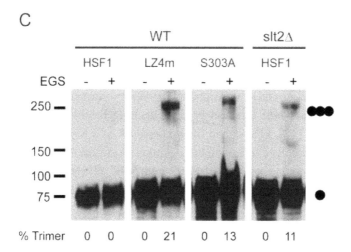

FIGURE 6: S303 phosphorylation of HSF1 in yeast is modulated by Slt2. (A) PS145 and LNY2 (slt2Δ) were transformed with a plasmid expressing wild-type HSF1 and were grown on galactose containing medium. Total protein extracts were analyzed for pS303, total HSF1 and Pgk1 by immunoblotting. (B) PS145 (WT) expressing wild-type HSF1 or the S303A mutant or LNY2 (slt2Δ) expressing wild-type HSF1 were grown in dextrose supplemented medium for 4 d. Growth was monitored by measuring O.D.600. (C) PS145 (WT) expressing wild-type HSF1, the LZ4m mutant or the S303A mutant and LNY2 (slt2Δ) expressing HSF1 were evaluated for HSF1 multimerization by EGS cross-linking, SDS-PAGE, and immunoblotting. The positions of molecular weight markers are indicated on the left and circles indicating the expected migration of HSF1 monomers and trimers are on the right. Levels of HSF1 trimer as percent of total HSF1 are shown below.

the HSF1 S303A and S303/307A mutant proteins that was not observed for wild-type HSF1 or the S307A mutant (Figure 7A), nor did we observe this change in mobility in the yeast system (Figure 1D). While the nature of this electrophoretic mobility shift is unknown, the HSF1 S303A and S303/S307A mutant alleles also exhibited lower steady state levels when exposed to MG132, suggesting that these proteins, despite having increased activity, might be less stable (Figure 7A, C). Because S303 phosphorylation has been proposed to promote HSF1 sumoylation in mammalian cells [16] it is possible that lack of sumoylation results in the altered electrophoretic mobility. Despite the fact that equal amounts of plasmid DNA were transfected for each mutant, we observed elevated steady state protein levels for the HSF1 S307A mutant (Figure 7A, C). While we have not definitively demonstrated that the S307A mutant protein has increased stability in comparison to wild-type HSF1, this finding correlates with the increased protein levels we observed for the HSF1 mutants expressed in yeast (Figure 1D, F) and will require further investigation. Interestingly, when we expressed the HSF1 S307A mutant in hsf1$^{-/-}$ cells we did not observe a reduction in S303 phosphorylation (Figure 7A, D) as was observed in yeast cells (Figure 1D, E) suggesting that priming requirements for S303 phosphorylation may change in different expression systems.

5.3 DISCUSSION

Mammalian HSF1 activity is regulated via complex regulatory mechanisms that include post-translation modifications as well as inter- and intra-molecular protein-protein interactions [2]. While our understanding of these regulatory mechanisms remains incomplete, earlier work has suggested that many of these mechanisms may be conserved in yeast [25], [26], [27], [28]. This is evident, in part, by repression of the human HSF1 protein when it is expressed in *S. cerevisiae* via coiled-coil domain and HSF1 loop interactions. In this report we show that evaluation of the mechanisms that regulate HSF1 activity in yeast via post-translational modifications can lead to important insights into the mechanisms that regulate HSF1 in mammalian cells.

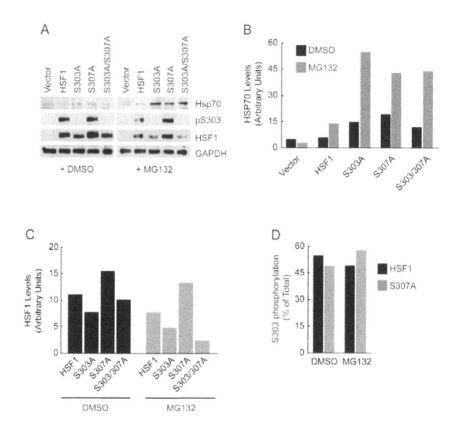

FIGURE 7: S303 and S307 repress HSF1 activity in hsf1$^{-/-}$ MEFs. (A) hsf1$^{-/-}$ MEFs were transfected with an empty vector or plasmids expressing wild-type HSF1 or the S303A, S307A or the S303/307A mutants. The transfected cells were treated with DMSO solvent or MG132 (10 µM) for 5 h. Total protein extracts were analyzed for Hsp70, pS303 and HSF1 by immunoblotting. GAPDH serves as a loading control. (B) Protein levels of Hsp70 were normalized to GAPDH, from panel A. (C) Protein levels of HSF1 were normalized to GAPDH, from panel A. (D) Levels of HSF1 phosphorylated at S303 were quantified and are shown as a percent of total HSF1, from panel A.

Previous experiments using HSF1 fusions with the constitutively bound Gal4 or lexA DNA-binding domains demonstrated that phosphorylation of S303 contributed to the repression of HSF1 transactivation [17], [18]. In this report we show that alanine substitution of S303, in the context of full length HSF1, also results in increased levels of trimerized HSF1 both in un-stressed yeast and in mammalian cells. This suggests that aside from repressing transactivation, S303 phosphorylation can also repress earlier points in the HSF1 activation pathway. Interestingly, we also show that repression of HSF1 activity through S303 phosphorylation may occur independent of K298 sumoylation in yeast, as arginine substitution of K298 does not promote HSF1 activation in yeast. It should be noted that not all of the mechanisms that regulate human HSF1 in mammalian cells are conserved in yeast. While human HSF1 is repressed in both yeast and mammalian cells through an amino-terminal coiled-coil as well as a carboxyl-terminal linker domain, the ability of wild type human HSF1 to respond to proteotoxic compounds or thermal stress, for example, appears to be strikingly absent in yeast [25], [27], [28]. Nevertheless, the ability of S303 phosphorylation to promote repression of human HSF1 in yeast independent of K298 sumoylation suggests that our understanding of the mechanisms by which S303 phosphorylation represses HSF1 activity remains incomplete. S303 and S307 are located in the regulatory domain of HSF1, a proposed binding site for the protein chaperone Hsp90 [41]. As such, it is tempting to speculate that phosphorylation of these residues might affect binding to Hsp90.

An understanding of how phosphorylation regulates HSF1 activity and what protein kinases phosphorylate HSF1 remains largely incomplete [20], [22]. Early reports showed that in vitro, HSF1 S307 phosphorylation acted as an essential priming event for S303 phosphorylation [22]. However, a subsequent report showed this priming event was not required in vivo and that HSF1 S303 phosphorylation occurred independent of S307 phosphorylation in K562 cells [16]. The work presented here using the yeast model system furthers our understanding of these regulatory mechanisms and may begin to clarify the conflicting mechanisms underlying S303 phosphorylation. Specifically, our data suggest that while phosphorylation of S303 can occur independently of S307 phosphorylation in both yeast and mammalian cells, S303 phosphorylation may be enhanced by

S307 phosphorylation in the non-native yeast system. While a mechanistic basis for this difference in the requirements for S303 phosphorylation remains unknown when HSF1 is expressed in yeast, structural differences could change the priming requirements for S303 phosphorylation. Such changes in HSF1 might occur due to different protein interactions and as such it is not surprising that in in vitro experiments, using only recombinant HSF1 protein, phosphorylation of S303 is fully dependent on S307 phosphorylation. However, further studies will be required to fully test these hypotheses.

Here, we demonstrate that in both yeast and mammalian cells phosphorylation of HSF1 S303 appears to occur independently of GSK3, previously thought to be the primary kinase responsible for S303 phosphorylation [20], [22]. Rather, as suggested by loss of function analysis, we propose that the MAPK Slt2 is one candidate that phosphorylates HSF1 at S303 in yeast though residual phosphorylation of HSF1 at S303 in an *slt2Δ* strain suggests that other MAPKs may also contribute to S303 phosphorylation. Differences in HSF1 structure between the in vivo and in vitro systems may also explain why different kinases can target S303 for phosphorylation under different conditions. We speculate that under some cellular conditions, for example physiological stress or different cell types, HSF1 structure may be altered, thereby shifting the S303-kinase specificity from a MAPK to GSK3. This might, in part, contribute to the complexity in identifying all of the mammalian kinases that phosphorylate S303. While GSK3 does not appear to phosphorylate HSF1 at S303 in vivo, data presented here nevertheless support a role for GSK3 as a repressor of HSF1 activity. It should be noted that several other serine residues in the HSF1 coding sequence, including S307, are located within putative GSK3 consensus sites [39].

The importance in understanding HSF1 regulation is underscored by recent findings showing that pharmacological activation of HSF1 can increase protein chaperone expression and ameliorate cytotoxicity in models of protein folding disease [28], [42], [43], [44], [45]. As such, it is important to further our understanding of the mechanisms that repress HSF1 activity as potential points of therapeutic intervention in disease. For example, our data has shown that the loss of S303-dependent HSF1 repression can lead to the accumulation of protein chaperones and as such could be efficacious in the treatment of protein folding diseases. In support of this possibility Rimoldi et al showed that over-expression of the HSF1 S303G mutant in HeLa cells reduced aggregation and inclusion formation of an aggregation prone Ataxin1-31Q mutant protein [46] In addition, Fujimoto et al showed that overexpression of a constitutively active HSF1 mutant lacking the regulatory domain, which includes S303 and S307, suppressed the aggregation and cytotoxicity of a mutant Huntingtin protein in both cell culture and mice [47]. Furthermore, Carmichael et al suggested that GSK3-inhibitors might prove useful in the treatment of polyQ-expansion diseases [48].

TABLE 1: Yeast strains used in this study.

Strain	Genotype
PS145	*MATa ade2-1 trp1-1 can1-100 leu2-3, 112 his3-11, 15 ura3-1 hsf1Δ::LEU2 Ycp50gal-y-HSF*
TYPH499	*MATa ura3-52 lys2-801 ade2-101 trp1-Δ63 his3-Δ200 leu2-Δ1*
BY4741	*MATa his3Δ1 leu2Δ0 met15Δ0 ura3Δ0*
LNY1	*MATa ade2-1 trp1-1 can1-100 leu2-3, 112 his3-11, 15 ura3-1 hsf1Δ::LEU2 Ycp50gal-yHSF rim11Δ::HIS3*
LNY2	*MATa ade2-1 trp1-1 can1-100 leu2-3, 112 his3-11, 15 ura3-1 hsf1Δ::LEU2 Ycp50gal-yHSF slt2Δ::HIS3*
LNY3	*MATa ura3-52 lys2-801 ade2-101 trp1-Δ63 his3-Δ200 leu2-Δ1 rim11Δ::TRP1 mck1Δ::HIS3 mrk1Δ::URA3 ygk3Δ::kanMX*

5.4 MATERIALS AND METHODS

5.4.1 YEAST STRAINS, PLASMIDS

S. cerevisiae strains used in this study are listed in Table 1. Yeast expression plasmids pRS424-GPD-HSF1 and pRS424-GPD-HSF1LZ4m were described previously [25]. Point mutations were introduced into the HSF1 coding sequence using the QuickChange Site-directed mutagenesis kit (Stratagene) and confirmed by DNA sequencing. YEp351-Slt2-FLAG was kindly provided by Dr. David E. Levin [49]. Mammalian expression plasmids were generated by subcloning the HSF1 open reading frame from yeast vectors into the mammalian vector pcDNA3.1.

5.4.2 CELL CULTURE MAINTENANCE, TRANSFECTION AND SIRNA

Mammalian cell lines used in the study were hsf1$^{-/-}$ MEF cells [30] and HeLa cells (ATCC, CCL-2). The MEF cells were maintained in DMEM supplemented with 10% fetal bovine serum (FBS), 0.1 mM nonessential amino acids, 100 U/ml penicillin/streptomycin and 55 µM 2-mercapto-ethanol. HeLa cells were maintained in DMEM supplemented with 10% FBS and 100 U/ml penicillin/streptomycin. MEF cells were transfected with HSF1 expressing plasmids using a Nucleofector (Lonza) and Nucleofector solution MEF2. siRNA was purchased from Dharmacon and 2 nmoles of each siRNA were transfected into HeLa cells using Dharmafect 1. Knock-down of proteins was assayed 72 h after siRNA transfection by immunoblot analysis.

5.4.3 COMPLEMENTATION ASSAYS

Growth curve experiments were carried out in 96-well plates as described previously [28]. For spot assays yeast cells were grown overnight in ga-lactose-containing medium to allow for expression of GAL1-yHSF and reseeded the following day at O.D.$_{600}$ = 0.2 and spotted on either galactose or dextrose supplemented growth media.

5.4.4 IMMUNOBLOT AND CROSSLINKING ANALYSIS

Protein extracts were generated from yeast cultures using glass bead lysis in cell lysis buffer (25 mM Tris, 150 mM NaCl, 1% Triton X-100, 0.1% SDS, 1 mM EDTA) supplemented with protease inhibitors (Roche) and Halt phosphate inhibitor cocktail (Thermo Scientific Pierce). Proteins extracts were generated from mammalian cell culture using cell lysis buffer supplemented with protease and phosphatase inhibitors. Protein concentrations were quantified using the BCA assay and 80–100 µg of total protein was resolved by SDS-PAGE and transferred to a nitrocellulose membrane. HSF1 oligomerization was assessed using the amine-specific cross-linker ethylene glycol bis-succinimidyl succinate (EGS) (Pierce). Crosslinking analysis were carried out as described previously [28]. Antibodies used in this study were anti-phospho-S303(pS303) (ab47369, Abcam), anti-HSF1 [28], anti-Pgk1, anti-FLAG (M2, Sigma), anti-Hsp70 (C92, Stressmarq), anti-β-catenin (6B3, Cell Signaling), anti-GAPDH (6C5, Ambion) and anti-GSK3α/β (D75D3, Cell Signaling). Quantification of immunoblot data was done using Photoshop.

REFERENCES

1. Morimoto RI (2008) Proteotoxic stress and inducible chaperone networks in neuro-degenerative disease and aging. Genes Dev 22: 1427–1438.
2. Akerfelt M, Morimoto RI, Sistonen L (2010) Heat shock factors: integrators of cell stress, development and lifespan. Nature reviews Molecular cell biology.
3. Hahn J-S, Hu Z, Thiele DJ, Iyer VR (2004) Genome-wide analysis of the biology of stress responses through heat shock transcription factor. Mol Cell Biol 24: 5249–5256.
4. Farkas T, Kutskova YA, Zimarino V (1998) Intramolecular repression of mouse heat shock factor 1. Mol Cell Biol 18: 906–918.
5. Orosz A, Wisniewski J, Wu C (1996) Regulation of Drosophila heat shock factor trimerization: global sequence requirements and independence of nuclear localization. Mol Cell Biol 16: 7018–7030.
6. Rabindran SK, Haroun RI, Clos J, Wisniewski J, Wu C (1993) Regulation of heat shock factor trimer formation: role of a conserved leucine zipper. Science 259: 230–234.
7. Abravaya K, Myers MP, Murphy SP, Morimoto RI (1992) The human heat shock protein hsp70 interacts with HSF, the transcription factor that regulates heat shock gene expression. Genes Dev 6: 1153–1164.
8. Baler R, Welch WJ, Voellmy R (1992) Heat shock gene regulation by nascent poly-peptides and denatured proteins: hsp70 as a potential autoregulatory factor. J Cell Biol 117: 1151–1159.
9. Shi Y, Mosser DD, Morimoto RI (1998) Molecular chaperones as HSF1-specific transcriptional repressors. Genes Dev 12: 654–666.
10. Zou J, Guo Y, Guettouche T, Smith DF, Voellmy R (1998) Repression of heat shock transcription factor HSF1 activation by HSP90 (HSP90 complex) that forms a stress-sensitive complex with HSF1. Cell 94: 471–480.
11. Sarge KD, Murphy SP, Morimoto RI (1993) Activation of heat shock gene transcription by heat shock factor 1 involves oligomerization, acquisition of DNA-binding activity, and nuclear localization and can occur in the absence of stress. Mol Cell Biol 13: 1392–1407.
12. Lee BS, Chen J, Angelidis C, Jurivich DA, Morimoto RI (1995) Pharmacological modulation of heat shock factor 1 by antiinflammatory drugs results in protection against stress-induced cellular damage. Proc Natl Acad Sci USA 92: 7207–7211.

13. Guettouche T, Boellmann F, Lane WS, Voellmy R (2005) Analysis of phosphorylation of human heat shock factor 1 in cells experiencing a stress. BMC Biochem 6: 4.

14. Holmberg CI, Hietakangas V, Mikhailov A, Rantanen JO, Kallio M, et al. (2001) Phosphorylation of serine 230 promotes inducible transcriptional activity of heat shock factor 1. EMBO J 20: 3800–3810.

15. Xia W, Voellmy R (1997) Hyperphosphorylation of heat shock transcription factor 1 is correlated with transcriptional competence and slow dissociation of active factor trimers. J Biol Chem 272: 4094–4102.

16. Hietakangas V, Ahlskog JK, Jakobsson AM, Hellesuo M, Sahlberg NM, et al. (2003) Phosphorylation of serine 303 is a prerequisite for the stress-inducible SUMO modification of heat shock factor 1. Mol Cell Biol 23: 2953–2968.

17. Kline MP, Morimoto RI (1997) Repression of the heat shock factor 1 transcriptional activation domain is modulated by constitutive phosphorylation. Mol Cell Biol 17: 2107–2115.

18. Knauf U, Newton E, Kyriakis J, Kingston R (1996) Repression of human heat shock factor 1 activity at control temperature by phosphorylation. Genes & Development 10: 2782–2793.

19. Westerheide SD, Anckar J, Stevens SM, Sistonen L, Morimoto RI (2009) Stress-Inducible Regulation of Heat Shock Factor 1 by the Deacetylase SIRT1. Science 323: 1063–1066.

20. Chu B, Zhong R, Soncin F, Stevenson MA, Calderwood SK (1998) Transcriptional activity of heat shock factor 1 at 37 degrees C is repressed through phosphorylation on two distinct serine residues by glycogen synthase kinase 3 and protein kinases Calpha and Czeta. J Biol Chem 273: 18640–18646.

21. Wang X, Khaleque MA, Zhao MJ, Zhong R, Gaestel M, et al. (2006) Phosphorylation of HSF1 by MAPK-activated protein kinase 2 on serine 121, inhibits transcriptional activity and promotes HSP90 binding. J Biol Chem 281: 782–791.

22. Chu B, Soncin F, Price BD, Stevenson MA, Calderwood SK (1996) Sequential phosphorylation by mitogen-activated protein kinase and glycogen synthase kinase 3 represses transcriptional activation by heat shock factor-1. J Biol Chem 271: 30847–30857.

23. Xia W, Guo Y, Vilaboa N, Zuo J, Voellmy R (1998) Transcriptional activation of heat shock factor HSF1 probed by phosphopeptide analysis of factor 32P-labeled in vivo. J Biol Chem 273: 8749–8755.

24. Hietakangas V, Anckar J, Blomster HA, Fujimoto M, Palvimo JJ, et al. (2006) PDSM, a motif for phosphorylation-dependent SUMO modification. Proc Natl Acad Sci USA 103: 45–50.

25. Liu XD, Liu PC, Santoro N, Thiele DJ (1997) Conservation of a stress response: human heat shock transcription factors functionally substitute for yeast HSF. EMBO J 16: 6466–6477.

26. Ahn SG, Liu PC, Klyachko K, Morimoto RI, Thiele DJ (2001) The loop domain of heat shock transcription factor 1 dictates DNA-binding specificity and responses to heat stress. Genes & Development 15: 2134–2145.

27. Liu PC, Thiele DJ (1999) Modulation of human heat shock factor trimerization by the linker domain. J Biol Chem 274: 17219–17225.

28. Neef DW, Turski ML, Thiele DJ (2010) Modulation of heat shock transcription factor 1 as a therapeutic target for small molecule intervention in neurodegenerative disease. PLoS Biol 8: e1000291.

29. Sorger PK, Pelham HR (1988) Yeast heat shock factor is an essential DNA-binding protein that exhibits temperature-dependent phosphorylation. Cell 54: 855–864.

30. McMillan DR, Xiao X, Shao L, Graves K, Benjamin IJ (1998) Targeted disruption of heat shock transcription factor 1 abolishes thermotolerance and protection against heat-inducible apoptosis. J Biol Chem 273: 7523–7528.

31. Xavier I, Mercier P, McLoughlin C, Ali A, Woodgett J, et al. (2000) Glycogen synthase kinase 3β negatively regulates both DNA-binding and transcriptional activities of heat shock factor 1. Journal of Biological Chemistry 275: 29147.

32. Kassir Y, Rubin-Bejerano I, Mandel-Gutfreund Y (2006) The Saccharomyces cerevisiae GSK-3 beta homologs. Curr Drug Targets 7: 1455–1465.

33. Coghlan MP, Culbert AA, Cross DA, Corcoran SL, Yates JW, et al. (2000) Selective small molecule inhibitors of glycogen synthase kinase-3 modulate glycogen metabolism and gene transcription. Chem Biol 7: 793–803.

34. Salic A, Lee E, Mayer L, Kirschner MW (2000) Control of beta-catenin stability: reconstitution of the cytoplasmic steps of the wnt pathway in Xenopus egg extracts. Mol Cell 5: 523–532.

35. Khaleque MA, Bharti A, Sawyer D, Gong J, Benjamin IJ, et al. (2005) Induction of heat shock proteins by heregulin beta1 leads to protection from apoptosis and anchorage-independent growth. Oncogene 24: 6564–6573.

36. Winzeler EA, Shoemaker DD, Astromoff A, Liang H, Anderson K, et al. (1999) Functional characterization of the *S. cerevisiae* genome by gene deletion and parallel analysis. Science 285: 901–906.

37. Mazzoni C, Zarov P, Rambourg A, Mann C (1993) The SLT2 (MPK1) MAP kinase homolog is involved in polarized cell growth in Saccharomyces cerevisiae. J Cell Biol 123: 1821–1833.

38. Hahn JS, Thiele DJ (2002) Regulation of the Saccharomyces cerevisiae Slt2 kinase pathway by the stress-inducible Sdp1 dual specificity phosphatase. J Biol Chem 277: 21278–21284.

39. Gnad F, Ren S, Cox J, Olsen JV, Macek B, et al. (2007) PHOSIDA (phosphorylation site database): management, structural and evolutionary investigation, and prediction of phosphosites. Genome Biol 8: R250.

40. Truman AW, Millson SH, Nuttall JM, King V, Mollapour M, et al. (2006) Expressed in the Yeast Saccharomyces cerevisiae, Human ERK5 Is a Client of the Hsp90 Chaperone That Complements Loss of the Slt2p (Mpk1p) Cell Integrity Stress-Activated Protein Kinase. Eukaryotic Cell 5: 1914–1924.

41. Guo Y, Guettouche T, Fenna M, Boellmann F, Pratt WB, et al. (2001) Evidence for a mechanism of repression of heat shock factor 1 transcriptional activity by a multichaperone complex. J Biol Chem 276: 45791–45799.

42. Fujikake N, Nagai Y, Popiel HA, Okamoto Y, Yamaguchi M, et al. (2008) Heat shock transcription factor 1-activating compounds suppress polyglutamine-induced neurodegeneration through induction of multiple molecular chaperones. J Biol Chem 283: 26188–26197.

43. Auluck P, Meulener M, Bonini N (2005) Mechanisms of Suppression of{alpha}-Synuclein Neurotoxicity by Geldanamycin in Drosophila. Journal of Biological Chemistry 280: 2873–2878.
44. Auluck PK, Bonini NM (2002) Pharmacological prevention of Parkinson disease in Drosophila. Nat Med 8: 1185–1186.
45. Hay DG, Sathasivam K, Tobaben S, Stahl B, Marber M, et al. (2004) Progressive decrease in chaperone protein levels in a mouse model of Huntington's disease and induction of stress proteins as a therapeutic approach. Hum Mol Genet 13: 1389–1405.
46. Rimoldi M, Servadio A, Zimarino V (2001) Analysis of heat shock transcription factor for suppression of polyglutamine toxicity. Brain Res Bull 56: 353–362.
47. Fujimoto M, Takaki E, Hayashi T, Kitaura Y, Tanaka Y, et al. (2005) Active HSF1 significantly suppresses polyglutamine aggregate formation in cellular and mouse models. J Biol Chem 280: 34908–34916.
48. Carmichael J, Sugars KL, Bao YP, Rubinsztein DC (2002) Glycogen synthase kinase-3beta inhibitors prevent cellular polyglutamine toxicity caused by the Huntington's disease mutation. J Biol Chem 277: 33791–33798.
49. Kim K-Y, Truman AW, Levin DE (2008) Yeast Mpk1 mitogen-activated protein kinase activates transcription through Swi4/Swi6 by a noncatalytic mechanism that requires upstream signal. Molecular and Cellular Biology 28: 2579–2589.

CHAPTER 6

ENVIRONMENTAL REGULATION OF PRIONS IN YEAST

LIMING LI AND ANTHONY S. KOWAL

6.1 THE YEAST PRION CONCEPT

The term prion, proteinaceus infectious particle, was first used to describe the causative agent of a group of mammalian neurodegenerative diseases known as transmissible spongiform encephalopathies (TSEs) [1]. The mammalian prion protein (PrP) can exist in either a normal cellular conformation, PrP^C, or in multiple misfolded pathogenic conformations, collectively called PrP^{Sc}. PrP^{Sc} is considered infectious because it can recruit and convert its normal isomer PrP^C to its pathogenic conformation. This "protein-only" concept of infectivity has gained general acceptance and has been extended to explain some unusual non-Mendelian genetic elements in the budding yeast *Saccharomyces cerevisiae*. In yeast, these factors are transmitted from mother to daughter cell as particular self-propagating protein conformations, and are thus referred to as yeast prions [2].

Yeast prions share many features with PrP^{Sc}: both are capable of perpetuating particular conformational changes, forming amyloid fibrils (ordered protein aggregates with cross-β sheet structure and filamentous morphology) under physiological conditions, and both can exist as multiple "strains" or variants. However, a number of fundamental differences

This chapter was originally published under the Creative Commons Attribution License. Li L and Kowal AS. Environmental Regulation of Prions in Yeast. PLoS Pathogens 8,11 (2012). doi:10.1371/ journal.ppat.1002973.

between them are worth noting. First, yeast prion proteins and PrP do not share a significant sequence similarity. Almost all yeast prion proteins contain a domain with an unusually high content of glutamine (Q) and asparagine (N) residues (~45%), whereas PrP does not have such a region. The Q/N-rich domains of yeast prion proteins, termed prion forming domains (PrDs), are modular and transferrable and essential for the formation and propagation of their corresponding prions. Second, whereas the normal function of PrP is unclear, yeast prion proteins are involved in a wide range of functions, from transcriptional and translational regulation to nitrogen metabolism. To date, PrP is the only prion protein identified in mammals, whereas at least 8 prions have been identified in yeast: [PSI+], [URE3], [PIN+], [SWI+], [OCT+], [MOT3], [ISP+], and [MOD+] [3], [4] (capital letters indicate that these genetic elements are dominant, and brackets signify non-Mendelian patterns of inheritance). Finally, while PrPSc is associated with human disease, yeast prions are not associated with disease per se, but manifest as dominant, cytoplasmically inherited phenotypes.

6.2 PROTEIN-BASED INFECTIVITY OF YEAST PRIONS

Yeast prions do not infect nonprion cells through simple cell–cell contact. For example, coculturing [PRION+] and [prion−] cells of the same mating type does not result in prion transmission. However, sexual crosses between [PRION+] and [prion−] cells yield diploids that are all [PRION+], and tetrads derived from [PRION+] diploids will give rise to spores that are all [PRION+] (Figure 1A). In contrast, a diploid from a similar cross of a nucleic acid–based mutant to a wild-type partner gives rise to meiotic progeny in a 2:2 ratio. This "protein-only" infectivity of yeast prions can be also demonstrated by cytoduction, a process in which the cytoplasmic but not the nuclear components are mixed between partners (Figure 1B). [PRION+] donor cells can pass the [PRION+] state to nonprion recipient haploid progeny without exchange of genetic information. This "gold-standard" assay has been used to confirm if a phenotypic trait is cytoplamically inherited; all known yeast prions are cytoducible due to their protein-based infectivity. Prion infectivity can also be demonstrated by transformation of prion fibrils (Figure 1C). Incubating naïve [*prion*−] cells with amyloid fibrils assembled in vitro from recombinant prion proteins can result in *de novo* formation of stable, transmissible prions in the recipient. The first successful studies to demonstrate fibril-based transformation were conducted using the well-studied prion [*PSI+*], a translation termination modifier [5], [6]. This method of transformation provides simple, direct confirmation that the amyloidsformed in vitro are able to self-propagate by converting endogenously produced protein isomers into the [*PRION+*] state.

6.3 AN INTERACTION BETWEEN YEAST PRIONS AND THE CELLULAR MACHINERY

While infectious prion amyloids can be formed in a test tube autocatalytically, prion formation and propagation inside a cell requiresa supporting cellular network; imbalance of this network often results in prion destabilization or loss. For example, inhibiting the activity of the protein deaggregase Hsp104, which normally fragments prion fibrils into transmissible seeds, blocks prion transmission from mother to daughter during cell division and results in the loss of all amyloid prions [7]. Further, the abundant yeast cytoplasmic chaperone (Hsp70-Ssa) collaborates with two groups of cochaperones—the Jprotein family members (e.g., Sis1) and the nucleotide-exchangefactors (e.g., Sse1)—to play a crucial role in maintaining yeast prions [8]. In prions that have been examined thus far, manipulating the function of Hsp70-Ssa or its cochaperones has been found to result in their destabilization or loss [8]. Other cellular factors that have been identified as supporting the prionogenic cellular network include components of the cytoskeleton, the endocytotic machinery, and the ubiquitin-proteasome system (UPS) [9], [10]. Remarkably, a single yeast cell can harbor multiple prion elementssimultaneously, but they do not simply coexist; they can promote or inhibit each other's appearance and maintenance. For example, the presence of [*PIN+*] or [*URE3*] can facilitate [*PSI+*] induction [11]. However, they have also been shown to have antagonizing effects [11], [12]. Therefore, stable prion transmission is a consequence not only of dynamic interactions between coexisting prions and their protein determinants but also of other cellular components.

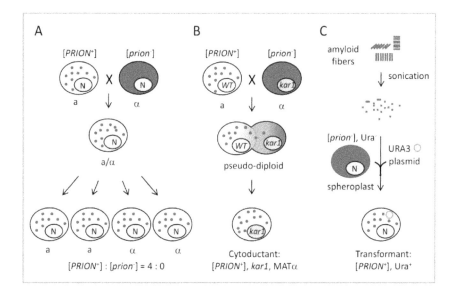

FIGURE 1: Yeast prions are "infectious." A) A sexual cross of [*PRION*+] and [*prion*−] cells of opposite mating types results in a [*PRION*+] diploid, which can give rise to fourspores that are all [*PRION*+] after sporulation. Note: in the case of weak [*PSI*+] and [*URE3*], [*PRION*+]×[*prion*−] crosses do not always give a 4:0 segregation in progeny. Some other random, non-Mendelian segregation ratios of progeny can be seen, such as 1:4, 1:3, 3:1, 4:0, as well as 2:2, due to their meiotic instabilities. B) Mating a [*PRION*+] donor with a [*prion*−] recipient carrying a kar1 mutation (which prevents nuclear fusion of the mating partners) will result in formation of a pseudodiploid carrying a mixed cytoplasm of the two mating partners. The pseudodiploid will give rise to haploid cytoductants containing either the donor or recipient nucleus. Shown is a cytoductant containing the recipient nucleus with a kar1 mutation. C) Transformation of [prion−] spheroplasts (yeast with cell wall removed) with amyloid fibers assembled from recombinant prion protein can result in *de novo* formation of heritable [*PRION*+] in the transformed cells. A URA3 plasmid (circle) was used as a selection marker for the transformation. Solid darker color indicates the soluble, diffused prion-determinant protein, whereas darker dots indicate the prion protein is in an aggregated prion conformation.

6.4 ENVIRONMENTAL REGULATION OF YEAST PRIONS

Prion proteins interact extensively with their cellular environments throughout the entire process of prion formation and propagation (Figure 2). Therefore any modulations that perturb this cellular interaction network will likely affect prionogenesis and prion stability. Intriguingly, supplementation of growth media with select chemical agents, such as the protein denaturant guanidine hydrochloride, the organic solvent dimethyl sulfoxide, alcohols, or potassium chloride salt, results in the loss or destabilization of the [*PSI+*] prion [13]. Thermal changes, treatments with antibiotics, or oxidative chemicals also have profound effects on [*PSI+*] propagation [13]–[15]. Prion *de novo* formation can be affected by environmental stresses as well. Mutations in heat-shock factor 1 (Hsf1), the master regulator of heat-response genes, drastically influence the frequency of [*PSI+*] induction. The observed effects of [*PSI+*] induction can be either an enhancement or inhibition, depending on the specific nature of the Hsf1 mutation [16]. In addition, data from an unbiased, high-throughput screen identified a group of stress-response proteins, including Msn2, a general stress-response regulator, and Hac1, a protein-unfolding response regulator,as modifiers of [*PSI+*] prionogenesis [17]. Mutants harboring deletion of *MSN2* or *HAC1*, or the exposure of wild-type cells to various extreme stressful conditions, drastically increased the frequency of [*PSI+*] induction [17]. Recent findings show that heatshock increases the synthesis of Lsb2, a short-lived protein facilitating [*PSI+*] *de novo* formation [9], suggesting another regulatory mechanism for the impact of environment on yeast prionogenesis.

6.5 POTENTIALLY DIVERSE ROLES FOR YEAST PRIONS IN EVOLUTION

Isogenic [*PRION+*] and [*prion−*] cells may exhibit completely different phenotypes under identical environmental conditions, but they can switch between these distinct phenotypic states spontaneously. It has been proposed that prion formation may be a mechanism to uncover otherwise hidden genetic variations to create new phenotypic traits, thus providing a

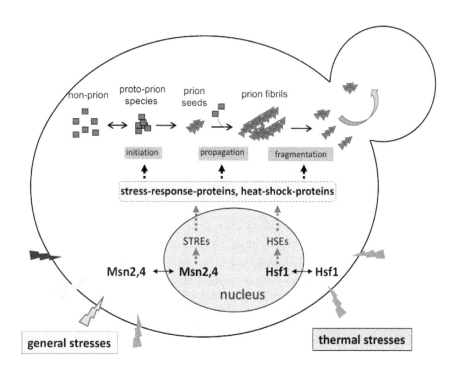

FIGURE 2: Environmental regulation of yeast prions. Prionogenesis is a multistep process in which the prion determinant protein undergoes changes in its secondary structure to form intermediate species and then prion (amyloid) fibrils; this process relies on other cellular machinery to drive these changes. Thermal stress results in the relocalization of heat-shock factor 1 (Hsf1) from the cytoplasm to the nucleus; here it binds to the heat-shockelement (HSEs) of heat-shock–protein genes, activating their transcription. Consequentially, a diverse group of heat-shock proteins (HSPs) are synthesized. Many HSPs (molecular chaperones) play important roles in prion formation and propagation, including Hsp104, Hsp70-Ssa, and Hsp40-Sis1. In a similar manner, general stresses including oxidative, osmotic, and heat stresses, activate a separate pathway in which Msn2,4 binds to the stress-response element (STREs) of stress-response genes, thereby activating their transcription. Some HSP genes also contain one or more STREs at their 5′-regulatory regions. Deletion of the *MSN2* gene results in a drastic increase of the frequency of [*PSI+*] formation, suggesting that some stress-response proteins are also involved in prion formation. However, the identity of the Msn2,4 targets that are involved in prionogenesis remain elusive. Note: for simplicity, only the two major stress-response pathways that are regulated by Hsf1 and Msn2,4 are shown.

means of rapid adaptive evolution [14], [18]. Indeed, the metastable nature of prion inheritance offers a potential for regulatory plasticity that cannot be readily achieved by nucleic acid mutation. Because prion-conferred phenotypic traits can be quickly spread between mating partners and progeny without altering the underlying nucleic acid sequence, prion-based inheritance might provide a rapid means to allow yeast to survive sudden undesirable environmental changes. That yeast prions and mammalian PrP[Sc] can exist as multiple heritable variants indicates the possibility of multilevel epigenetic regulation. Additionally, in its aggregated conformation, a prion protein may sequester other important cellular factors, causing, in effect, a multigene-knockdown phenotype. Lastly, since a single yeast cell can harbor multiple prion elements simultaneously, it is possible that different prion combinations might provide additional phenotypic diversity.

Indeed, it has been hypothesized that the [PSI+] prion aids the response of yeast to environmental changes in order to produce a number of new, temporary phenotypic traits [14]. Remarkably, some [PSI+]-mediated epigenetic traits can be fixed permanently in progeny through one-step outcross to become [PSI+] independent [18]. While it remains controversial whether the presence of a prion is beneficial to yeast [19], recent studies provide evidence to support the hypothesis that the prions provide a fitness advantage. For example, the recently discovered prion [MOD+] confers a gain-of-function resistance to antifungal agents [4]. Upon application of antifungal drugs, [MOD+] prion conversion increases, suggesting that de novo prion appearance is effected by selective pressure [4]. Crucially, yeast prions are not an artifact of laboratory manipulation; a recent study found several yeast prions ([PSI+], [PIN+], and [MOT3+]) in a number of wild strains [20], indicating that these prions arise from some selective pressure under natural conditions. Collectively, prion-mediated heritable conformational alterations potentiate evolutionary changes.

6.6 CONCLUSIONS

Although yeast prions are not associated with distinct human diseases, results from yeast prion research during the last two decades have provided invaluable information about protein misfolding, aggregation, and protein-based heredity and infectivity. The fact that multiple prions have been identified in yeast thus far (with additional promising prion candidates) suggests that their occurrence is a ubiquitous, natural biological phenomenon that deserves our understanding and further research efforts. Due to its simplicity and amenability to genetic and cell biological manipulation, yeast will remain a powerful model organism for prion research.

REFERENCES

1. Prusiner SB (1982) Novel proteinaceous infectious particles cause scrapie. Science 216: 136–144.
2. Wickner RB (1994) [URE3] as an altered Ure2 protein: evidence for a prion analog in *Saccharomyces cerevisiae*. Science 264: 566–569.
3. Crow ET, Li L (2011) Newly identified prions in budding yeast, and their possible functions. Semin Cell Dev Biol 22: 452–459.
4. Suzuki G, Shimazu N, Tanaka M (2012) A yeast prion, Mod5, promotes acquired drug resistance and cell survival under environmental stress. Science 336: 355–359.
5. King CY, Diaz-Avalos R (2004) Protein-only transmission of three yeast prion strains. Nature 428: 319–323.
6. Tanaka M, Chien P, Naber N, Cooke R, Weissman JS (2004) Conformational variations in an infectious protein determine prion strain differences. Nature 428: 323–328.
7. Tuite MF, Serio TR (2010) The prion hypothesis: from biological anomaly to basic regulatory mechanism. Nat Rev Mol Cell Biol 11: 823–833.
8. Liebman SW, Chernoff YO (2012) Prions in yeast. Genetics 191: 1041–1072.
9. Chernova TA, Romanyuk AV, Karpova TS, Shanks JR, Ali M, et al. (2011) Prion induction by the short-lived, stress-induced protein Lsb2 is regulated by ubiquitination and association with the actin cytoskeleton. Mol Cell 43: 242–252.
10. Ganusova EE, Ozolins LN, Bhagat S, Newnam GP, Wegrzyn RD, et al. (2006) Modulation of prion formation, aggregation, and toxicity by the actin cytoskeleton in yeast. Mol Cell Biol 26: 617–629.
11. Derkatch IL, Liebman SW (2007) Prion-prion interactions. Prion 1: 161–169.
12. Schwimmer C, Masison DC (2002) Antagonistic interactions between yeast [PSI(+)] and [URE3] prions and curing of [URE3] by Hsp70 protein chaperone Ssa1p but not by Ssa2p. Mol Cell Biol 22: 3590–3598.
13. Tuite MF, Mundy CR, Cox BS (1981) Agents that cause a high frequency of genetic change from [PSI+] to [psi−] in *Saccharomyces cerevisiae*. Genetics 98: 691–711.
14. True HL, Lindquist SL (2000) A yeast prion provides a mechanism for genetic variation and phenotypic diversity. Nature 407: 477–483.
15. Newnam GP, Birchmore JL, Chernoff YO (2011) Destabilization and recovery of a yeast prion after mild heat shock. J Mol Biol 408: 432–448.

16. Park KW, Hahn JS, Fan Q, Thiele DJ, Li L (2006) *De novo* appearance and "strain" formation of yeast prion [PSI+] are regulated by the heat-shock transcription factor. Genetics 173: 35–47.
17. Tyedmers J, Madariaga ML, Lindquist S (2008) Prion switching in response to environmental stress. PLoS Biol 6: e294 doi:10.1371/journal.pbio.0060294.
18. True HL, Berlin I, Lindquist SL (2004) Epigenetic regulation of translation reveals hidden genetic variation to produce complex traits. Nature 431: 184–187.
19. Nakayashiki T, Kurtzman CP, Edskes HK, Wickner RB (2005) Yeast prions [URE3] and [PSI+] are diseases. Proc Natl Acad Sci U S A 102: 10575–10580.
20. Halfmann R, Jarosz DF, Jones SK, Chang A, Lancaster AK, et al. (2012) Prions are a common mechanism for phenotypic inheritance in wild yeasts. Nature 482: 363–368.

CHAPTER 7

HETEROLOGOUS GLN/ASN-RICH PROTEINS IMPEDE THE PROPAGATION OF YEAST PRIONS BY ALTERING CHAPERONE AVAILABILITY

ZI YANG, JOO Y. HONG, IRINA L. DERKATCH, AND SUSAN W. LIEBMAN

7.1 INTRODUCTION

The infectivity of transmissible spongiform encephalopathies (TSEs) was explained by the prion hypothesis proposing that the inheritance of biological information can be achieved by self-propagating conformational changes in the prion protein PrP [1]. The prion list has since been extended to include protein-based genetic elements found in fungi [2]. The best-studied yeast prions [PSI+], [PIN+] (often called [RNQ+]) and [URE3] are, respectively, self-propagating conformations of: Sup35, a translation termination factor; Rnq1, a protein of unknown function; and Ure2, a nitrogen catabolism repression regulator [3]–[5]. Other recently discovered yeast prions include [SWI+], [OCT+], [ISP+], [MOT3+] and [MOD5] [6], [7]. The propagation of most [8]–[10], but not all [6] yeast prions is driven by their Q/N-rich prion domains that have the propensity to form aggregates

This chapter was originally published under the Creative Commons Attribution License. Yang Z, Hong JY, Derkatch IL, and Liebman SW. Heterologous Gln/Asn-Rich Proteins Impede the Propagation of Yeast Prions by Altering Chaperone Availability. PLoS Genetics 9,1 (2013). doi:10.1371/journal. pgen.1003236.

in vivo and assemble into self-seeding, β-sheet-rich amyloid fibers in vitro 11,12.

Prion propagation involves templated conversion of soluble protein into the prion state [13]. In vitro data show that amyloid fibers grow by recruiting protein monomers to fiber ends [14]. In addition, prion propagation requires fibers to be fragmented to create new ends for conversion and to allow efficient transmission of seeds to daughter cells [15]. Failure at any of these steps would lead to loss (curing) of the prion.

The Hsp104 chaperone is required for the propagation of yeast prions, and its elimination leads to prion loss [16], [17]. One role of Hsp104 in prion propagation is to shear prion aggregates [16], [18]–[23]. Overexpression of Hsp104 also cures cells of [PSI+] [16]. The mechanism appears to be more complex than simple over-shearing [24]–[28]. A recent study indicates that Hsp104 overexpression displaces the Hsp70 chaperone Ssa1 from binding to [PSI+] aggregates. Since Ssa1 is required for Hsp104 shearing activity, this inhibits shearing of [PSI+] aggregates leading to loss of [PSI+] [29].

Modulation of levels and mutations in Hsp70 chaperones and their co-chaperones have various effects on [PSI+] and [URE3] [30], [31]. For example, Ssa1/2 in excess cures [PSI+] [32] and overexpression of Ssa1 cures [URE3] [33]. Depletion of Sis1, an Hsp40 J-protein co-chaperone of Hsp70 specifically cures cells of [PSI+], [PIN+], [URE3] and [SWI+] and leads to an increase in the size of SDS-resistant Sup35 polymers derived from [PSI +] aggregates [34]–[37]. It has been suggested that Ssa1/2 and Sis1 recruit [PSI +] aggregates to Hsp104 for fragmentation, and that prion stability and propagation are mediated by the chaperone composition of prion aggregates [38]–[40]. Overexpression of the Hsp70 nucleotide exchange factor Sse1 or the Hsp40 chaperone Ydj1 cures [URE3] [41], [42].

When aggregated in the [PSI+] state, Sup35's participation in translation termination is greatly reduced. This leads to increased read-through of stop codons, including the ade1-14 nonsense allele that can be readily monitored by a red/white color assay [16], [43]–[45]. [PSI+] prion variants or strains manifest a range of distinct prion conformations that differ in levels of Sup35 aggregation and, consequently, in the frequency of stop-codon read-through. Since weak [PSI+] variants cause less read-through of stop codons, are less mitotically stable and contain more of the soluble non-prion form of Sup35 than strong [*PSI+*] variants, the color assay allows their distinction by the degree of red pigment accumulated [44], [46]–[49]. Also, these [*PSI+*] variants differ in the size of their SDS-resistant Sup35 polymers [20].

When full length Sup35, or just its Q/N-rich prion domain, is transiently overproduced in [*psi−*] cells, [*PSI+*] is induced to appear, presumably because the excess protein increases the chance that it will form a prion seed [2], [44], [50], [51]. However, efficient *de novo* induction of [*PSI+*] requires the presence of a heterologous prion, e.g. [*PIN+*] [3], [51], [52]. Heritable variants of the [*PIN+*] prion have been distinguished by their efficiency in inducing [*PSI+*] with the inducing efficiency gradually decreasing from very high to high to medium to low [*PIN+*] [53].

In a screen of a high-copy yeast library for genes that enhance [*PSI+*] induction in the absence of [*PIN+*], an excess of any of 11 Q/N-rich proteins was found to promote *de novo* [*PSI+*] appearance upon the overexpression of the prion domain of Sup35 [3]. Similarly, aggregation-prone polyQ sequences could substitute for [*PIN+*] in the case of [*PSI+*] induction, and [*PIN+*] also facilitated the aggregation of proteins with extended polyQ stretches [54]–[56]. It was proposed that the aggregates formed by these Q/N-rich proteins provide a nidus for the formation of the first [*PSI+*] seeds, which then promote Sup35's rapid aggregation. This cross-seeding model postulates a direct interaction between Q/N-rich domains of a newly forming prion and preexisting heterologous prion or prion-like aggregates [3], [54], [57], [58].

Several studies have indicated that heterologous prions or prion proteins can also inhibit prion propagation. For example, some [*PIN+*] variants impede the inheritance of [*PSI+*] [32], [59] and [*PSI+*] and [*URE3*] slightly destabilize each other [33], [53]. Also, overexpression of the Ure2 prion domain or several other fragments of Ure2 cures [*URE3*] [60], and overexpression of some Rnq1 fragments encompassing the Q/N-rich C-terminal domain but lacking the N-terminus is inhibitory to [*PSI+*] and [*URE3*] propagation in the presence of [*PIN+*] [61]. Finally, overexpression of rnq1 N-terminal mutants causes enlargement of [*PSI+*] aggregates leading to loss of [*PSI+*] [62]. The molecular basis of these antagonistic interactions is unknown.

Here we report that overexpression of a number of Q/N-rich proteins can impede the propagation of the Q/N-rich prions, [*PSI+*] and [*URE3*].

Our studies reveal a physical interaction between two such heterologous Q/N-rich protein aggregates and Hsp104. This hinders the availability of Hsp104 to shear prion aggregates, thereby inhibiting prion propagation. In contrast another overexpressed Q/N-rich protein does not sequester Hsp104, but rather appears to cure [*PSI*+] by increasing the level of Hsp104.

7.2 RESULTS

7.2.1 OVEREXPRESSION OF SOME Q/N RICH PROTEINS THAT ELIMINATE THE [PIN+] REQUIREMENT FOR THE INDUCTION OF [PSI+] ALSO DESTABILIZE PRE-EXISTING PRIONS

In an unsaturated genetic screen for overexpressed proteins that cure cells of [*PSI*+], the most efficient curing was observed in the presence of a plasmid encoding a Q/N-rich portion of the CYC8 gene. Strikingly, CYC8 was one of the 11 genes we previously uncovered in a screen for genes that in high copy substitute for the [*PIN*+] requirement for the *de novo* induction of [*PSI*+] [3]. Therefore we asked if overproduction of the other proteins identified in the [*PSI*+] induction screen would also destabilize pre-existing [*PSI*+]. Of the 11 chromosomal DNA fragments, 8 (*STE18, YCK1, PIN2, URE2, PIN3, NEW1, NUP116* and *LSM4*) encode full-length proteins with Q/N-rich domains, and another 3 encode partial genes: the C-terminal Q/N-rich domains of *PIN4* and *CYC8*, and the N-terminal Q/N-rich domain of *SWI1*, respectively, called here *PIN4C, CYC8C* and *SWI1N*.

Weak (w) [*PSI*+][*PIN*+] was transformed with the 11 multicopy plasmids with the *URA3* and *leu2-d* markers [63], and encoding the above mentioned Q/N-rich proteins and protein fragments. Transformants grown on leucineless media that amplified the plasmids to a high-copy number were then examined for the presence of [*PSI*+] using the color assay. This assay is based on the accumulation of a red pigment in ade1 mutants and the requirement of the Sup35 protein for proper termination at stop codons: cells in which much of the Sup35 release factor is sequestered into [*PSI+*] aggregates are unable to efficiently terminate translation at the premature stop codon in *ade1-14*, and some full-length Ade1 is synthesized despite the mutation. Thus, *ade1-14* cells that give rise to white or pink colonies are ⌊*PSI+*⌋, while those that grow into red colonies are [*psi−*].

A high proportion of red colonies indicated that amplification of plasmids encoding Pin4C, Cyc8C, Yck1 and Ste18, but not Pin2, Pin3, Ure2 and New1, caused efficient loss of w[*PSI+*] (Figure 1A). Representative red colonies were confirmed to be [*psi−*], because they exhibited diffuse fluorescence after being crossed to a [*psi−*] tester strain carrying Sup35NM-GFP [64], [65]. Furthermore, the resulting [*psi−*] state remained unchanged after elimination of the library plasmids, confirming that [*PSI+*] was indeed lost, not just transiently inhibited. This [*PSI+*] loss is not caused by a growth advantage of [*psi−*] over [*PSI+*] cells when the Q/N-rich domains are overexpressed (Figure S1). Overexpression of Swi1N, Nup116 and Lsm4 caused growth inhibition in w[*PSI+*][*PIN+*] cells, thus impeding analysis of curing of [*PSI+*] by those proteins.

Curiously, the overexpressed plasmids (*PIN4C, CYC8C, STE18, YCK1*) that caused the most efficient curing of w[*PSI+*] also caused efficient induction of [*PSI+*] (Figure 1B). The efficiencies of [*PSI+*] induction were examined in a [*psi−*][*pin−*] *rnq1Δ* strain by assaying read-through of the premature stop codon in the ade1-14 allele, which was detected as growth on SD-Ade (see Materials and Methods).

7.2.2 [URE3] IS ALSO DESTABILIZED BY HIGH COPY PLASMIDS ENCODING Q/N-RICH DOMAINS

We next examined if the Q/N-rich proteins could cure cells of another Q/N-rich prion, [*URE3*]. A [*URE3*][*PIN+*][*psi−*] derivative of 74-D694 with *SUP35* endogenously tagged with GFP was used. In this background, cells are light red in the absence of [*URE3*], but become dark red when they are [*URE3*] [66].

Overproduced Pin4C and Ste18, which caused efficient curing of w[*PSI+*], caused 58% and 49% loss of [*URE3*] respectively (Figure 1C), indicating that they each have a destabilizing effect on different Q/N-rich

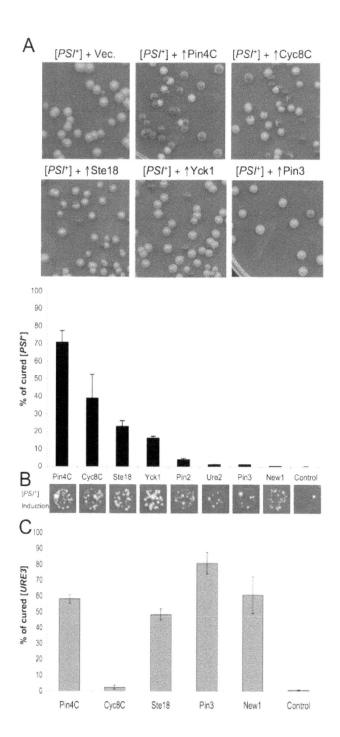

FIGURE 1: High copy plasmids that enhance induction of [*PSI+*] also cure pre-existing [*PSI+*] and [*URE3*]. (A) Overexpressed Q/N-rich proteins cause loss of [*PSI+*]. Weak (w) [*PSI+*][*PIN+*] (L1758) was transformed with plasmids encoding the indicated Q/N-rich proteins or fragments (*PIN4C* and *CYC8C*), or with the empty vector pHR81. Representative images of transformants plated on YPD following amplification of plasmids on SD-Leu are shown (upper). The efficiency of curing (lower) was determined as the percentage of red colonies indicative of [*psi−*] among ~1100 colonies. (B) Induction of [*PSI+*] in a rnq1Δ::HIS3 [*psi−*][*pin−*] 74-D694 strain (L3125) carrying the same plasmids as in (A). [*PSI+*] was induced by overexpression of *SUP35NM-GFP* from *pCUP1-SUP35NM::GFP-TRP*1 in 50 µM Cu²⁺ following library plasmid amplification on SD-Leu. Shown is growth on SD-Ade, which indicates the presence of [*PSI+*]: spots are representative of three repeated experiments. The Ade+ colonies were verified to be [*PSI+*] by visualization of Sup35NM-GFP dots. (C) Overexpressed Q/N-rich domains cause [*URE3*] curing. The [*URE3*] derivative of 74D-694 [*PIN+*][*psi−*] expressing the GFP tagged endogenous Sup35 (L3154), was transformed with high copy plasmids encoding the Q/N-rich proteins or protein fragments, or with the empty vector pHR81. The percentage of cured [ure-o] cells among ~1000 colonies was determined using the color assay described in Materials and Methods. Error bars show standard error of the mean.

prions. However, overproduced Cyc8C, another efficient [*PSI+*] curer, destabilized [*URE3*] only slightly, while Pin3 and New1, which did not have a significant effect on w[*PSI+*] propagation, caused 81% and 61% loss of [*URE3*] respectively (Figure 1C). These distinctions in the ability of different Q/N-rich proteins to cure cells of [*PSI+*] and [*URE3*] imply that they are curing prions via distinct mechanisms.

7.2.3 MEDIUM AND VERY HIGH [PIN+], BUT NOT HIGH [PIN+], VARIANTS ARE DESTABILIZED BY OVEREXPRESSION OF PIN4C

Pin4C was chosen for further investigation of how overexpressed Q/N-rich domains cure cells of prions, since it was the most efficient in curing both w[*PSI+*] and [*URE3*]. We thus asked if overexpression of Pin4C could affect propagation of [*PIN+*] and found that overexpression of Pin4C caused 30% loss of medium [*PIN+*], 8% loss of very high [*PIN+*], but had no effect on high [*PIN+*] (see Materials and Methods). All [*PIN+*] strains used in the study of curing of [*PSI+*] and [*URE3*] in this paper were high [*PIN+*].

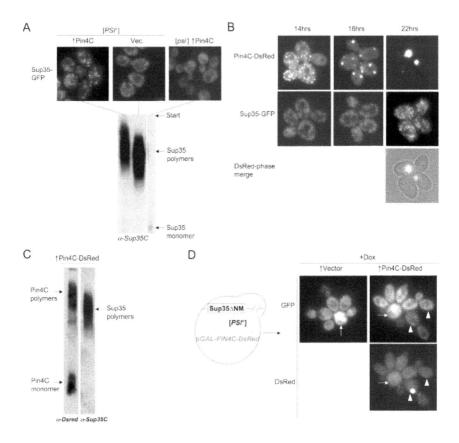

FIGURE 2: Pin4C overexpression leads to larger [*PSI+*] aggregates. (A) Sup35-GFP aggregates become larger upon overexpression of Pin4C. Upper panels: Representative GFP images of strong [*PSI+*][*PIN+*] *SUP35-GFP* expressing strains (GF657) after an overnight induction of *PIN4C*; empty vector control was incubated in galactose medium for the same amount of time; [*psi−*] culture shown is after 4-days of induction of *PIN4C*. Lower panels: Lysates prepared from cultures shown above were treated with 2% SDS at room temperature and analyzed for the presence of Sup35 by SDD-AGE with anti-Sup35C-GFP. (B) Overexpression of Pin4C-DsRed leads to the formation of single dot-like aggregates concomitant with enlargement of Sup35-GFP aggregates. Representative DsRed and GFP images are of strong [*PSI+*][*PIN+*] SUP35-GFP (GF657) cells after induction of pHR81*GAL-PIN4C-DsRED* for the times indicated. Cells were grown in liquid plasmid selective galactose media. (C) Pin4C-DsRed aggregates are composed of SDS-resistant polymers. Lysates from cultures where ~60% of cells contain large single Pin4-DsRed dots were treated with 2% SDS at room temperature and analyzed by SDD-AGE. The blot was probed with anti-DsRed antibody (left lane), stripped and re-probed with anti-Sup35C antibody (right lane). (D) Increased size of visible [*PSI+*] aggregates requires

continuous synthesis of the Sup35 protein. Single cells of strong [PSI+][PIN+] expressing SUP35ΔNM at its endogenous locus and harboring extrachromosomal pTET-SUP35-GFP (L3126) were transformed with pHR81GAL-PIN4C-DsRED, or vector, pHR81GAL-DsRED. Cells were grown in 2% raffinose +2% galactose +0.025 μg/ml doxcycline for 6 hrs, which induced PIN4C-DsRED and allowed the Sup35-GFP level to be close to the normal Sup35 level. Single cells with diffuse DsRed were then micromanipulated and grown for 18 hrs dividing ~3 times on 2% raffinose +2% galactose +10 μg/ml doxcycline medium where new synthesis of Sup35-GFP is repressed. GFP and DsRed images of a representative part of a growing microcolony are shown. The arrows point to mother cells, and the arrowheads point to two daughter cells with and without a large DsRed dot.

7.2.4 [PSI+] AGGREGATES INCREASE IN SIZE UPON OVEREXPRESSION OF PIN4C

SUP35-GFP strains expressing GFP tagged Sup35 in the original chromosomal location under the control of the SUP35 promoter were employed to allow for real-time visualization of [PSI+] aggregates as the cells were cured of [PSI+] by Pin4C overexpression. In [PSI+] cells Sup35-GFP accumulates in numerous small cytoplasmic foci. It was previously shown that this SUP35-GFP construct is functional, i.e. it can replace the essential Sup35 protein and can stably propagate strong [PSI+] [21], [67]. We proceeded with strong [PSI+] because w[PSI+] was somewhat unstable in the presence of the Sup35-GFP replacement. Thus, all [PSI+] strains used in the rest of the paper were strong [PSI+].

To tightly control Pin4C overexpression, a plasmid bearing PIN4C driven by the inducible GAL promoter was introduced into a strong [PSI+] [PIN+] SUP35-GFP strain (GF657). Overexpression of PIN4C on this leu2-d amplified plasmid (pHR81GAL-PIN4C) caused a 99% loss of strong [PSI+] in this [PIN+] strain with endogenously tagged SUP35-GFP. Likewise, strong [PSI+] was also efficiently cured (45% loss) by overexpressing Pin4C (using pHR81H-PIN4C) in 74D-694 with untagged Sup35. Consistent with our previous findings that high [PIN+] was not cured by overexpression of Pin4C, we found that [PIN+] was still maintained in derivatives of 74D-694 that were cured of strong [PSI+] as a result of Pin4C overexpression.

Changes in the state of strong [*PSI+*] caused by excess Pin4C were monitored over time. After overnight overexpression of Pin4C, the Sup35-GFP foci became brighter, bigger and more distinct in ~80% of the cells in comparison with the numerous tiny Sup35-GFP foci formed in control cells lacking the Pin4C plasmid (Figure 2A). Likewise, Pin4C overexpression caused the size distribution of SDS-resistant [*PSI+*] polymers to shift dramatically to larger complexes (Figure 2A). These enlarged Sup35-GFP foci that formed following overnight expression of Pin4C were still capable of propagating [*PSI+*] if expression of Pin4C was turned off. However, when Pin4C was expressed for another 4 days, Sup35-GFP became diffuse and [*PSI+*] was lost (Figure S2).

7.2.5 OVEREXPRESSED PIN4C FORMS AMYLOID-LIKE AGGREGATES, WHICH DO NOT COLOCALIZE WITH [PSI+] AGGREGATES

As expected of a [*PSI+*]-promoting Q/N-rich protein, overexpression of *PIN4C* leads to formation of Pin4C aggregates. When *PIN4C*-DsRED is expressed from a multicopy plasmid under the control of the *GAL* promoter in strong [*PSI+*][*PIN+*] cells, Pin4C-DsRed aggregates first appear as multiple cytosolic puncta after 14 hrs of overexpression. By 22 hours the Pin4C-DsRed usually forms one large focus per cell (Figure 2B). The formation of the large Pin4C-DsRed dot was always accompanied by the appearance of large Sup35-GFP foci within the cell, but they did not colocalize (Figure 2B). Furthermore, although Pin4C aggregates have the SDS-resistant characteristic of amyloid aggregates (Figure 2C), the sizes of Sup35 and Pin4C SDS-resistant polymers were not identical (Figure 2C), suggesting that Pin4C is not a component of SDS-resistant [*PSI+*] polymers.

7.2.6 PIN4C-INDUCED INCREASE OF [PSI+] AGGREGATE SIZE REQUIRES CONTINUOUS SYNTHESIS OF SUP35

The large Sup35 aggregates that appeared in the presence of excess Pin4C could have been formed by the simple association of existing [*PSI+*]

aggregates, or by enlargement of individual aggregates, e.g. due to reduced shearing of growing amyloid fibers. The first possibility is modeled on our previous finding that overexpressed Sup35 causes pre-existing
[*PSI+*] aggregates to coalesce into larger particles [68]. Thus we reasoned
that overproduced Pin4C might "glue" existing [*PSI+*] aggregates together through heterologous Q/N-rich domain interactions.

To test this, we examined the appearance of pre-existing [*PSI+*] aggregates when Pin4C was overexpressed. Pre-existing aggregates were made
of protein encoded by extrachromosomal *SUP35-GFP* controlled by the
repressible TETr promoter in a strain lacking the endogenous *SUP35* prion
domain (*SUP35ΔNM*). After 6 hrs of expression of the *GAL* controlled
PIN4C-DsRED (i.e. before diffuse Pin4-DsRed formed big foci and before any changes in [*PSI+*] aggregates could be noted in previous experiments), single cells were micromanipulated onto solid galactose medium
containing doxycycline, where expression of Sup35-GFP was repressed.
Later, in cells where large Pin4-DsRed foci appeared, the Sup35-GFP aggregates were still tiny, just as in cells with diffuse Pin4-DsRed examined
at the same time (Figure 2D). The faint appearance of the GFP aggregates was due to lack of newly synthesized Sup35-GFP. The absence of
large Sup35-GFP foci even in cells with large Pin4C-DsRed aggregates,
suggests that the large Sup35-GFP foci seen in the presence of continued
Sup35-GFP synthesis are not formed by coalescence of previously formed
[*PSI+*] aggregates. Rather, [*PSI+*] aggregates appear to become larger by
continuously incorporating newly synthesized Sup35-GFP upon overexpression of Pin4C.

7.2.7 PIN4C OVEREXPRESSION REDUCES SUP35-GFP AGGREGATE MOBILITY AND TRANSMISSION TO DAUGHTER CELLS

The dynamics of Sup35-GFP in dividing [*PSI+*] cells upon Pin4C overexpression was probed using fluorescence recovery after photobleaching (FRAP). The rate of transfer of Sup35-GFP from mother to daughter
cells was examined by first completely photobleaching daughter cells and
then measuring the fluorescence recovery of the daughter (Figure 3A). As

shown previously [69], [70], soluble Sup35-GFP in [*psi*−] cells was much more mobile than predominantly aggregated Sup35 in [*PSI*+] cells; the average half-time for recovery in isogenic [*psi*−] versus [*PSI*+] daughters was, respectively, 7 s versus 63 s (Figure 3B, 3C and 3E).

The fluorescence recovery measured following Pin4C overexpression in 12 [*PSI*+] cells containing large Sup35-GFP foci indicated that the population of Pin4C expressing cells is heterogeneous. Three cells exhibited almost no recovery, indicating a major defect in transmission of Sup35-GFP (Figure 3D (I) and 3E). In another 4 cells, Sup35-GFP only recovered to 67% of the intensity observed prior to photobleaching, with a half-time of 126 s that is twice as long as in [*PSI*+] without Pin4C overexpression (Figure 3D (II) and 3E). In one cell 98% recovery was completed with a half-time of 243 s (data not shown). Yet, in 3 cells, fluorescence recovered to 100% with half-times similar to that in [*PSI*+] cells without Pin4C (Figure 3D (III) and 3E). We also observed one cell (data not shown) exhibiting 83% recovery with a half-time of only 17.79 s.

The slow flux of Sup35-GFP in 8 of these mother-daughter pairs indicates that Sup35-GFP often becomes extremely immobile following overexpression of Pin4C, which is consistent with an increase in Sup35 aggregate size and a reduction in the segregation of prion seeds to daughters. However, the presence of cells with normal recovery suggests that at least in some cells, in addition to the large Sup35-GFP foci, there were still small prion seeds available to be transmitted to daughters. Finally, the existence of cells with very fast flow of Sup35-GFP from mother to daughter cells indicates a high level of soluble Sup35 that might be already inefficiently sequestered by the few large Sup35 aggregates still remaining in the mother cell. Differences in the rates of Sup35-GFP transfer among individual mother-daughter pairs suggest that overexpressed Pin4C creates heterogeneity in the properties of prion aggregates during the curing process.

7.2.8 MICROCOLONIES OVEREXPRESSING PIN4C SHOW PROGRESSIVE LOSS OF [PSI+]

To further assess how the appearance of large immobile Sup35 aggregates caused by excess Pin4C correlates with the loss of [*PSI*+], individual cells

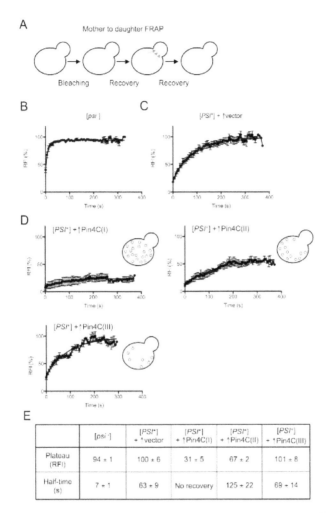

FIGURE 3: Overexpressed Pin4C reduces the transmission of Sup35-GFP from mother to daughter cells. (A) Diagram of experiment. Fluorescence in daughter cells was photobleached and the time course of fluorescence recovery of the daughter cells was measured as described previously [70]. (B–D) Quantitative FRAP analysis of Sup35-GFP in 10 [psi−][pin−] (GF658) cells (B), 9 strong [PSI+][PIN+] (GF657) cells harboring the pHR81GAL vector (C), and 12 strong [PSI+][PIN+] (GF657) cells harboring pHR81GAL-*PIN4C* and containing enlarged Sup35-GFP foci following overnight Pin4C overexpression (D). The relative fluorescence intensity (RFI) of the bleached daughter cell was determined every 5 s after completion of photobleaching and normalization. Error bars indicate the standard error of the mean. RFI in (D) represents the average of 3 cells (I), 4 cells (II) and 3 cells (III). The analysis of two more cells is not shown. (E) The recovery plateau level and half-time from the curves in B–D are listed.

of the [*PSI+*] Sup35-GFP strain carrying pHR81GAL-PIN4C-DsRED were micromanipulated and grown on 2% raffinose + 2% galactose plates where the DsRed tagged Pin4C was expressed. As the cells divided we monitored the changes of Sup35-GFP distribution in the cells within the microcolonies. The outer edges of microcolonies with a single layer of cells were imaged since the central portion of the microcolony included multiple layers of overlapping cells. In the edge of one sector (Figure 4, upper panel), Sup35-GFP remained in the multiple tiny foci seen in [*PSI+*] cells prior to Pin4C induction. But in the edge of another sector (Figure 4, lower panel), Sup35-GFP foci increased in size and were reduced in number progressively in dividing cells, which eventually segregated out [*psi−*] cells (also see Figure S3). Different phenotypes observed in different sectors may be due to differences of PIN4C plasmid copy number within individual cells.

7.2.9 CELL DIVISION IS REQUIRED FOR OVEREXPRESSION OF PIN4C TO CURE [PSI+]

Previous studies showed that cell division was essential for the loss of the [*PSI+*] prion in GuHCl-treated cells [71]. To test if cell division was required for the overexpression of Pin4C to cure [*PSI+*], loss of [*PSI+*] was compared in a *MATa* strong [*PSI+*] *SUP35-GFP* strain overexpressing Pin4C in the presence and absence of growth arrest induced by α-factor. Pin4C was overexpressed in liquid galactose for 40 hrs, and 50 μM α-factor was added at this stage, i.e. when Sup35-GFP aggregates were larger and fewer in number, but before the emergence of any diffuse [*psi−*] cells. After overexpressing Pin4C for another 16 hrs, cells were plated on YPD to score for [*PSI+*] loss. The α-factor arrest caused an 88% decrease in colony-forming units, CFUs. Cultures whose growth was arrested vs. not arrested respectively showed, 1% vs. 59% loss of [*PSI+*] (Figure 5). Reduced loss of [*PSI+*] during the α-factor arrest indicates that cell division is required for curing of [*PSI+*] by overexpressed Pin4C.

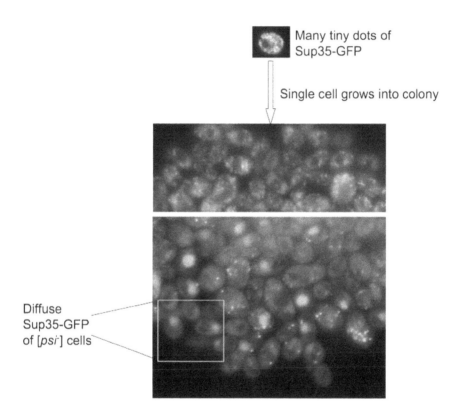

Many tiny dots of
Sup35-GFP

Single cell grows into colony

Diffuse
Sup35-GFP
of [psiˉ] cells

FIGURE 4: Microcolonies overexpressing Pin4C-DsRed show progressive loss of [*PSI*+]. Single strong [*PSI*+][*PIN*+] *SUP35-GFP* expressing (GF657) cells carrying pHR81*GAL-PIN4C-DsRED* were micromanipulated and grown on 2% raffinose +2% galactose to induce Pin4C-DsRed for ~24 hrs. Portions of a microcolony are shown as the merge of GFP and DsRed images.

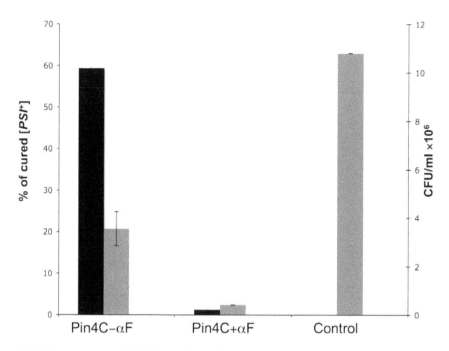

FIGURE 5: Curing of [*PSI*+] by Pin4C depends on cell division. Three *MATa* strong [*PSI*+][*PIN*+] *SUP35-GFP* (GF845) transformants harboring pHR81*GAL-PIN4C* were grown in liquid galactose media for 40 hrs, when Sup35-GFP foci became larger in size and fewer in number. Then cells were transferred to fresh galactose medium with or without the addition of 50 μM a-factor for another 16 hrs. Three transformants harboring empty vector pHR81-*GAL* were transferred to galactose medium without the addition of 50 μM a-factor as control. Samples were taken, diluted and plated on YPD where the percentage of red cured [*PSI*+] among ~850 colonies was scored (black bars). There was no [*PSI*+] loss in the control. The number of CFUs is shown by gray bars. Error bars show standard error of the mean.

7.2.10 OVEREXPRESSION OF PIN4C DOES NOT CHANGE CHAPERONE LEVELS

Since [*PSI*+] propagation is sensitive to optimal levels of chaperones such as Hsp104, Ssa1/2, Ssb1, Sse1 and Sis1 [72]–[74], it seemed possible that excess Pin4C caused a change in levels of chaperones which then led to loss of [*PSI*+]. However, Pin4C overexpression did not cause a significant alteration in levels of Ssa1/2, Ssb1, Sse1 and Sis1 (Figure S4A).

More thorough analysis revealed that the level of Hsp104 in cells following Pin4C overexpression was reduced to 83% of that in cultures not overproducing Pin4C (Figure S4B and Table S1). Because previous studies showed that a heterozygous disruption of *HSP104* has no effect on [*PSI+*] propagation [75], and that loss of [*PSI+*] is only initiated when the Hsp104 levels drop well below 50% of the normal level [76], it appeared unlikely that the slight decrease in Hsp104 level induced by excess Pin4C would cause [*PSI+*] loss. Indeed, a heterozygous disruption of *HSP104* did not facilitate curing of [*PSI+*] in our strains (Figure S4C).

7.2.11 OVEREXPRESSED PIN4C TITRATES HSP104-GFP AWAY FROM THE CYTOPLASM

The increased size of Sup35 polymers seen upon Pin4C overexpression was similar to that seen upon inhibition of Hsp104 due to a block of prion fragmentation [20]. Thus we considered the possibility that excess Pin4C cures [*PSI+*] by titrating Hsp104 away. Since [*PSI+*] aggregates were found to associate with Hsp104 [38], it seemed possible that Pin4C aggregates also harbor Hsp104.

To visualize the distribution of Hsp104, we used the Hsp104-GFP strain from the endogenously GFP-tagged yeast library [77]. In unstressed cells, Hsp104-GFP is observed as diffuse GFP or occasionally tiny foci with diffuse background (Figure 6A). However, after 16 hrs of induction of untagged Pin4C, Hsp104-GFP coalesced into one large aggregate per cell. Such large Hsp104-GFP aggregates were never found in control cells without Pin4C overexpression. When Pin4C tagged with DsRed was used, the big Pin4C-DsRed focus colocalized with the coalesced Hsp104-GFP (Figure 6A). Furthermore, Hsp104 was co-immunocaptured with large Pin4C-DsRed aggregates in strong [*PSI+*][*PIN+*] cells (Figure 6B). The sequestration of Hsp104 by overexpressed Pin4C reduced the cytoplasmic level of Hsp104 to about 33% of its normal level (Figure 7) which could have inhibited Hsp104 from shearing [*PSI+*] aggregates and producing new seeds for prion propagation.

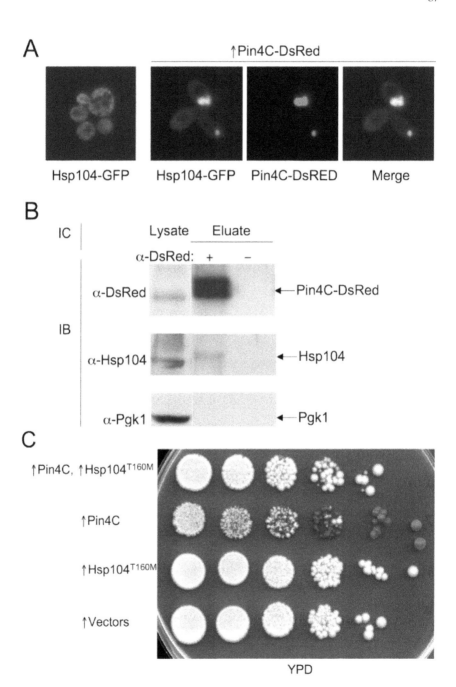

A

↑Pin4C-DsRed

Hsp104-GFP   Hsp104-GFP   Pin4C-DsRED   Merge

B

IC

Lysate   Eluate

α-DsRed: +   −

IB

α-DsRed   ←Pin4C-DsRed

α-Hsp104   ←Hsp104

α-Pgk1   ←Pgk1

C

↑Pin4C, ↑Hsp104^{T160M}

↑Pin4C

↑Hsp104^{T160M}

↑Vectors

YPD

FIGURE 6: Overexpressed Pin4C sequesters Hsp104 from the cytoplasm. (A) Overexpressed Pin4C sequestered Hsp104-GFP to colocalize with Pin4C-DsRed aggregates. Representative images of cells with endogenous Hsp104 tagged with GFP without or with 16 hrs of induction of pHR81*GAL-PIN4C-DsRED* are shown. (B) Overexpressed Pin4C binds to Hsp104. The interaction between Hsp104 and overexpressed Pin4C was assayed in strong [*PSI+*][*PIN+*] (GF657) following overnight overexpression of Pin4C-DsRed. 60 µg of total protein was loaded as the "lysate". The Pin4C-DsRed complex immunocaptured with anti-DsRed from 500 µg of total protein was loaded as "eluate". The same membrane was immunoblotted (IB) with anti-DsRed, then with anti-Hsp104, and re-probed with anti-Pgk1 as a control. No co-immunocapture of endogenous Pgk1 with Pin4C was detected, implying that Hsp104 was specifically immunocaptured with the Pin4C complex. The slightly slower migration of Hsp104 in the "eluate" relative to its migration in the "lysate" is probably due to the different buffers used during immunocapture. (C) Overexpression of Hsp104^{T160M} suppresses curing of strong [*PSI+*] by Pin4C. Transformants with pHR81*GAL-PIN4C* and pRS413*GAL-HSP104T160M* (↑Pin4C, ↑Hsp104^{T160M}); or with pHR81*GAL-PIN4C* and empty vector pRS413*GAL* (↑Pin4C); or with pRS413*GAL-HSP104*^{T160M} and pHR81*GAL* (↑Hsp104^{T160M}); or with both empty vectors pHR81*GAL* and pRS413*GAL* (↑vectors) were selected on plasmid selective glucose medium, replica-plated to plasmid selective inducing galactose medium, and then 10-fold serially diluted (10^5→10^0 cells from left to right) and spotted onto glucose YPD medium where expression of Pin4C and Hsp104^{T160M} is turned off. There was no growth inhibition in cells overexpressing Pin4C and Hsp104^{T160M} compared to those overexpressing Pin4C alone when spotted on a galactose plate (Figure S5). [*PSI+*] loss was scored by the appearance of red [*psi−*] colonies.

7.2.12 CURING OF [PSI+] BY PIN4C IS RESCUED BY ELEVATED LEVELS OF HSP104

If Pin4C sequestration of Hsp104 causes curing of [*PSI+*], elevation of Hsp104 levels should antagonize this. Unlike wild-type Hsp104, over-expression of Hsp104^{T160M} does not cure cells of [*PSI+*]. Also, when expressed at the normal level, Hsp104^{T160M} maintains [*PSI+*] [26]. Therefore we overexpressed the Hsp104^{T160M} mutant allele. Excess Hsp104^{T160M} did not reduce Pin4C aggregation (data not shown), however it reduced the efficiency with which overexpressed Pin4C caused loss of [*PSI+*] (Figure 6C).

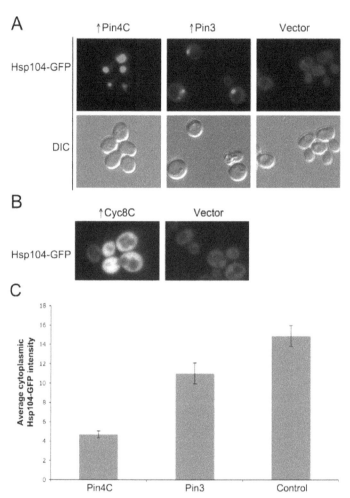

FIGURE 7: Effects of overexpressed Pin3 and Cyc8C on Hsp104. (A) Overexpressed Pin3 caused endogenous Hsp104-GFP to coalesce into big aggregates. Representative images of Hsp104-GFP cells after induction of pHR81*GAL-PIN4C* or pHR81*GAL-PIN3* for 16 hrs are shown. All the images shown were taken in the same exposure time. (B) Hsp104-GFP became brighter upon overexpression of Cyc8C. Representative images of Hsp104-GFP cells after 16 hrs induction of Cyc8C from pHR81 based high copy vectors with the insert encoding *CYC8C* or the empty vector on SD-Leu liquid medium. All the images shown were taken in the same exposure time. (C) Overexpressed Pin3 caused the sequestration of Hsp104 less effectively than Pin4C. The bar graphs represent the average of the mean fluorescence signal intensity in cytoplasmic regions devoid of aggregates and excluding the vacuole in 22 Hsp104-GFP cells. Error bars indicate the standard error of the mean. P<0.0001 for comparisons of fluorescence intensity in cells with overexpressed Pin4C or Pin3 and vector control.

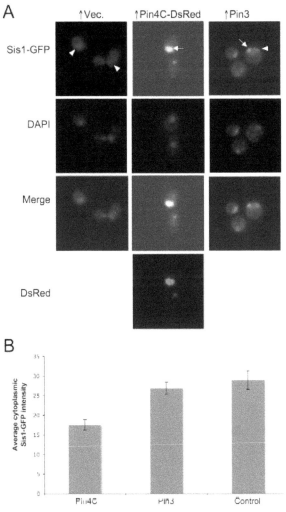

FIGURE 8: Overexpressed Pin4C or Pin3 caused Sis1 to coalesce.: (A) Overexpressed Pin4C or Pin3 causes Sis1-GFP to aggregate. Representative images of endogenous Sis1 tagged with GFP in the absence or presence of 16 hrs of induction of pHR81*GAL-PIN4C-DsRED* or pHR81*GAL-PIN3* are shown. Fixed cells were permeabilized and stained with DAPI. The arrowhead points to nuclear Sis1-GFP signals, and the arrow points to aggregated Sis1-GFP that is not nuclear. (B) Overexpressed Pin3 did not cause effective sequestration of Sis1 from the cytoplasm. The bar graphs represent the average of the mean fluorescence signal intensity in cytoplasmic regions devoid of aggregates and excluding the vacuole in 20 Sis1-GFP cells. Error bars indicate the standard error of the mean. P<0.0001 for comparisons of fluorescence intensity in cells with overexpressed Pin4C or Pin3 and vector control.

7.2.13 INCREASED LEVELS OF SIS1 PREVENT AGGREGATION OF PIN4C AND REDUCE THE ABILITY OF PIN4C TO CURE [PSI+]

Sis1 is a chaperone involved in cleaving [*PSI*+] aggregates and generating new prion seeds. It was hypothesized to recruit Hsp104 to the sites of prion aggregation [39]. We observed that overexpressed Pin4C also sequestered Sis1. As described previously [78], without Pin4C overexpression most Sis1-GFP was found in the nucleus (Figure 8A). Upon Pin4C-DsRed over-expression, much of the Sis1-GFP colocalized with cytoplasmic Pin4C-DsRed aggregates, and the amount of Sis1-GFP remaining in the nucleus was significantly reduced (Figure 8). Thus we asked if increased levels of Sis1 would affect curing of [*PSI*+] by overexpression of Pin4C. Indeed, we found that the loss of [*PSI*+] by overexpression of Pin4C was sig-nificantly reduced by overexpression of Sis1 (Figure 9A). Furthermore, in cells with excess Sis1, overproduced Pin4-DsRed accumulated in several small foci and did not form the huge single focus observed in the absence of excess Sis1. Also, in cells with excess Sis1, Sup35-GFP remained in multiple tiny foci, that did not enlarge upon Pin4C overexpression (Figure 9B). Thus, overproduced Sis1 prevents overexpressed Pin4C from form-ing big foci and reduces the formation of large Sup35 aggregates, which may cause decreased [*PSI*+] curing by Pin4C.

7.2.14 EFFECTS OF OVEREXPRESSED PIN3 AND CYC8C ON CHAPERONES

To investigate if titrating Hsp104 is a general mechanism by which the heterologous Q/N-rich proteins cure prions, we visualized Hsp104-GFP in cells overexpressing Pin3 or Cyc8C. Like Pin4C, overexpressed Pin3 caused Hsp104-GFP to coalesce into large aggregates and reduced the lev-el of diffuse cytoplasmic Hsp104-GFP fluorescence relative to empty vec-tor controls. However, overexpressed Pin3 also caused a slight increase in the cellular levels of Hsp104, Sse1, Ydj1 and Sis1 (Figure S6A). The com-bined result of these two effects was that the Hsp104 cytoplasmic level was about 74% of that seen in cells without Pin4C overexpression (Figure 7C). Likewise, although overexpressed Pin3 caused Sis1-GFP to coalesce

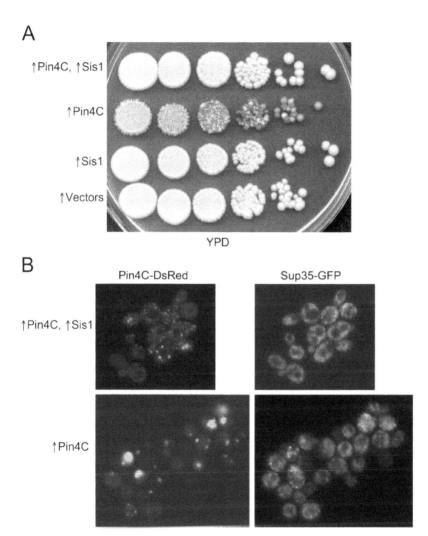

FIGURE 9: Overexpressed Sis1 reduces elimination of [*PSI+*] by excess Pin4C. (A) Increased levels of Sis1 inhibit [*PSI+*] curing by excess Pin4C. Strong [*PSI+*][*PIN+*] (GF657) cells co-transformed with pHR81*GAL-PIN4C* and pYES3*GAL-SIS1* (↑Pin4C, ↑Sis1); or with pHR81*GAL-PIN4C* and empty vector pYES3*GAL* (↑Pin4C); or with pYES3*GAL-SIS1* and empty vector pHR81*GAL* (↑Sis1); or with two empty vectors pHR81*GAL* and pYES3*GAL* (↑vectors), were examined as described above (see Figure 6C). (B) Overexpressed Sis1 prevents overproduced Pin4C from forming large aggregates. Representative fluorescent images of strong [*PSI+*][*PIN+*] (GF657) cells harboring pHR81*GAL-PIN4C-DsRED* and pYES3*GAL-SIS1* (↑Pin4C, ↑Sis1), or pHR81*GAL-PIN4C-DsRED* and empty vector pYES3*GAL* (↑Pin4C) after overnight induction in liquid galactose are shown

(Figure 8A), the levels of Sis1 that remained in the cytoplasm were similar to controls (Figure 8B). In contrast to overexpressed Pin4C or Pin3, over-expressed Cyc8C caused an 8-fold increase in the Hsp104 level (Figure S6B), and did not sequester Hsp104 (Figure 7B). Overexpressed Cyc8C also caused a slight increase in the Sis1 level (Figure S6B).

7.2.15 OVEREXPRESSED GPG1 MAY TITRATE CHAPERONES AWAY FROM THE CYTOPLASM

We next investigate if non-Q/N-rich aggregates that cure [PSI+] also se-quester chaperones. Gpg1 is a mimic of a G protein γ subunit. Like over-expressed Pin4C, overexpressed Gpg1 formed aggregates, had reduced curing efficiency when Hsp104 was overexpressed, but did not affect the cellular levels of Hsp104 [79]. Previous work visualizing Gpg1 cur-ing of [PSI+] aggregates was complicated by the use of overexpressed Sup35NM-GFP [79]. When we examined the effect of excess Gpg1 on flu-orescent aggregates in [PSI+] cells with endogenous Sup35 tagged with GFP, we found that the fluorescent dots got larger and fewer in number (Figure 10A), just as seen when Pin4C was overexpressed. Furthermore, like excess Pin4C, excess Gpg1 caused endogenous Hsp104 tagged with GFP to aggregate into foci. However, despite this aggregation we did not detect any reduction in the intensity of cytoplasmic Hsp104-GFP com-pared to the vector control (Figure 10B).

7.3 DISCUSSION

Surprisingly, several factors that enhance prion induction also cause prion destabilization: [PIN+] both facilitates [PSI+] appearance and destabi-lizes [PSI+] [51], [59]; overexpression of the Ure2 prion domain both induces de novo [URE3] appearance and cures [URE3] [60]; overexpres-sion of the same chaperones both enhance de novo [URE3] generation and destabilize existing [URE3] [80]. We now report that overexpression of many of the 11 Q/N-rich proteins, which in high copy substitute for the [PIN+] requirement for [PSI+] induction, also destabilize pre-existing

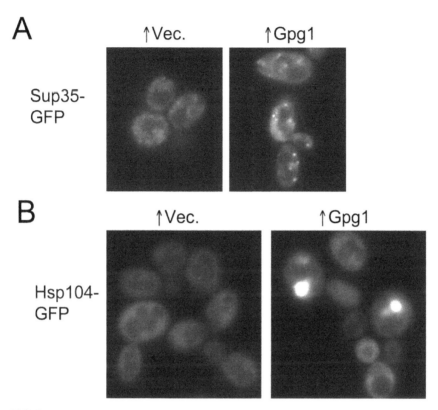

FIGURE 10: Overexpression of the non-Q/N-rich protein Gpg1 leads to larger [PSI +] aggregates and causes Hsp104 to aggregate. (A) Overexpressed Gpg1 leads to larger [*PSI+*] aggregates. Representative GFP images of strong [*PSI+*][*PIN+*] *SUP35-GFP* expressing cells (GF657) transformed with pRS316-*GAL1-GPG1* or empty vector control were of cells incubated overnight in galactose medium for the same amount of time. (B) Overexpressed Gpg1 caused the aggregation of Hsp104-GFP. Representative images of cells with endogenous Hsp104 tagged with GFP with 16 hrs of induction of pRS316-*GAL1-GPG1* or empty vector as control are shown.

[*PSI+*] or [*URE3*] (Figure 1), indicating that Q/N-rich proteins can also both enhance prion appearance and impair propagation of existing prions.

Our finding that the high copy plasmids encoding 11 Q/N-rich domains can promote the *de novo* induction of [*PSI+*] in an *rnq1Δ* strain, establishes that the induction of [*PSI+*] does not first require the appearance of [*PIN+*]. Rather, this suggests a direct interaction between heterologous Q/N-rich proteins and Sup35, and, as originally hypothesized [3], and as indicated by earlier in vitro studies [54], [58], that the Q/N-rich proteins themselves are likely substituting for [PIN+] as the cross-seeds.

Since certain [PIN+] variants (e.g. low [PIN+], medium [PIN+] and very high [PIN+] but not high [PIN+]) destabilize [PSI+], curing of [PSI+] by Q/N-rich proteins could result from changing the [PSI+]-inducing high [PIN+] into a destabilizing [PIN+]. However, this seems unlikely because the effects of the destabilizing [PIN+]s are limited to weak but not strong [PSI+] [59], while overexpressed Pin4C also cures strong [PSI+]. Furthermore, overexpressed Q/N-rich proteins caused loss of weak [PSI+] even without [PIN+] (data not shown).

Other previously hypothesized mechanisms of prion curing by Q/N-rich proteins are that the overexpressed Q/N-rich domain of Ure2 inhibits prion fiber growth leading to curing of [URE3], either by incorporating into the growing tip of the [URE3] seed thereby blocking or "capping" its growth [60], [81], [82], or by sequestering Ure2 preventing it from joining prion fibers [81], [82]. However, this mechanism implies a very efficient interaction between the curing protein and the prion-forming protein, which is likely only in the case of very high homology.

Our studies indicate that overexpression of Pin4C allows [PSI+] aggregates to continuously incorporate soluble Sup35, but prevents proper fragmentation (Figure 2, Figure 3, Figure 4, Figure 5). Thus, overproduced Pin4C causes [PSI+] loss via a defect in breakage of growing [PSI+] fibers and transmission of prion seeds. This is quite different from either the capping or sequestration of its own protein models. Since prion fragmentation is crucially dependent on chaperones, particularly on Hsp104, overexpressed Pin4C could affect the function of Hsp104 on [PSI+]. One possibility is that Pin4C coats [PSI+] aggregates through the interaction of their Q/N-rich domains and thus shields Sup35 polymers from the shearing activity of chaperones. However, we did not detect co-localization of Sup35-GFP and Pin4C-DsRed foci (Figure 2B).

Although there is no significant alteration in total levels of chaperones (Figure S4), excess Pin4C sequesters Hsp104 and Sis1 (Figure 6, Figure 7, Figure 8), reducing their availability to shear [PSI+] aggregates. Although depletion of Sis1 only causes delayed and gradual loss of [PSI+] [35], the reduced availability of Sis1 may enhance the effect of Hsp104 sequestration. Furthermore, although excess Sis1 prevented curing of [PSI+] by

Pin4C overexpression, excess Sis1 also decreases Pin4C aggregation (Figure 9).

We also observed that different Q/N-rich proteins had distinct effects toward the yeast Q/N-rich prions, [*PSI+*] and [*URE3*] (Figure 1). For example, our data is consistent with the recent study showing that overproduction of New1 does not cure [*PSI+*] [83], but we further uncover that overexpressed New1 does cause efficient curing of [*URE3*]. Likewise, Pin3 cures [*URE3*] but not [*PSI+*]. One explanation could be that these proteins sequester Hsp104 less efficiently than the proteins that cure both prions such as Pin4C. Indeed, we found that overexpressed Pin3 reduced the cytoplasmic level of Hsp104 less effectively than Pin4C (Figure 7 and Figure S6A). Since more than 50% of the cellular Hsp104 remained diffuse when Pin3 was overexpressed, curing of [*PSI+*] would not be expected [76]. However, since there are fewer prion seeds in [*URE3*] than weak [*PSI+*] cells [35], this slight reduction in soluble Hsp104 might be sufficient to cure [*URE3*]. Indeed, previous studies showed that different prions and different prion variants have different susceptibilities towards chaperone activities [37], [69], [84]. Although overexpressed Pin3 sequestered Sis1 (Figure 8A), there was no significant difference in the fluorescence intensity of diffuse Sis1-GFP in the cytoplasm with Pin3 overexpression compared to the control without Pin3 overexpression (Figure 8B), suggesting that sequestration of Sis1 by excess Pin3 does not significantly contribute to curing of [*URE3*]. There was also a slight increase in the total cellular levels of Sse1 and Ydj1, which could contribute to the Pin3 curing.

Surprisingly, overexpression of Cyc8C cures [*PSI+*] but not [*URE3*] (Figure 1). This cannot be explained by sequestration of Hsp104. Indeed, overexpression of Cyc8C did not cause aggregation of Hsp104 (Figure 7B), but rather increased the Hsp104 level 8-fold while having no effect on Ssa1 (Figure S6B). Since overexpression of Hsp104 cures [*PSI+*] but not [*URE3*] [16], [42], this provides a plausible mechanism. We also observed a slight increase in the Sis1 level by overexpressed Cyc8C (Figure S6B). It was previously shown that overproduced Sis1 enhances curing of [*PSI+*] by overexpressed Hsp104 [80], therefore overexpressed Cyc8C may cure [*PSI+*] through additive effects of the increased level of Hsp104 and Sis1.

Our findings may provide an explanation for previous observations that overexpressed proteins lead to curing of prions. Overexpression of the

Rnq1Δ100 protein (the Q/N-rich C-terminal domain) eliminates [*PSI+*] and [*URE3*] in the presence of [*PIN+*] [61]. Also mutations in the non-Q/N-rich domains of *RNQ1* cause an increase in the size of Sup35 aggregates leading to curing [62]. Both of these phenomena could be because the Rnq1 fragment or mutants form [*PIN+*]-dependent aggregates that sequester Hsp104 and/or other chaperones and reduce their availability to aid [*PSI+*] propagation.

The chaperone titration curing mechanism may also be applicable to non-Q/N-rich aggregates. Indeed, a similar mechanism could explain prion curing caused by overexpression of the non-Q/N-rich protein Gpg1 (Figure 10). Although Gpg1 sequestered Hsp104, we could not detect a reduction in cytoplasmic Hsp104. Nonetheless, other chaperones might also be sequestered by Gpg1 aggregates leading to reduced prion shearing and prion loss.

A recent study also indicates that the cellular localization of chaperones can have a direct impact prion propagation. Indeed, this appears to explain the long-standing mystery of why overexpression of Hsp104 cures cells of [*PSI+*] but not of other prions [16], [36], [42], [51]. Hsp104 overexpression was shown to inhibit shearing of [*PSI+*] aggregates because the excess of Hsp104 displaced the Hsp70 chaperone Ssa1 from the [*PSI+*] aggregate. Hsp104 does not bind to other prions in the absence of Ssa1, consistent with the absence of curing [29].

Other studies also indicate that titration of cellular proteins by amyloid aggregates can have a profound effect on the cell. Indeed, sequestration of essential proteins by amyloid aggregates was previously shown to cause prion toxicity. The large Sup35 aggregate that forms when Sup35 is overexpressed in [*PSI+*] cells sequesters the essential Sup35 binding partner Sup45, resulting in death [68]. Likewise, the large Rnq1 aggregates that form when Rnq1 is overexpressed in [*PIN+*] cells sequester the core spindle pole body component Spc42 causing toxicity [85]. Also, polyglutamine (polyQ) aggregates sequester essential endocytic components such as Sla2 [86] and endoplasmic reticulum associated degradation proteins in [*PIN+*] dependent toxicity [87], and sequester Sup35 in [*PSI+*]-dependent polyQ toxicity [88].

Our results establish sequestration of specific chaperones by overexpressed proteins as a general mechanism to alter the cellular localization of chaperones and therefore inhibit prion propagation. Similar mechanisms could influence phenotypic variation by regulating the balance of chaperones needed for prion propagation, in response to environmental stimuli. Since biochemical pathways controlling prion formation and/or maintenance appear to be conserved from yeast to mammals, titration of chaperones via heterologous Q/N-rich aggregates might provide a new approach to prion and amyloid disease intervention.

7.4 MATERIALS AND METHODS

7.4.1 STRAINS, MEDIA, AND PLASMIDS

All strains used are described in Table 1. GF657 and GF658, respectively, are strong [PSI+][PIN+] and [psi−][pin−] versions of 74-D694 with endogenous SUP35 replaced with SUP35-GFP (SY80 and SY84 from T. R. Serio) [21]. Unless otherwise stated, all strains used in the study are high [PIN+] [53]. L3079 was constructed by disrupting chromosomal RNQ1 of GF657 with Δrnq1::HIS3. L3126 was constructed by transforming GF657 with exogenous pTET'-SUP35-GFP to maintain [PSI+] and then replacing chromosomal SUP35-GFP at its genomic locus with SUP35ΔNM by integration and excision using MluI digested pEMBL-Δ3ATG [89]. This construct was verified by detection of SUP35ΔNM (deletion of amino acid residues 1–253) on a western blot. A [psi−][pin−] version of strain 64-D697 transformed with pCUP1-SUP35NM::GFP-TRP1 was crossed to strains to confirm their [PSI+] state which was indicated by the appearance of fluorescent foci. GF657 was cured of [PSI+] by overexpression of Pin4C to generate L3107, and then grown on ethidium bromide [90] to generate a [rho−] petite version L3116. GF827 was grown on 5 mM guanidine hydrochloride to cure [PSI+] [91] and [PIN+] [52] to generate L3152. Cytoduction from donor L3152 into recipient L3116 gave rise to the 74D-694 [URE3][PIN+][psi−] derivative, L3154. The [rho+] cytoductants were confirmed by their inability to grow on medium lacking histidine. GF708 transformed with pURE2-URE2N::GFP-HIS3 was crossed to strains to score for their [URE3] state which was indicated by the appearance of fluorescent foci.

TABLE 1: Yeast strains.

Strain	Genotype	References
64-D697	*MATa ade1-14 ura3-52 leu2-3, 112 trp1-289 lys9-A21 [psi-][pin-]*	[51]
74D-694	*MATa ade1-14 leu2-3, 112 his3-Δ200 trp1-289 ura3-52*	[50]
L1749	*MATa ade1-14 leu2-3, 112 his3-Δ200 trp1-289 ura3-52 high [PIN+] [psi-]*	[59]
L1758	*MATa ade1-14 leu2-3, 112 his3-Δ200 trp1-289 ura3-52 weak [PSI+] high [PIN+]*	[50]
L1762	*MATa ade1-14 leu2-3, 112 his3-Δ200 trp1-289 ura3-52 strong [PSI+] high [PIN+]*	[51]
L1945	*MATa ade1-14 leu2-3, 112 his3-Δ200 trp1-289 ura3-52 medium [PIN+][psi-]*	[59]
L1953	*MATa ade1-14 leu2-3, 112 his3-Δ200 trp1-289 ura3-52 very high [PIN+][psi-]*	[59]
L3079	*MATα ade1-14 leu2-3, 112 his3-Δ200 trp1-289 ura3-52 rnq1Δ:: HIS3 SUP35-GFP strong [PSI+]*	this study
L3107	*MATα ade1-14 leu2-3, 112 his3-Δ200 trp1-289 ura3-52 SUP35-GFP high [PIN+][psi-]*	this study
L3116	*MATα ade1-14 leu2-3, 112 his3-Δ200 trp1-289 ura3-52 SUP35-GFP high [PIN+][psi-][rho-]*	this study
L3125	*MATa ade1-14 leu2-3, 112 his3-Δ200 trp1-289 ura3-52 rnq1Δ::HIS3 [psi-][pin-]*	this study
L3126	*MATα ade1-14 leu2-3, 112 his3-Δ200 trp1-289 ura3-52 SUP35ΔNM pTETrSUP35-GFP strong [PSI+] high [PIN+]*	this study
L3152	*MATa, ura3, leu2, trp1, pDAL5::ADE2 pDAL5::CAN1 kar1 [URE3] [psi-][pin-]*	this study
L3154	*MATα ade1-14 leu2-3, 112 his3-Δ200 trp1-289 ura3-52 SUP35-GFP high [PIN+][URE3][psi-]*	this study
GF657	*MATα ade1-14 leu2-3, 112 his3-Δ200 trp1-289 ura3-52 SUP35-GFP strong [PSI+] high [PIN+]*	SY80 [21]
GF658	*MATα ade1-14 leu2-3, 112 his3-Δ200 trp1-289 ura3-52 SUP35-GFP [psi-][pin-]*	SY84 [21]
GF708	*MATa ade1-14 his3-Δ200 ura3-52 leu2-3, 112 trp1Δ lys2 [psi-][pin-]*	Cured version of GT81-1C [97]
GF827	*MATa, ura3, leu2, trp1, pDAL5::ADE2 pDAL5::CAN1 kar1 [URE2] [PSI+][PIN+]*	BY241 [98]
GF844	*MATα ade1-14 leu2-3, 112 his3-Δ200 trp1-289 ura3-52 HSP104::LEU2 SUP35-GFP [psi-]*	SY97 [75]
GF845	*MATa ade1-14 leu2-3, 112 his3-Δ200 trp1-289 ura3-52 SUP35-GFP strong [PSI+]*	SY81 [21]
GF886	*MATa his3-Δ1 leu2Δ0 met15Δ0 ura3Δ0 HSP104-GFP:: HIS3 [PIN+]*	77
GF894	*MATa his3-Δ1 leu2Δ0 met15Δ0 ura3Δ0 SIS1-GFP::HIS3 [PIN+]*	77

Standard yeast media were used [92]. For overexpression of library high copy plasmids [93] transformants were selected on plasmid selective synthetic media with dextrose lacking uracil (SD-Ura) and then spread on synthetic media lacking both uracil and leucine (SD-Ura-Leu) to amplify the copy number of leu2-d bearing plasmids about 100 fold [63]. For Pin4C overexpression from pHR81*GAL-PIN4C* in liquid medium, cultures were grown in 2% raffinose synthetic media lacking uracil for ~8 hrs and then transferred to 2% raffinose + 2% galactose media lacking uracil and leucine (SGal-Ura-Leu) for ~16 hrs. Transformants carrying double plasmids were selected on SD-Ura-Trp and replicated to SGal-Ura-Trp-Leu to induce overexpression of the GAL controlled genes on both plasmids.

7.4.2 PLASMIDS

pCUP1-SUP35NM::GFP-TRP1 carries fusions of amino acids 1–254 of Sup35 and GFP under the *CUP1* promoter [3]. *pURE2-URE2N::GFP-HIS3* (a kind gift from R. B. Wickner) carries fusions of amino acids 1–65 of Ure2 and GFP under the *URE2* promoter. The high copy genomic library used in our earlier study was constructed in the pHR81 (2 μ *URA3 leu2-d*) vector [3], [93]. The *PIN4C ClaI-XhoII* fragment isolated from the *PIN4* library clone #277 [3] was Klenow-filled and cloned into pHR81H (2 μ *HIS3 leu2-d*, with the *URA3* marker of pHR81 exchanged for the *HIS3* marker) vector at the blunt-ended BamHI site to generate pHR81H-*PIN4C* (2 μ *HIS3 leu2-d*). The *GAL* promoter isolated from pRS316-*GAL1* (a kind gift from A. Bretscher) on the *XhoI* fragment was filled in and cloned into the unique pHR81 *BamH*I site to create pHR81*GAL*. The *PIN4C* fragment isolated from the *PIN4* library clone #277 [3] as an *XhoII* fragment was filled in and cloned into pHR81*GAL* at the *BamHI* site to generate pHR-81*GAL-PIN4C*. *DsRED* as a *NdeI-NotI* fragment digested from pDsRed-Monomer-N1 vector (Clontech) was filled in and cloned into pHR81*GAL* at the *BamHI* site to produce pHR81*GAL-DsRED*. The *PIN4C* fragment was PCR amplified using primers with *BamHI* linkers and subcloned into pHR81*GAL-DsRED* at the *BamHI* site to produce pHR81*GAL-PIN4C-DsRED*. The PCR primers were P1 (5′-ggctcgagtggatcggcgggaggaaatt-

gaaag-3′) and P2 (5′agtggatcctcgagtggtacctctagaagtatataataccatagattc-3′).
The *SUP35-GFP* fragment isolated as a *SacI-BamHI* fragment from p1744
(SB238, kindly provided by T. R. Serio) was subcloned into p*TETr* vector
p1331 (a kind gift of pCM184 from E. Herrero) to create p*TETr-SUP35-
GFP*. Plasmid p*GAL-HSP104*^{T160M} is a kind gift from D. C. Masison [26].
Plasmid p*YES3GAL-SIS1* is a kind gift from S. L. Lindquist [94]. *GPG1*
isolated from p*GPD-GPG1* (a kind gift from Y. Nakamura and H. Kurahashi) [79] on the *BamHI-XhoI* fragment was filled in and cloned into the
unique pRS316-*GAL1 BamHI* site to create pRS316-*GAL1-GPG1*.

7.4.3 SCORING FOR LOSS OF PRIONS

A weak [*PSI+*][*PIN+*] variant of 74-D694 (L1758) was transformed with
the pHR81 based high copy vectors with inserts encoding any of 9 Q/N-
rich domains: *PIN4C* clone #277, *CYC8C* clone #151, *STE18* clone #299,
YCK1 clone #103, *PIN2* clone #222, *URE2* clone #155, *PIN3* clone #80,
NEW1 clone #39 and *LSM4* clone #288, whose overexpression substitutes
for [*PIN+*] in [*PSI+*] induction [3]. For each clone, three transformants
selected on SD-Ura were spread on SD-Ura-Leu where the library plasmids were amplified to high copy number because of their poorly expressed leu2-d allele [63]. Three equal size colonies were resuspended in
water and spread on YPD where the percentage of red [*psi−*] colonies was
scored. A similar method was used to score for loss of strong [*PSI+*] in the
high [*PIN+*] variant of 74-D694 (L1762) by overexpression of pHR81H-
PIN4C (2 μ *HIS3 leu2-d*) among ~1500 colonies.

In order to confirm that red colonies were [*psi−*], randomly chosen
red colonies for each clone that lost the plasmid were crossed to [*psi−*]
64-D697 harboring p*CUP1-SUP35NM::GFP-TRP1*. Expression of the
Sup35NM-GFP fusion in resulting diploids induced on SD-Ura-Trp containing 50 μM Cu2+, decorates prion aggregates as GFP foci in [*PSI+*]
while remaining diffuse in [*psi−*] due to the absence of aggregates. To
confirm the [*URE3*] prion state, randomly chosen L3154 dark and light
red colonies for each clone were crossed to GF708 harboring p*URE2-
URE2N::GFP-HIS3*. The resulting diploids made from dark red colonies
produced Ure2-GFP foci confirming that they were [*URE3*], and diploids

made from light red colonies gave rise to diffuse fluorescence, confirming that they were [ure-o].

Strong [*PSI*+][*PIN*+] SUP35-GFP strain (GF657) was transformed with the pHR81*GAL* empty vector or pHR81*GAL-PIN4C*. Three transformants selected on SD-Ura were spread on SGal-Ura-Leu to overexpress Pin4C. Three equal size colonies were resuspended in water and spread on YPD where the percentage of red [*psi*−] colonies was scored. The efficiency of curing was determined as the percentage of red colonies indicative of [*psi*−] among ~2500 colonies. Red colonies were confirmed to be [*psi*−] by their diffuse Sup35-GFP.

Medium [*PIN*+][*psi*−] (L1945), high [*PIN*+][*psi*−] (L1749) and very high [*PIN*+][*psi*−] (L1953) strains were transformed with pHR81 empty vector or pHR81-*PIN4C*. Two transformants selected on SD-Ura were spread on SD-Ura-Leu to overexpress Pin4C. Three equal size colonies of each transformant were resuspended in water and spread on YPD. Ten colonies from each clone were crossed to [*pin*−] 64-D697 harboring pCUP1-RNQ1::GFP-TRP1 and diffused Rnq1-GFP was scored as [*pin*−].

7.4.4 ASSAYS FOR INDUCTION OF [PSI+]

An *rnq1Δ::HIS3* [*psi*−][*pin*−] 74-D694 strain bearing any of the 9 genomic library plasmids tested for curing of [PSI +] was transformed with pCUP1-SUP35NM::GFP-TRP1. Transformants selected on SD-Ura-Trp medium were diluted, and about $4×10^4$ cells were spotted on SD-Ura-Leu-Trp medium to amplify the *leu2-d* library plasmids followed by replica-plating on SD-Ura-Leu-Trp containing 50 μM Cu^{2+} to induce [*PSI*+]. Then cells were replica-plated on SD-Ade and scored for [*PSI*+]. The resulting Ade+ colonies were verified to be [*PSI*+] by the formation of GFP dots following expression of pCUP1-SUP35NM::GFP-TRP1.

7.4.5 FLUORESCENCE MICROSCOPY

Yeast expressing Sup35-GFP and Pin4C-DsRed were imaged with a Zeiss Axioskop 2 microscope. For time lapse experiments, single cells were micromanipulated onto a 2% agar patch, then covered with a coverslip and placed on 2% raffinose + 2% galactose plates to allow cell growth. The patch with the coverslip in place was then transferred to a glass slide to image the microcolony and was returned to the plate for further growth. For colocalization experiments, yeast expressing Hsp104-GFP were grown overnight in 2% glucose and then induced in 2% raffinose + 2% galactose liquid medium to overexpress Pin4-DsRed.

7.4.6 FRAP

Fluorescence recovery after photobleaching (FRAP) was performed on a Zeiss LSM510 Axiovert confocal microscope. Mothers with buds smaller than 2.5 μm in diameter were selected in GF657 where Sup35-GFP foci became larger following Pin4 overexpression. Buds were completely bleached with a 488-nm laser at 100% power. After photobleaching, single scan images were collected every 5s with 3% laser and 5× zoom power. The pinhole was fully open to allow complete bleaching and to yield enough signal for fluorescent recovery.

Relative fluorescence intensity (RFI) was determined by $RFI = ((Ne_t - Ne_{min}/N1_t)/(Ne_0 - Ne_{min}/N1_0)) \times 100$, where Ne_t is the average intensity of the bleached bud at time t and $N1_t$ is the average intensity of its non-bleached mother cell at the corresponding time used to compensate for loss in total fluorescence [95]. Ne_0 and $N1_0$ represent average intensities of the bleached bud and its non-bleached mother respectively before photobleaching. Ne_{min} is the minimum fluorescence intensity of the bud seen. The half-time that indicates the speed of mobility and the plateau level of recovery were measured by curve fitting the RFI data to a one-phase exponential association alogorithm with GraphPad Prism.

7.4.7 QUANTIFICATION OF FLUORESCENCE INTENSITY

Fluorescent images were acquired using the same exposure time for all the samples. Using ImageJ, cytoplasmic regions devoid of aggregates and the vacuole were selected with the "brush" selection tool. The mean fluorescence intensity in the selected area was quantified using the "measure" function. The area around the cell was selected as the background. The data for each cell was obtained by calculating the mean fluorescence in the cytoplasm subtracted by that in the background.

7.4.8 A-FACTOR ARREST

Cell growth was arrested by the addition of 50 μM of the yeast mating pheromone α-factor. α-factor peptide (Trp-His-Trp-Leu-Gln-Leu-Lys-Pro-Gly-Gln-Pro-Met-Tyr)was from GenScript.

7.4.9 PREPARATION AND ANALYSIS OF YEAST CELL LYSATES

Cells overexpressing *PIN4* or *PIN4-DsRED* were grown in 150 ml of 2% raffinose +2% galactose media to an A_{600} OD of 0.3–0.8, where 80% of cells contained larger Sup35-GFP foci. Lysates were prepared as described [32]. For chaperone analysis, equal amounts of total proteins in precleard lysates were analyzed by Western blot using previously described antibodies [38]. Monoclonal anti-Pgk1 antibody was from Invitrogen. For semi-denaturing detergent agarose gel electrophoresis (SDD-AGE), 50–80 μg of total protein in precleared lysates were incubated for 7 min in sample buffer with 2% SDS at room temperature and resolved on 1.5% agarose gels [96].

7.4.10 IMMUNOCAPTURE OF CELL LYSATES ON MAGNETIC BEADS

Immunocapture experiments were essentially as described [38] with the following changes: 750 μl of a higher salt lysis buffer [LB2: 40 mM Tris-HCl (pH 7.5), 150 mM KCl, 5 mM $MgCl_2$, 10% glycerol] was used; 500 μl lysates of 0.5–1.0 mg/ml proteins were incubated with 3 μl of α-DsRed antibody for 2 hrs on ice; samples were mixed with 50 μl magnetic beads with immobilized G protein (Miltenyi Biotec) and incubated on ice for 30 min. Finally, beads were washed with 1.0 ml of each of the following solutions at 4°C in the following order to remove nonspecifically bound proteins: LB2 with 1% Triton X-100; LB2, 210 mM KCl, 1% Triton X-100; LB2 with 1% Triton X-100; LB2; LB1 [40 mM Tris-HCl (pH 7.5), 50 mM KCl, 5 mM MgCl2, 5% glycerol]; 20 mM Tris-HCl (pH 7.6). α-DsRed for immunocapturing was monoclonal antibody from Clontech, and α-DsRed for detection was a polyclonal antibody from Santa Cruz Biotechnology.

REFERENCES

1. Prusiner SB (1982) Novel proteinaceous infectious particles cause scrapie. Science 216: 136–144.
2. Wickner RB (1994) [*URE3*] as an altered URE2 protein: evidence for a prion analog in Saccharomyces cerevisiae. Science 264: 566–569.
3. Derkatch IL, Bradley ME, Hong JY, Liebman SW (2001) Prions affect the appearance of other prions: the story of [*PIN+*]. Cell 106: 171–182.
4. Sondheimer N, Lindquist S (2000) Rnq1: an epigenetic modifier of protein function in yeast. Mol Cell 5: 163–172.
5. Wickner RB, Masison DC, Edskes HK (1995) [*PSI+*] and [*URE3*] as yeast prions. Yeast 11: 1671–1685.
6. Suzuki G, Shimazu N, Tanaka M (2012) A yeast prion, Mod5, promotes acquired drug resistance and cell survival under environmental stress. Science 336: 355–359.
7. Crow ET, Li L (2011) Newly identified prions in budding yeast, and their possible functions. Semin Cell Dev Biol 22: 452–459.
8. DePace AH, Santoso A, Hillner P, Weissman JS (1998) A critical role for amino-terminal glutamine/asparagine repeats in the formation and propagation of a yeast prion. Cell 93: 1241–1252.
9. Halfmann R, Alberti S, Krishnan R, Lyle N, O'Donnell CW, et al. (2011) Opposing effects of glutamine and asparagine govern prion formation by intrinsically disordered proteins. Mol Cell 43: 72–84.
10. Ross ED, Toombs JA (2010) The effects of amino acid composition on yeast prion formation and prion domain interactions. Prion 4: 60–65.
11. Glover JR, Kowal AS, Schirmer EC, Patino MM, Liu JJ, et al. (1997) Self-seeded fibers formed by Sup35, the protein determinant of [*PSI+*], a heritable prion-like factor of S. cerevisiae. Cell 89: 811–819.
12. King CY, Tittmann P, Gross H, Gebert R, Aebi M, et al. (1997) Prion-inducing domain 2–114 of yeast Sup35 protein transforms in vitro into amyloid-like filaments. Proc Natl Acad Sci U S A 94: 6618–6622.
13. Serio TR, AG C, AS K, GJ S, JJ M, et al. (2000) Nucleated Conformational Conversion and the Replication of Conformational Information by a Prion Determinant. Science 289: 1317–1321.

14. Collins SR, Douglass A, Vale RD, Weissman JS (2004) Mechanism of Prion Propagation: Amyloid Growth Occurs by Monomer Addition. PLoS Biol 2: e321 doi:10.1371/journal.pbio.0020321.

15. Inoue Y (2009) Life cycle of yeast prions: propagation mediated by amyloid fibrils. Protein Pept Lett 16: 271–276.

16. Chernoff YO, Lindquist SL, Ono B, Inge-Vechtomov SG, Liebman SW (1995) Role of the chaperone protein Hsp104 in propagation of the yeast prion-like factor [PSI+]. Science 268: 880–884.

17. Haslberger T, Bukau B, Mogk A (2010) Towards a unifying mechanism for ClpB/Hsp104-mediated protein disaggregation and prion propagation. Biochem Cell Biol 88: 63–75.

18. Cox B, Ness F, Tuite M (2003) Analysis of the generation and segregation of propagons: entities that propagate the [PSI+] prion in yeast. Genetics 165: 23–33.

19. Eaglestone SS, Ruddock LW, Cox BS, Tuite MF (2000) Guanidine hydrochloride blocks a critical step in the propagation of the prion-like determinant [PSI+] of Saccharomyces cerevisiae. Proc Natl Acad Sci U S A 97: 240–244.

20. Kryndushkin DS, Alexandrov IM, Ter-Avanesyan MD, Kushnirov VV (2003) Yeast [PSI+] prion aggregates are formed by small Sup35 polymers fragmented by Hsp104. J Biol Chem 278: 49636–49643.

21. Satpute-Krishnan P, Langseth SX, Serio TR (2007) Hsp104-dependent remodeling of prion complexes mediates protein-only inheritance. PLoS Biol 5: e24 doi:10.1371/journal.pbio.0050024.

22. Shorter J, Lindquist S (2004) Hsp104 catalyzes formation and elimination of self-replicating Sup35 prion conformers. Science 304: 1793–1797.

23. Tessarz P, Mogk A, Bukau B (2008) Substrate threading through the central pore of the Hsp104 chaperone as a common mechanism for protein disaggregation and prion propagation. Mol Microbiol 68: 87–97.

24. Helsen CW, Glover JR (2012) A new perspective on Hsp104-mediated propagation and curing of the yeast prion [PSI+]. Prion 6: 234–239.

25. Romanova NV, Chernoff YO (2009) Hsp104 and prion propagation. Protein Pept Lett 16: 598–605.

26. Hung GC, Masison DC (2006) N-terminal domain of yeast Hsp104 chaperone is dispensable for thermotolerance and prion propagation but necessary for curing prions by Hsp104 overexpression. Genetics 173: 611–620.

27. Moosavi B, Wongwigkarn J, Tuite MF (2010) Hsp70/Hsp90 co-chaperones are required for efficient Hsp104-mediated elimination of the yeast [PSI+] prion but not for prion propagation. Yeast 27: 167–179.

28. Helsen CW, Glover JR (2012) Insight into molecular basis of curing of [PSI+] prion by overexpression of 104-kDa heat shock protein (Hsp104). J Biol Chem 287: 542–556.

29. Winkler J, Tyedmers J, Bukau B, Mogk A (2012) Hsp70 targets Hsp100 chaperones to substrates for protein disaggregation and prion fragmentation. J Cell Biol 198: 387–404.

30. Masison DC, Kirkland PA, Sharma D (2009) Influence of Hsp70s and their regulators on yeast prion propagation. Prion 3: 65–73.

31. Perrett S, Jones GW (2008) Insights into the mechanism of prion propagation. Curr Opin Struct Biol 18: 52–59.

32. Mathur V, Hong JY, Liebman SW (2009) Ssa1 overexpression and [*PIN+*] variants cure [*PSI+*] by dilution of aggregates. J Mol Biol 390: 155–167.

33. Schwimmer C, Masison DC (2002) Antagonistic interactions between yeast [*PSI+*] and [*URE3*] prions and curing of [*URE3*] by Hsp70 protein chaperone Ssa1p but not by Ssa2p. Mol Cell Biol 22: 3590–3598.

34. Aron R, Higurashi T, Sahi C, Craig EA (2007) J-protein co-chaperone Sis1 required for generation of [*RNQ+*] seeds necessary for prion propagation. Embo J 26: 3794–3803.

35. Higurashi T, Hines JK, Sahi C, Aron R, Craig EA (2008) Specificity of the J-protein Sis1 in the propagation of 3 yeast prions. Proc Natl Acad Sci U S A 105: 16596–16601.

36. Sondheimer N LN, Craig EA, Lindquist S (2001) The role of Sis1 in the maintenance of the [*RNQ+*] prion. EMBO J 20: 2435–2442.

37. Hines JK, Li X, Du Z, Higurashi T, Li L, et al. (2011) [SWI], the prion formed by the chromatin remodeling factor Swi1, is highly sensitive to alterations in Hsp70 chaperone system activity. PLoS Genet 7: e1001309 doi:10.1371/journal.pgen.1001309.

38. Bagriantsev SN, Gracheva EO, Richmond JE, Liebman SW (2008) Variant-specific [*PSI+*] Infection Is Transmitted by Sup35 Polymers within [*PSI+*] Aggregates with Heterogeneous Protein Composition. Mol Biol Cell 19: 2433–2443.

39. Tipton KA, Verges KJ, Weissman JS (2008) In vivo monitoring of the prion replication cycle reveals a critical role for Sis1 in delivering substrates to Hsp104. Mol Cell 32: 584–591.

40. Winkler J, Tyedmers J, Bukau B, Mogk A (2012) Chaperone networks in protein disaggregation and prion propagation. J Struct Biol 179: 152–160.

41. Kryndushkin D, Wickner RB (2007) Nucleotide exchange factors for Hsp70s are required for [*URE3*] prion propagation in Saccharomyces cerevisiae. Mol Biol Cell 18: 2149–2154.

42. Moriyama H, Edskes HK, Wickner RB (2000) [*URE3*] prion propagation in Saccharomyces cerevisiae: requirement for chaperone Hsp104 and curing by overexpressed chaperone Ydj1p. Mol Cell Biol 20: 8916–8922.

43. Cox BS (1965) [*PSI+*], a cytoplasmic suppressor of super-suppressor in yeast. Heredity 20: 505–521.

44. Derkatch IL, Chernoff YO, Kushnirov VV, Inge-Vechtomov SG, Liebman SW (1996) Genesis and variability of [*PSI+*] prion factors in Saccharomyces cerevisiae. Genetics 144: 1375–1386.

45. Liebman SW, Derkatch IL (1999) The yeast [*PSI+*] prion: making sense of nonsense. J Biol Chem 274: 1181–1184.

46. Klng CY (2001) Supporting the structural basis of prion strains· induction and identification of [*PSI+*] variants. J Mol Biol 307: 1247–1260.

47. Kochneva-Pervukhova NV, Chechenova MB, Valouev IA, Kushnirov VV, Smirnov VN, et al. (2001) [*PSI+*] prion generation in yeast: characterization of the 'strain' difference. Yeast 18: 489–497.

48. Zhou P, Derkatch IL, Uptain SM, Patino MM, Lindquist S, et al. (1999) The yeast non-Mendelian factor [*ETA+*] is a variant of [*PSI+*], a prion-like form of release factor eRF3. Embo J 18: 1182–1191.

49. Tanaka M, Chien P, Naber N, Cooke R, Weissman JS (2004) Conformational variations in an infectious protein determine prion strain differences. Nature 428: 323–328.

50. Chernoff YO, Derkach IL, Inge-Vechtomov SG (1993) Multicopy SUP35 gene induces de-novo appearance of psi-like factors in the yeast Saccharomyces cerevisiae. Curr Genet 24: 268–270.

51. Derkatch IL, Bradley ME, Zhou P, Chernoff YO, Liebman SW (1997) Genetic and environmental factors affecting the *de novo* appearance of the [*PSI+*] prion in Saccharomyces cerevisiae. Genetics 147: 507–519.

52. Derkatch IL, Bradley ME, Masse SV, Zadorsky SP, Polozkov GV, et al. (2000) Dependence and independence of [*PSI+*] and [*PIN+*]: a two-prion system in yeast? Embo J 19: 1942–1952.

53. Bradley ME, Edskes HK, Hong JY, Wickner RB, Liebman SW (2002) Interactions among prions and prion "strains" in yeast. Proc Natl Acad Sci U S A 99 (Suppl 4) 16392–16399.

54. Derkatch IL, Uptain SM, Outeiro TF, Krishnan R, Lindquist SL, et al. (2004) Effects of Q/N-rich, polyQ, and non-polyQ amyloids on the *de novo* formation of the [*PSI+*] prion in yeast and aggregation of Sup35 in vitro. Proc Natl Acad Sci U S A 101: 12934–12939.

55. Meriin AB, Zhang X, He X, Newnam GP, Chernoff YO, et al. (2002) Huntington toxicity in yeast model depends on polyglutamine aggregation mediated by a prion-like protein Rnq1. J Cell Biol 157: 997–1004.

56. Osherovich LZ, Weissman JS (2001) Multiple Gln/Asn-rich prion domains confer susceptibility to induction of the yeast [*PSI+*] prion. Cell 106: 183–194.

57. Choe YJ, Ryu Y, Kim HJ, Seok YJ (2009) Increased [*PSI+*] appearance by fusion of Rnq1 with the prion domain of Sup35 in Saccharomyces cerevisiae. Eukaryot Cell 8: 968–976.

58. Vitrenko YA, Gracheva EO, Richmond JE, Liebman SW (2007) Visualization of aggregation of the Rnq1 prion domain and cross-seeding interactions with Sup35NM. J Biol Chem 282: 1779–1787.

59. Bradley ME, Liebman SW (2003) Destabilizing interactions among [*PSI+*] and [*PIN+*] yeast prion variants. Genetics 165. 1675–1685.

60. Edskes HK, Gray VT, Wickner RB (1999) The [*URE3*] prion is an aggregated form of Ure2p that can be cured by overexpression of Ure2p fragments. Proc Natl Acad Sci U S A 96: 1498–1503.

61. Kurahashi H, Ishiwata M, Shibata S, Nakamura Y (2008) A regulatory role of the Rnq1 nonprion domain for prion propagation and polyglutamine aggregates. Mol Cell Biol 28: 3313–3323.

62. Kurahashi H, Pack CG, Shibata S, Oishi K, Sako Y, et al. (2011) [*PSI+*] aggregate enlargement in rnq1 nonprion domain mutants, leading to a loss of prion in yeast. Genes Cells 16: 576–589.

63. Erhart E, Hollenberg CP (1983) The presence of a defective LEU2 gene on 2 mu DNA recombinant plasmids of Saccharomyces cerevisiae is responsible for curing and high copy number. J Bacteriol 156: 625–635.

64. Patino MM, Liu JJ, Glover JR, Lindquist S (1996) Support for the prion hypothesis for inheritance of a phenotypic trait in yeast. Science 273: 622–626.

65. Zhou P, Derkatch IL, Liebman SW (2001) The relationship between visible intracellular aggregates that appear after overexpression of Sup35 and the yeast prion-like elements [PSI+] and [PIN+]. Mol Microbiol 39: 37–46.

66. Hong JY, Mathur V, Liebman SW (2011) A new colour assay for [URE3] prion in a genetic background used to score for the [PSI+] prion. Yeast 28: 555–560.

67. Satpute-Krishnan P, Serio TR (2005) Prion protein remodelling confers an immediate phenotypic switch. Nature 437: 262–265.

68. Vishveshwara N, Bradley ME, Liebman SW (2009) Sequestration of essential proteins causes prion associated toxicity in yeast. Mol Microbiol 73: 1101–1114.

69. Derdowski A, Sindi SS, Klaips CL, DiSalvo S, Serio TR (2011) A size threshold limits prion transmission and establishes phenotypic diversity. Science 330: 680–683.

70. Kawai-Noma S, Pack CG, Tsuji T, Kinjo M, Taguchi H (2009) Single mother-daughter pair analysis to clarify the diffusion properties of yeast prion Sup35 in guanidine-HCl-treated [PSI+] cells. Genes Cells 14: 1045–1054.

71. Byrne LJ, Cox BS, Cole DJ, Ridout MS, Morgan BJ, et al. (2007) Cell division is essential for elimination of the yeast [PSI+] prion by guanidine hydrochloride. Proc Natl Acad Sci U S A 104: 11688–11693.

72. Chernoff YO (2007) Stress and prions: lessons from the yeast model. FEBS Lett 581: 3695–3701.

73. Fan Q, Park KW, Du Z, Morano KA, Li L (2007) The role of Sse1 in the de novo formation and variant determination of the [PSI+] prion. Genetics 177: 1583–1593.

74. Sadlish H, Rampelt H, Shorter J, Wegrzyn RD, Andreasson C, et al. (2008) Hsp110 chaperones regulate prion formation and propagation in S. cerevisiae by two discrete activities. PLoS ONE 3: e1763 doi:10.1371/journal.pone.0001763.

75. Disalvo S, Derdowski A, Pezza JA, Serio TR (2011) Dominant prion mutants induce curing through pathways that promote chaperone-mediated disaggregation. Nat Struct Mol Biol 18: 486–492.

76. Wegrzyn RD, Bapat K, Newnam GP, Zink AD, Chernoff YO (2001) Mechanism of prion loss after Hsp104 inactivation in yeast. Mol Cell Biol 21: 4656–4669.

77. Huh WK, Falvo JV, Gerke LC, Carroll AS, Howson RW, et al. (2003) Global analysis of protein localization in budding yeast. Nature 425: 686–691.

78. Douglas PM, Summers DW, Ren HY, Cyr DM (2009) Reciprocal efficiency of RNQ1 and polyglutamine detoxification in the cytosol and nucleus. Mol Biol Cell 20: 4162–4173.

79. Ishiwata M, Kurahashi H, Nakamura Y (2009) A G-protein gamma subunit mimic is a general antagonist of prion propagation in Saccharomyces cerevisiae. Proc Natl Acad Sci U S A 106: 791–796.

80. Kryndushkin DS, Engel A, Edskes HK, Wickner RB (2011) Molecular Chaperone Hsp104 Can Promote Yeast Prion Generation. Genetics 188: 339–348.

81. Edskes HK, Wickner RB (2002) Conservation of a portion of the S. cerevisiae Ure2p prion domain that interacts with the full-length protein. Proc Natl Acad Sci U S A 99 (Suppl 4) 16384–16391.

82. Ripaud L, Maillet L, Cullin C (2003) The mechanisms of [*URE3*] prion elimination demonstrate that large aggregates of Ure2p are dead-end products. Embo J 22: 5251–5259.

83. Inoue Y, Kawai-Noma S, Koike-Takeshita A, Taguchi H, Yoshida M (2011) Yeast prion protein New1 can break Sup35 amyloid fibrils into fragments in an ATP-dependent manner. Genes Cells 16: 545–556.

84. Hines JK, Higurashi T, Srinivasan M, Craig EA (2011) Influence of prion variant and yeast strain variation on prion-molecular chaperone requirements. Prion 5: 238–244.

85. Treusch S, Lindquist S (2012) An intrinsically disordered yeast prion arrests the cell cycle by sequestering a spindle pole body component. J Cell Biol 197: 369–379.

86. Meriin ABZX, Alexandrov IM, Salnikova AB, Ter-Avanesian MD, Chernoff YO, Sherman MY (2007) Endocytosis machinery is involved in aggregation of proteins with expanded polyglutamine domains. FASEB J 21: 1915–1925.

87. Duennwald ML, Lindquist S (2008) Impaired ERAD and ER stress are early and specific events in polyglutamine toxicity. Genes Dev 22: 3308–3319.

88. Gong H, Romanova NV, Allen KD, Chandramowlishwaran P, Gokhale K, et al. (2012) Polyglutamine toxicity is controlled by prion composition and gene dosage in yeast. PLoS Genet 8: e1002634 doi:10.1371/journal.pgen.1002634.

89. Ter-Avanesyan MD, Kushnirov VV, Dagkesamanskaya AR, Didichenko SA, Chernoff YO, et al. (1993) Deletion analysis of the SUP35 gene of the yeast Saccharomyces cerevisiae reveals two non-overlapping functional regions in the encoded protein. Mol Microbiol 7: 683–692.

90. Goldring ES, Grossman LI, Krupnick D, Cryer DR, Marmur J (1970) The petite mutation in yeast. Loss of mitochondrial deoxyribonucleic acid during induction of petites with ethidium bromide. J Mol Biol 52: 323–335.

91. Tuite MF, Mundy CR, Cox BS (1981) Agents that cause a high frequency of genetic change from [*PSI+*] to [*psi−*] in Saccharomyces cerevisiae. Genetics 98: 691–711.

92. Sherman F, Fink, G R & Hicks, J B. (1986) Methods in Yeast Genetics;Sherman F, Fink, G. R. & Hicks, J. B., editor. Plainview, New York: Cold Spring Harbor Lab.

93. Nehlin JO, Carlberg M, Ronne H (1989) Yeast galactose permease is related to yeast and mammalian glucose transporters. Gene 85: 313–319.

94. Douglas PM, Treusch S, Ren HY, Halfmann R, Duennwald ML, et al. (2008) Chaperone-dependent amyloid assembly protects cells from prion toxicity. Proc Natl Acad Sci U S A 105: 7206–7211.

95. Lippincott-Schwartz J, Snapp E, Kenworthy A (2001) Studying protein dynamics in living cells. Nat Rev Mol Cell Biol 2: 444–456.

96. Bagriantsev S, Liebman SW (2004) Specificity of prion assembly in vivo. [*PSI+*] and [*PIN+*] form separate structures in yeast. J Biol Chem 279: 51042–51048.

97. Chernoff YO NG, Kumar J, Allen K, Zink AD (1999) Evidence for a protein mutator in yeast: role of the Hsp70-related chaperone ssb in formation, stability, and toxicity of the [*PSI+*] prion. Mol Cell Biol 19: 8103–8112.

98. Brachmann A, Baxa U, Wickner RB (2005) Prion generation in vitro: amyloid of Ure2p is infectious. Embo J 24: 3082–3092.

This article has supplemental information that is not featured in this version of the text. To view these files, please visit the original version of the article as cited in the beginning of this chapter.

PART III

GENE EXPRESSION
AND TRAIT LOCUS ANALYSIS

CHAPTER 8

STRATEGY OF TRANSCRIPTION REGULATION IN THE BUDDING YEAST

SAGI LEVY, JAN IHMELS, MIRI CARMI, ADINA WEINBERGER, GILGI FRIEDLANDER, AND NAAMA BARKAI

8.1 INTRODUCTION

Cellular functionality is tightly coupled to the external environment. The type of nutrients available defines the internal metabolic flow, while their abundance often limits the rate of biomass production and energy available for growth. An abundance of toxins impede upon various aspects of cellular machinery, including metabolic capacity, protein stability or DNA integrity. Over evolutionary time scales, cells may encounter virtually endless environmental states at widely different frequencies. Maintaining optimal functionality in the presence of such external variability is a central evolutionary constraint.

Gene expression plays a central role in the adaptation to changing conditions. Studies in the budding yeast *S. cerevisiae*, for example, have shown that cellular transcription program is dramatically modified by changes in nutrient availability, growth conditions, temperature, and a variety of other environmental condition tested.

This chapter was originally published under the Creative Commons Attribution License. Levy S, Ihmels J, Carmi M, Weinberger A, Friedlander G, and Barkai N. Strategy of Transcription Regulation in the Budding Yeast. PLoS ONE 2,2 (2007). doi:10.1371/journal.pone.0000250.

How do cells coordinate their gene expression with varying environments? One strategy is to use feedback mechanisms, which directly link gene expression with internal needs. Within such strategy, internal variables, such as the rate of biomass production or the internal pools of nutrients or energy, feedbacks to properly tune gene expression with the corresponding functional needs. Cells then respond directly to the relevant functional parameter that needs to be monitored, regardless of the specific external perturbation that may have altered this parameter. The primary advantage of such internal feedback strategy is the capacity to ensure optimality of response under a wide diversity of external conditions. It may limit, however, the speed of the response since environmental changes will alter internal characters (e.g. growth rate) only after a delay (e.g. when the intracellular pool of nutrient is diminished).

An alternative strategy for coordinating gene expression with physiological requirements is to infer the expected physiological state from the external environment. This can be done using evolutionary-tuned signaling pathways. For example, in yeast, the TOR and the PKA pathways sense the source and availability of carbon and nitrogen and regulate the expression levels of multiple gene groups [1]–[3]. In principle, the information received by those signaling pathways could coordinate gene expression with the expected growth rate at the specific level of nutrients available. Such a strategy can be used to optimize the speed of the response. It is likely, however, to limit the range of environments that can be properly recognized. In particular, it will hinder optimal response to newly encountered situations.

The need to coordinate gene expression with internal state is best exemplified in the case of ribosomal biogenesis. In rapidly growing yeast cells, the synthesis of ribosomes accounts for the cell's single largest expenditure of biosynthesis energy. With the need to produce ~2000 ribosomes every minute, 60% of total transcription is devoted to ribosomal RNA. Similarly, 50% of RNA polymerase II transcription, as well as 90% of mRNA splicing, are devoted to ribosomal protein [4]. Over 150 non-ribosomal genes are involved in various aspects of ribosomal RNA processing or transport [5]. Tight coordination of ribosomal biosynthesis with the requirement for protein translation is thus needed to maintain efficient utilization of biosynthesis resources. Indeed, the number of ribosomes in yeast cells appears to be linked to cell size and growth rate [6], [7]. Similarly, tumor cells often express high levels of ribosome and other translation related factors, in accordance with their elevated growth rates [8], [9].

In bacteria, the rate of ribosome synthesis increases approximately with the square of the growth rate [10]. Underlying this growth-dependent control are well-studied examples of regulation by an internal-feedback. The purine nucleotide (ATP and GTP), whose concentration reflect the nutritional state as well as the translational activity of the cell, play a key role in this regulation. In fact, purine NTP levels directly regulates rRNA transcription [11]–[14], and thus determine the rate of ribosomal biogenesis. An additional layer of regulation by internal feedback is provided by the stringent response, which is induced by uncharged tRNA. Uncharged tRNAs accumulate when the internal level of amino-acids is not sufficiently to support the rate of protein synthesis. In those instances, the stringent response feedbacks to repress the transcription of genes associated with the translational apparatus, including tRNAs, rRNAs, ribosomal proteins, translational factors, and synthetases [15]–[17].

Growth-rate dependent control of gene expression was implied also in the regulation of gene expression in the budding yeast [18]. The dependence of gene expression on growth rate was characterized most comprehensively using continuous chemostat cultures growing at steady state [18]. A large multitude of genes exhibited an expression pattern that was strongly correlated with growth rate. In particular, ribosomal biogenesis gene expression increased with increasing growth rate, whereas the expression of stress-related genes decreased. Based on the genomic distribution of growth-correlated genes, it was suggested that growth-rate dependence is achieved through replication-mediated changes in chromatin modification [18]. In the context of the internal vs. external regulatory strategies discussed above, however, the interpretation of the observed correlations is difficult, since changes in growth rate necessarily involved change in the amount of the limiting factor (glucose) in the medium. Gene expression was measured at steady state, and, at least formally, it may be that expression levels were tuned by the external glucose concentration, according to the evolutionary-expected growth rate, rather than as feedback by the growth-rate itself.

More generally, since under most conditions the actual and environmentally-expected states are compatible, it is difficult to discern whether gene expression is regulated predominantly by an internal feedback-mechanism, or whether it is tuned by an environmental signal based on its expected influence on the internal state. To try and overcome this limitation, we undertook two approaches. First, we set to measure gene expression under conditions that decouple the perceived environmental signal from its actual effect on cell growth. This was done by focusing on a specific mutant (*adh1*) which grows better on glycerol than on glucose. Under such conditions, environmental cues dominate, leading to increased level of ribosomal biogenesis gene expression in slowly-growing cells. Second, we examined the correlation between gene expression and growth rate during the dynamic response of chemostat grown culture to environmental changes. The focus on the dynamics, rather than the steady state behavior, allowed us to examine the relative timing of the two responses. Ribosome biogenesis gene expression responded rapidly to changes in the environments but was rather oblivious to longer-term changes in growth rate. Lack of correlation between growth rate and ribosomal biogenesis gene expression was also observed in a compendium of 196 deletion mutants, for which both growth rates and genome-wide expression profiles were reported [19]. Taken together, our results suggest that the apparent coupling of translation-related gene expression and growth rate is not causal, but reflects the evolutionary fine-tuning of signal transduction mechanisms. The capacity to recognize and prepare to conditions that may alter changes in growth rate was probably a major selection force during yeast evolution.

8.2 RESULTS

8.2.1 RIBOSOMAL BIOGENESIS GENE EXPRESSION IS HIGHLY RESPONSIVE TO CHANGING ENVIRONMENTS

We analyzed the expression levels of genes involved in ribosomal biogenesis using an annotated database of over 1500 genome-wide

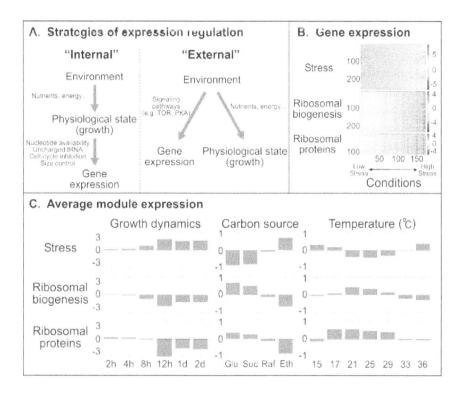

FIGURE 1: (a) Environmental versus growth dependent modulation of gene expression, Gene expression can be modulated by direct signaling pathways which sense the environmental cues. Alternatively, it can be regulated by internal cues which are coupled directly to cell growth. In this study we characterize the relative contribution of external vs. internal sensing to the regulation of gene expression. (b,c) Expression profile of ESR modules during growth at different environmental conditions (reported in [25]; see table S1 for the list of genes). (b) The colormap represents the $\log_2$-expression-ratio of the respective gene module. The conditions are sorted by the average expression level of the stress module. (c) Average $\log_2$-expression-ratio of the respective gene module under the indicated condition. Stress-induced genes are activated, while ribosomal biogenesis and RP genes are repressed under conditions associated with slow growth (e.g. non-optimal temperatures, non-fermentable carbon source, or stationary phase).

expression profiles in *S. cerevisiae* [20]. Our previous studies had identified two groups of co-regulated genes involved in making the ribosome [20]–[22]. The first group includes genes that code for the ribosomal proteins (RP) themselves. The second group is composed of genes that assist in the proper assembly of the ribosome, such as genes involved in the processing of rRNA, and is denoted here as ribosomal biogenesis gene module.

As described previously, genes of both groups are co-expressed under a large number of conditions, and display a strong inverse correlation with genes induced as part of the general stress response [23]. This correlation pattern was particularly evident in the prototypic responses, termed environmental stress response (ESR), to a variety of environmental stresses (including heat-shock, peroxide shock or high osmolarity) [24], [25]. The ESR is characterized by the repression of ribosomal proteins and ribosomal biogenesis genes together with the induction of a common set of stress-related genes (Fig 1b). We note that although both RP and ribosomal biogenesis genes were repressed by environmental stresses, their responses were characterized by different kinetics (rapid repression of ribosomal biogenesis genes followed by a slower repression of RP genes) leading to their identification as two separate transcription modules (i.e. sets of coordinately expressed genes) [25].

Gene expression in unperturbed cells was also reported [25] (Fig. 1c). The expression of both RP and ribosomal biogenesis genes was maximal during logarithmic growth, and decreased as the cultures approached saturation. As expected, the level of ribosomal biogenesis gene expression was also correlated with the efficiency by which the corresponding carbon is utilized: expression was high in cells grown in a fermentable carbon source (glucose), whereas low expression was observed during growth in non-fermentable carbon sources such as ethanol. Similar differences in the level of ribosomal biogenesis genes were observed also when comparing steady state growth at different temperatures. In all the above conditions, the stress-induced part of the ESR exhibited high expression levels in conditions where ribosomal biogenesis genes expression was low, and low expression levels in conditions where ribosomal biogenesis gene expression was high. Although growth rates were not reported, the gene expression pattern seems consistent with the general notion that the expression of ribosomal biogenesis gene increases with growth rate, whereas the expression of stress genes decreases.

8.2.2 GENE EXPRESSION IN THE PRESENCE OF INTERNAL VS. EXTERNAL CONFLICT

To examine whether the apparent correlation between the level of ribosomal biogenesis gene expression and growth rate reflects a feedback-mediated regulation that is sensitive to the growth rate itself, we sought a situation where the recognized environmental signal as perceived by the cell is separated from its actual effect on growth. Wild-type yeast cells grow most rapidly when provided with glucose as a carbon source. We considered a mutant strain deleted of the gene *ADH1*, which codes for an enzyme responsible for ethanol production during glucose fermentation. This strain cannot ferment glucose, and consequently grows faster on non-fermentable carbon sources (e.g. glycerol) than on glucose (Fig. 2a, S1).

We reasoned that comparing the gene expression pattern of *ADH1* cells grown in glucose vs. glycerol would allow us to distinguish between internal vs. external strategies of gene expression regulation. If internal, growth-rate dependent feedback dominates, ribosomal biogenesis gene expression will be higher in glycerol than in glucose, in accordance with the higher growth rate of the cells. Importantly, this prediction holds regardless of the mechanism which hinders *ADH1* growth in glucose. Alternatively, if signaling by the external environment dominates, ribosomal biogenesis gene expression will be higher in glucose, in accordance with the evolutionary expectation of higher growth rate of wild-type cells in glucose.

Using full-genome microarrays, we compared the expression profile of log-phase *adh1* mutant cells grown on glycerol vs glucose medium (Fig. 2b). The expression of ribosomal biogenesis and RP genes was higher in glucose, although the growth rate of the cells was lower. Transfer of cells from glucose to glycerol led to a corresponding repression of ribosomal biogenesis genes (Fig. 2c). Similarly, the expression of stress genes was higher in glycerol (Fig. 2b), and was induced upon transfer from glucose to glycerol (Fig. 2c). Taken together, it appears that the expression of ribosomal biogenesis genes, as well as that of stress genes, was tuned to the environment rather than to the rate of cellular growth.

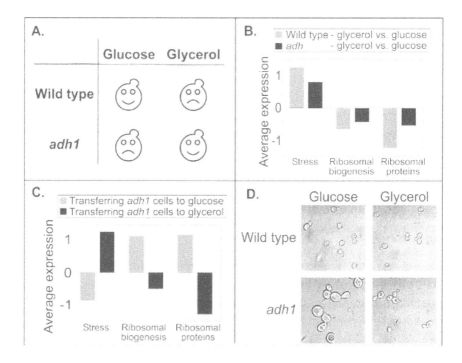

FIGURE 2: Ribosomal gene expression under "confusing" conditions. (a) Glucose is the preferred carbon source for wild-type yeast cells. In contrast, adh1 mutant cells grow faster when glycerol is available (See also Fig S1). (b) Wild-type and mutant cells were grown to log phase on media containing glucose or glycerol as the only carbon source (Methods). The expression profile on glycerol vs. glucose was quantified with microarrays. The bars represent the average $\log_2$-ratio-expression of the ESR modules. Notably, both wild type and mutant cells activate the ESR program. Stress genes are induced while ribosomal biogenesis genes and RP are repressed. (c) Response of *adh1* mutants to change in carbon source. Overnight cultures of *adh1* cells were transferred both to glycerol and to fresh glucose media. The bars represent the average expression after 30 minutes with respect to the initial cell culture expression ($\log_2$-ratios). As expected, the ESR is repressed upon transfer to fresh glucose medium. However, upon transfer to glycerol, the favorable carbon source of these mutant cells, the ESR is activated. Specifically, stress genes are induced while ribosomal biogenesis and RP genes are repressed. (d) Wild-type and *adh1* mutant cells were monitored upon transfer to glycerol versus their transfer to fresh glucose. As previously reported, the size of slow growing wild type cells (glycerol) is smaller than fast growing ones (glucose) [26]. Interestingly, *adh1* mutants display an opposite behavior. Specifically, slow growing *adh1* mutants (glucose) are appreciably larger than the fast growing ones (glycerol). Thus, it appears that the cell size may also be tuned to the environmental signals rather than to cell division.

Interestingly, cell size was also tuned to the environmental signal rather then the growth rate. In *S. cerevisiae*, poor media and low growth rates correspond to smaller cell size. We observed, however, that *adh1* cells growing in glucose were larger than *adh1* cells growing in glycerol (Fig. 2d). Moreover, *adh1* cells growing on glucose were significantly larger than wild-type cells in glucose. This abnormal size may reflect the imbalance between their (high) ribosomal content and their (slow) growth rate, in consistence with the observed impact of ribosomal content on cell size [26].

8.2.3 TEMPORAL ADAPTATION TO PERTURBATION IN CONTINUOUS CULTURES

Our results suggest that upon conflicting signals, when the expected growth rate at a particular environment is different from the actual rate of growth, the environmental signal dominates. Next we asked whether growth-dependent signals play an important role under non-conflicting situations. As we argue above, however, in the absence of conflict the internal physiological state (e.g. growth rate) is compatible with the environmentally-expected one, making it is difficult to disentangle the signals received from the external environment from these that are generated as feedback from internal processes. For examples, changing the steady state growth rate in continuous cultures necessarily involves a change in the amount of limiting factor. It is thus hard to discern whether the changes in gene expression result from the change in growth rate itself, or whether they were generated as a response to the change in the abundance of the limiting factor.

We reasoned that characterizing temporal kinetics, rather than steady state behavior, may overcome this limitation. Consider an environmental perturbation that modulates both the gene expression and the growth rate. At steady state, these changes are likely to be coordinated. By following the temporal evolution of both responses, however, we can conclude whether the change in growth rate followed or precede the changes in gene expression. For example, if internal feedback mechanism is at function, any change in growth rate will be followed by a change in gene expression.

To define a well-controlled environment where the kinetics of both gene expression and growth rate can be monitored, we grew cells in continuous cultures in a chemostat (Methods). Cells were grown to steady state, and were subsequently subjected to perturbations. Microarrays were used to follow their genome-wide transcription response to each perturbation at subsequent time points following the perturbation. Notably, the changes in gene expression changes are analyzed relative to the gene expression in the unperturbed (steady state) culture, and thus reflect only the response to environmental cues. In parallel, we measured also the cells growth rate, by following the changes in their optical density (Fig. 3a, 3b, S2). At steady state, the optical density remains constant in time, and the rate of cell growth is set by the dilution rate of the chemostat (0.2 hr−1 in our experiment) [27]. Since the cells are continuously diluted, and dilution rate remains fixed throughout the experiment, any change in growth rate will be reflected practically immediate by a change in the optical density.

TABLE 1: Environmental cues imposed on steady state growing yeast cells.

Limiting factor	signal	Final conc./Temperature in the chemostat	Notes
Histidine	Clotrimazole	10 μM	Sigma C6019. Dissolved in DMSO
	DTT	6 mM	Sigma D9779
	NaCl	0.31 M	
	Heat shock	Shift from 30°C to 37°C for 5 hours	
	Histidine	+2 mg/l	Double the limiting factor level
Glucose	H_2O_2	0.6 mM	
	DTT	6 mM	Sigma D9779
	NaCl	0.4 M	
	Heat shock	Shift from 30°C to 39°C for 5 hours	
	Glucose	+0.2 g/L	Double the limiting factor level

Since changes in growth rate necessarily involve changes in other factors, such as abundance or level of nutrients or drugs in the media, it was important to diversify the type of perturbations, so that a general correlation

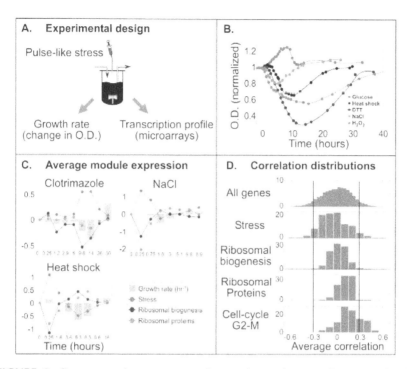

FIGURE 3: Gene expression versus growth rates in continuous cultures. (a) Steady state growing cells were subjected to different environmental stresses. Growth rate was calculated from the measured optical density. Simultaneously, the profile of gene expression was quantified with microarrays. In our study we looked for possible correlations between growth and gene expression (see Methods for more details). (b) Cells grown at steady state in a glucose-limited chemostat were subjected to 5 different environmental cues. We examined 4 stress pulses: H_2O_2 (0.6mM), DTT (6mM), NaCl (0.4M), heat shock (39°C for 5 hours), and a stress-relieving pulse of glucose (0.2 g/l). The Y-axis is normalized such that the steady state O.D. (at t<0) is 1. The markers designate O.D. sampling time, and solid lines are spline interpolations of the data. The perturbations in histidine-limited chemostat are shown in figure S2. (c) Dynamics of the average log_2-ratio-expression for three ESR gene modules. The DTT, NaCl and heat shock pulses were given during glucose depletion, and the clotrimazole perturbation was given in histidine limitation. Bars designate the cells growth rate, which is normalized relative to the steady state growth rate (see Fig. S3 for other perturbations dynamics). (d) Each gene was assigned a growth correlation score (GCR), quantifying the correlation between its expression level and growth rate, averaged over all perturbations (Methods). The GCR distribution of several gene modules is shown. The X-axis represents the calculated average correlation, and Y-axis represents the fraction of genes within the module. The two vertical lines designate the threshold for the 5% of genes with the highest or lowest GCR. Notably, the distribution for the ribosomal biogenesis genes lies between the thresholds, while the cell-cycle G2-M gene group is enriched with positive GCR genes.

with growth-rate could be inferred. We considered four environmental stresses (Heat shock, high osmolarity, peroxide, DTT), a drug (clotrimazole), and supplementation of the limiting factor (histidine or glucose, respectively) (Table 1). Moreover, the experiments were performed using two chemostat cultures, growing in different limiting nutrients (histidine and glucose). Taken together, we analyzed 10 time course experiments, with a total of 83 arrays.

All perturbations led to a significant change in culture density, ranging from 15% increase in biomass (addition of histidine to the histidine-limited chemostat) to 70% reduction in biomass (addition of DTT to the glucose-limited chemostat) (Fig. 3c, S3). Significant changes in cell growth were observed about ~1 hour after the onset of the perturbation. The rate of growth kept changing for long time periods, adapting back on a scale of 10–30 hours, depending on the experiment. Typically, an 'over-shoot' was observed before reaching steady state, where cells grew faster than the dilution rate (Fig. 3c, S3).

8.2.4 TEMPORAL KINETICS OF RIBOSOMAL BIOGENESIS GENE EXPRESSION VS. GROWTH RATE

Changes in gene expression were observed immediately following the perturbation (Figs. 3c, S3). The gene expression response was similar to that characterized in batch cultures: environmental stresses led to a strong induction of stress-induced genes and the repression of ribosomal biogenesis genes. Also as expected, the addition of glucose to a glucose-limited chemostat led to a decrease in the ESR genes, and increase in RP and ribosomal biogenesis genes (Fig. S3).

Maximal change in gene expression was typically observed already at the first time point examined (10–20 minutes after the onset of the perturbation). In most perturbations, growth rate was hardly altered at that time. Thus, the initial gene expression response appears to be dominated by direct environmental signals, and not by growth-rate dependent feedbacks.

The strong initial response to environmental changes may mask further growth-dependent signal. We asked whether growth-rate effects may dominate during the recovery from perturbation. The environmental perturbations were given in a pulse-like addition to the media, and were washed out of the media on a time scale of ~5 hours (corresponding to the dilution rate). The long-time temporal recovery from perturbation varied between experiments, but no consistent correlation between growth rate and ribosomal biogenesis gene expression was observed. During the re- covery from heat-shock, for example, slow growth was associated with high levels of ribosomal biogenesis genes (Fig. 3c). During the recovery from DTT, temporal changes in growth rate seem to follow gene expres- sion, rather then vice-versa: gene expression adapted to near its steady state level at 8.6 hours, a time where growth rate was minimal (Fig. 3c). Similarly, steady state-levels of gene expression were observed at 27 hours, a time when growth rate was still much higher than its steady state value. In a large number of perturbations, such as NaCl, gene expression was rela- tively constant during the long-term adaptation, although significant chang- es in growth rate were observed (Fig. 3c). Some correlation between gene expression and growth rate was observed during the response to clotrima- zole (Fig. 3c). However, even in this case, the large increase in growth rate (over-shoot) during the final adaptation back to steady state (at 31 hours) was not accompanied by a corresponding change in gene expression.

To explore the link between ESR gene expression and growth rate more systematically, we measured the correlation between the average ex- pression of each gene group and growth rate (Table S2). This was done separately for each perturbation and then averaged over all perturbations tested. As expected from the analysis of individual perturbations, no sig- nificant correlation was observed for either of the groups considered (RP, ribosomal biogenesis genes and stress genes) (Table S2).

Taken together, our results indicate that environmental signal dominate over possible growth-rate dependent feedback in the modulation of gene expression during response to perturbation.

8.2.5 RIBOSOMAL BIOGENESIS GENE EXPRESSION VS. GROWTH RATES IN MUTANT CELLS

An additional means to decouple the growth rate from the environmental signal is by genetic mutations. We have asked whether growth-rate dependent feedback on gene expression can be recognized under such conditions. Gene deletion may impact gene expression through a multitude of effects, depending on its precise function. However, if a direct growth-rate dependent feedback significantly contributes to the regulation of gene expression, it would lead to an apparent correlation between expression levels and growth rates when a large number of unrelated mutants are examined.

We considered the previously published compendium of deletion profiles, characterizing the transcription profiles of 270 deletion mutants [19]. Growth rates of 196 of these deletion strains were measured in an independent experiment, allowing us to examine the correlation of gene expression with cell growth rate. Consistent with our results above, no correlation was found between the expression levels of the ribosomal biogenesis genes and the reported growth rates ($r = 0.03$). A small inverse correlation was observed with the expression of the stress genes ($r = -0.19$), but this correlation was only marginally significant and considerably lower than those observed with the highest correlated gene groups (e.g. cell cycle group with $r = 0.58$, see also Table S2).

Interestingly, a strong, and highly significant correlation was observed between the RP gene expression and the growth rate ($r = 0.53$). This is in contrast to the negligible correlation observed in the chemostat experiment. The decoupling of RP gene expression from ribosomal biogenesis gene expression is also unique to this dataset, as in most other datasets expression of these groups is tightly correlated. A possible explanation is that the mutant strains have undergone further genetic adaptation (secondary mutations c.f. Ref [28]) that adjusted the level of ribosomal proteins with growth rates [29]. Further experiments are required to examine this possibility.

8.2.6 IDENTIFYING GROWTH-CORRELATED GENES

Taken together, it appears that growth rate per-se has a minimal impact on the expression levels of genes involved in ribosomal biogenesis. To examine whether other gene groups display a tighter link to growth rate, we extended the analysis described above to all yeast genes. Specifically, we assigned each gene a growth correlation score (GCR), quantifying the

Gene group	No. of genes	Continuous culture		Compendium data	
		Enrichment [%]	P-value	Enrichment [%]	P-value
Cell-cycle G2/M	56	25	$4\cdot10^{-7}$	29	$7\cdot10^{-9}$
Cell-cycle M/G1	63	37	$< 10^{-15}$	8	0.21
Cell-cycle G1	280	5	0.65	26	$< 10^{-15}$
Histones	43	12	0.06	44	$< 10^{-15}$
Ribosomal proteins	117	1	1	58	$< 10^{-15}$
Mitochondrial ribosomal proteins	100	26	$2\cdot10^{-12}$	1	0.99
Phospholipids metabolism	22	23	$4\cdot10^{-3}$	55	$8\cdot10^{-11}$
Oxidative phosphorylation	51	29	10^{-8}	0	1
Cell wall	24	4	0.71	42	$9\cdot10^{-8}$
Ribosomal biogenesis	**231**	**0.4**	**1**	**2**	**0.998**
PAU	64	19	$6\cdot10^{-5}$	48	$< 10^{-15}$
Calcium calmodulin	33	30	$3\cdot10^{-6}$	24	$2\cdot10^{-4}$
Amino-acid biosynthesis	303	15	0.66	31	$< 10^{-15}$
Iron transport	54	0	1	48	$< 10^{-15}$
Phosphate and iron utilization	42	0	1	24	$3\cdot10^{-5}$
TCA cycle	10	10	0.40	50	$6\cdot10^{-5}$
Ty retro-transposons	61	5	0.59	16	$8\cdot10^{-4}$
Peroxide shock	25	20	$7\cdot10^{-3}$	28	$2\cdot10^{-4}$
Stress	255	4	0.90	9	$4\cdot10^{-3}$
Gluconeogenesis	30	10	0.19	13	$6\cdot10^{-2}$

FIGURE 4: Enrichment of positive and negative correlated genes with previously defined gene modules. The table summarizes the analysis which spanned over hundreds of gene modules. For each gene group and experiment we calculated the enrichment with the positive (darker gray) and negative (lighter gray) correlated genes. Note that none of the gene modules were enriched with both positive and negative correlated genes and only the highest score is displayed. Less significant enrichments are shown in black (P-value>10^{-3}) (see also Tables S3).

correlation between its expression level and growth rate, averaged over all perturbations (Methods). A positive GCR indicates a correlation with growth rate, whereas a negative GCR indicates an inverse correlation. We focused on the 5% of genes with the highest or lowest GCR, respectively and characterized their properties by analyzing enrichment within different co-expressed gene groups defined previously [20], [22] (Fig. 3d). To examine the generality of the results, we repeated the same analysis also for the compendium of deletion mutants, using the available gene expression and growth rate of the strains, as described above. The results are summarized in Figure 4.

The group of genes that are positively correlated with growth was enriched primarily with cell cycle genes. In the chemostat experiment, enrichment was found primarily for genes regulated during the G2/M and M/G1 transition (p-value of $4?10^{-7}$ and $<10^{-15}$, respectively), whereas in the compendium data enrichment was found for the G2/M transition as well as G1/S transition (p-values of $7?10^{-9}$ and $<10^{-15}$, respectively). Other groups that were linked to growth included Mitochondrial ribosomal proteins and oxidative phosphorylation (chemostat experiment), and also gene groups related to histones, ribosomal proteins, phospholipids metabolism and cell wall (deletion mutants). In both experiments, the group of genes that were negatively correlated with growth was enriched with calcium calmodulin related genes (Figure 4).

8.3 DISCUSSION

8.3.1 SUMMARY

Gene expression is responsive to both external and internal cues. In this study, we wished to characterize the relative importance of these two modes of signaling to the regulation of genes involved in ribosomal biogenesis. Specifically, we sought to characterize the potential importance of internal feedback mechanisms that sense and respond to the cellular physiological state. Such growth-dependent regulation is well characterized in bacteria leading to direct dependency of the rate of ribosomal biogenesis on cellular growth rate. This feedback is mediated, at least in part, by the nucleotide dependency of rRNA transcription, and by the accumulation of uncharged tRNA, through the stringent response [10], [13], [15], [30].

Our results indicate that in yeast, growth-related signals do not contribute significantly to the expression of ribosomal biogenesis genes. Several observations support this conclusion. First, under "confusion" situation, where the environmental signal and actual growth rate conflict, gene expression appears to be linked to the environment and not to the growth rate. Second, even in the absence of conflicting signals, the linkage of ribosomal biogenesis gene expression to growth rate relies on the availability of a proper environmental signal. Thus, no correlation was observed during the long-term recovery of chemostat-grown cells, when large variations in growth rate were observed, and we did not detect situations where ribosomal biogenesis gene expression was clearly dictated by growth rate. Similarly, no linkage between these factors was observed in a compendium of deletion mutants, where both gene expression and growth rate were modified not by changing an environment but rather by deleting specific genes.

We identified several gene groups which displayed tight correlation to growth rate. With the exception of the cell cycle genes, most groups identified were specific to a particular experimental condition (see also Table S3). It may be that correlation with growth is observed only for genes involved in rate limiting processes; For example, gluconeogenesis genes displayed a correlation with growth in the glucose-limited chemostat (Table S3). The correlation of amino-acid biosynthesis and phosphate utilization groups with growth rate in the compendium of deletion mutant may reflect limitation of the media used.

In conclusion, it appears that yeast cells control the level of ribosome biogenesis gene expression primarily by responding to environmental signals. The apparent correlation between gene expression and growth rate, observed under steady state conditions, is likely to reflect the evolutionary fine-tuning of these signals with the expected growth rate, rather than direct growth-dependent signals. This strategy might compensate the exact tuning of gene expression with growth rate, particularly under conditions that are rarely encountered over evolutionary times. On the other hand, it allows for rapid response and preparation to anticipated changes that will impact the growth rate. Our results suggest that the capacity to anticipate and prepare for environmentally-mediated changes in growth rate, before such changes actually occur, presented a major selection force during yeast evolution.

8.4 METHODS

8.4.1 MEDIA AND STRAINS

For the limiting histidine and glucose chemostats we used the following two strains, respectively: W303-1 bar1::P$_{ADH1}$-CFP-*KANMX* (*MATa, leu2-Δ1, trp1-Δ63, his3-Δ200, ura3-52, ade2-101*) and BY4741 (Euroscarf; *MATa, his3-Δ1, leu2-Δ0, met15-Δ0, ura3-Δ0*). For the mutant experiments we used BY4741 *adh1::KANMX* cells (Euroscarf), and wild-type BY4741 cells were used as a control.

For the *adh1* mutant experiment, YPD and YP-glycerol (2%) media were used. Histidine limited chemostat medium consisted of bacto-yeast nitrogen base without amino acids and with ammonium sulfate (0.67%), glucose (2%) and a drop-out mix (0.2%) of the following combination: L-Histidine (1.3 mg), Adenine sulfate (20 mg), Uracil (20 mg), L-Tryptophan (20 mg), L-Leucine (100 mg), L-Arginine (20 mg), L-Methionine (20 mg), L-Tyrosine (30 mg), L-Isoleucine (30 mg), L-Lysine (30 mg), L-Phenylalanine (50 mg), L-Glutamic acid (100 mg), L-Aspartic acid (100 mg), L-Valine (150 mg), L-Threonine (200 mg), L-Serine (400 mg). The final concentration of histidine in this medium is 2 mg/l. Limited glucose medium consisted of bacto-yeast nitrogen base without amino acids and with ammonium sulfate (0.67%), glucose (0.02%) and a drop-out mix (0.2%) as mentioned above but with 20 mg/l of L-histidine. Media for chemostat experiments included also 20 µl/liter of antifoam (Sigma A5758).

Importantly, media composition was chosen such that each chemostat has a single limiting factor. Specifically, the response of the yeast culture to addition of different medium components was tested to assure that the cells are sensitive only to changes in histidine or glucose concentrations.

8.4.2 ADH1 MUTANT EXPERIMENT

adh1 deletion strain and its isogenic wild type strain BY4741 were used to generate genome wide expression profiles. For the first experiment (Fig. 2B), cultures were grown in YPD for 8 hours and then washed and diluted into either YPD or YP-Glycerol. The cells were harvested after 4-5 duplications, upon reaching the concentration of ~$2*10^6$ cells/ml (see also Fig. S1). For the second experiment (Fig. 2C), *adh1* strain was grown on YPD to O.D. 0.4, and then washed and diluted into either YPD or YP-Glycerol. Cells were harvested after 30 minutes of growth and compared to samples of the initial overnight culture (at time t = 0).

8.4.3 CONTINUOUS CULTURE EXPERIMENTS

Continuous cultures were grown in BioFlo 110 fermentors (New Brunswick Scientific) in the following conditions: dilution rate = 0.2 h^{-1}, fermentor working volume = 750ml, temperature = 30°C, agitation speed = 500 RPM, air flux = 1 LPM, and the dissolved oxygen level was always above 70% of the saturation level. The pH of the media was titrated with NaOH to values of 5 and 5.3–5.6 for the histidine and glucose limited chemostats, respectively. After reaching steady state, cells were subjected to 10 different environmental cues (Table 1). Cell samples were harvested at various times during each perturbation. As a control, batch culture cells were grown in similar media and harvested at late logarithmic phase (Note that patterns of gene expression in steady-state closely resemble those of corresponding batch cultures just before they exhaust the nutrient [31]).

8.4.4 MICROARRAY HYBRIDIZATION AND DATA QUANTIFICATION

Samples were collected at various time points and RNA was isolated in order to generate cDNA, which was then labeled by the indirect amino-allyl method. Cy5-and Cy3-labeled cDNA was hybridized onto microarrays of ORFs representing the complete *S. cerevisiae* genome (University Health Network, Ontario), using cDNA from untreated cultures as reference. The arrays were scanned, and expression data was extracted by image analysis. Expression ratios were $\log_2$ transformed, corrected for technical biases and normalized (see also Text S1).

8.4.5 MODULE IDENTIFICATION

Transcription modules were identified by applying the Iterative Signature Algorithm to the full expression dataset comprising all conditions [20], [21]. The Iterative Signature Algorithm is an iterative extension of the Signature Algorithm, designed for a global decomposition of large-scale expression matrices into transcription modules, in the absence of a-priory information. Modules are fixed-points of the Signature Algorithm, and are identified in a heuristic search by iterating from a large ensemble of randomly composed input sets until convergence is reached. The gene threshold of the Signature Algorithm determines the resolution of the analysis. The full set of transcription modules used in this work is available in the Supplementary Materials and further details are presented at the following URL: http://barkai-serv.weizmann.ac.il/Modules.

8.4.6 CORRELATION BETWEEN GROWTH RATE AND GENE EXPRESSION

The measured O.D. was interpolated using the Matlab csaps function (Fig. 3b, S2). The cells growth rate (μ) was calculated using the chemostat cell density equation: $\mu = (\acute{n}/n) + D$, where n is the cell density (interpolated O.D) and D is the dilution rate [27]. Correlations between growth and gene expression were calculated using the Matlab corrcoef function. Each gene was assigned a growth correlation score (GCR), quantifying the correlation between its expression level and growth rate, averaged over all perturbations. Sorting the genes by their GCR revealed two gene groups: the group of 5% most positive correlated genes (positive GCR) and 5% most negatively correlated genes (negative GCR). In order to reduce noise, we neglected genes that were not affected by the environmental perturbations (see Text S1).

Our study focused on finding the enrichment of positive GCR and negative GCR genes in different co-regulated gene groups, which were previously defined by the signature algorithm [20], [21]. The analysis was unbiased and done over all gene modules within all threshold levels. The P-value for each group enrichment with positive and negative GCR was obtained by calculating the chance of getting at least such an enrichment by chance. Specifically:

$$P \ value = 1 - \sum_{i=1}^{X-1} H(i, K, N, M) \qquad \text{where}$$

$$H(X, K, N, M) \equiv \frac{\binom{K}{X} \cdot \binom{M-K}{N-X}}{\binom{M}{N}}$$

is the hyper-geometrical distribution designating the probability for enrichment of X of a possible K correlated genes from a group with N genes (module size) without replacement from a pool of M genes.

Gene groups were sorted both by their P-value for enrichment with positive GCR, and by their P-value for enrichment with negative GCR. The analysis was done for the continuous culture experiments considering all 10 perturbations (Figure 4). The analysis was done also for each of the two limiting factor experiments separately (only the 5 perturbations in histidine or glucose depletion, Table S3). The same calculation was repeated for the compendium data. Additional figures and tables are available in the Supplementary Material.

8.4.7 ACCESSION NUMBERS

All raw data is available through Gene Expression Omnibus (GEO) accession number GSE6302, at the following website: http://www.ncbi.nlm. nih.gov/geo.

REFERENCES

1.   Powers T, Walter P (1999) Regulation of ribosome biogenesis by the rapamycin-sensitive TOR-signaling pathway in Saccharomyces cerevisiae. Mol Biol Cell 10: 987–1000. doi: 10.1091/mbc.10.4.987.
2.   Martin DE, Soulard A, Hall MN (2004) TOR regulates ribosomal protein gene expression via PKA and the Forkhead transcription factor FHL1. Cell 119: 969–979. doi: 10.1016/j.cell.2004.11.047.
3.   Zurita-Martinez SA, Cardenas ME (2005) Tor and cyclic AMP-protein kinase A: two parallel pathways regulating expression of genes required for cell growth. Eukaryot Cell 4: 63–71. doi: 10.1128/EC.4.1.63-71.2005.
4.   Warner JR (1999) The economics of ribosome biosynthesis in yeast. Trends Biochem Sci 24: 437–440. doi: 10.1016/S0968-0004(99)01460-7.
5.   Fatica A, Tollervey D (2002) Making ribosomes. Curr Opin Cell Biol 14: 313–318. doi: 10.1016/S0955-0674(02)00336-8.
6.   Moss T (2004) At the crossroads of growth control; making ribosomal RNA. Curr Opin Genet Dev 14: 210–217. doi: 10.1016/j.gde.2004.02.005.
7.   Jorgensen P, Rupes I, Sharom JR, Schneper L, Broach JR, et al. (2004) A dynamic transcriptional network communicates growth potential to ribosome synthesis and critical cell size. Genes Dev 18: 2491–2505. doi: 10.1101/gad.1228804.
8.   Ruggero D, Pandolfi PP (2003) Does the ribosome translate cancer? Nat Rev Cancer 3: 179–192. doi: 10.1038/nrc1015.
9.   Wang H, Zhao LN, Li KZ, Ling R, Li XJ, et al. (2006) Overexpression of ribosomal protein L15 is associated with cell proliferation in gastric cancer. BMC Cancer 6: 91. doi: 10.1186/1471-2407-6-91.
10. Gourse RL, Gaal T, Bartlett MS, Appleman JA, Ross W (1996) rRNA transcription and growth rate-dependent regulation of ribosome synthesis in Escherichia coli. Annu Rev Microbiol 50: 645–677. doi: 10.1146/annurev.micro.50.1.645.
11. Gaal T, Bartlett MS, Ross W, Turnbough CL Jr., Gourse RL (1997) Transcription regulation by initiating NTP concentration: rRNA synthesis in bacteria. Science 278: 2092–2097. doi: 10.1126/science.278.5346.2092.
12. Paul BJ, Ross W, Gaal T, Gourse RL (2004) rRNA transcription in Escherichia coli. Annu Rev Genet 38: 749–770. doi: 10.1146/annurev.genet.38.072902.091347.
13. Schneider DA, Gaal T, Gourse RL (2002) NTP-sensing by rRNA promoters in Escherichia coli is direct. Proc Natl Acad Sci U S A 99: 8602–8607. doi: 10.1073/pnas.132285199.

14. Walker KA, Mallik P, Pratt TS, Osuna R (2004) The Escherichia coli Fis promoter is regulated by changes in the levels of its transcription initiation nucleotide CTP. J Biol Chem 279: 50818–50828. doi: 10.1074/jbc.M406285200.

15. Chatterji D, Ojha AK (2001) Revisiting the stringent response, ppGpp and starvation signaling. Curr Opin Microbiol 4: 160–165. doi: 10.1016/S1369-5274(00)00182-X.

16. Jain V, Kumar M, Chatterji D (2006) ppGpp: stringent response and survival. J Microbiol 44: 1–10.

17. Wendrich TM, Blaha G, Wilson DN, Marahiel MA, Nierhaus KH (2002) Dissection of the mechanism for the stringent factor RelA. Mol Cell 10: 779–788. doi: 10.1016/S1097-2765(02)00656-1.

18. Regenberg B, Grotkjaer T, Winther O, Fausboll A, Akesson M, et al. (2006) Growth-rate regulated genes have profound impact on interpretation of transcriptome profiling in Saccharomyces cerevisiae. Genome Biol 7: R107. doi: 10.1186/gb-2006-7-11-r107.

19. Hughes TR, Marton MJ, Jones AR, Roberts CJ, Stoughton R, et al. (2000) Functional discovery via a compendium of expression profiles. Cell 102: 109–126. doi: 10.1016/S0092-8674(00)00015-5.

20. Ihmels J, Friedlander G, Bergmann S, Sarig O, Ziv Y, et al. (2002) Revealing modular organization in the yeast transcriptional network. Nat Genet 31: 370–377. doi: 10.1038/ng941.

21. Bergmann S, Ihmels J, Barkai N (2003) Iterative signature algorithm for the analysis of large-scale gene expression data. Phys Rev E Stat Nonlin Soft Matter Phys 67: 031902. doi: 10.1103/PhysRevE.67.031902.

22. Ihmels J, Bergmann S, Barkai N (2004) Defining transcription modules using large-scale gene expression data. Bioinformatics 20: 1993–2003. doi: 10.1093/bioinformatics/bth166.

23. Ihmels J, Bergmann S, Gerami-Nejad M, Yanai I, McClellan M, et al. (2005) Rewiring of the yeast transcriptional network through the evolution of motif usage. Science 309: 938–940. doi: 10.1126/science.1113833.

24. Causton HC, Ren B, Koh SS, Harbison CT, Kanin E, et al. (2001) Remodeling of yeast genome expression in response to environmental changes. Mol Biol Cell 12: 323–337. doi: 10.1091/mbc.12.2.323.

25. Gasch AP, Spellman PT, Kao CM, Carmel-Harel O, Eisen MB, et al. (2000) Genomic expression programs in the response of yeast cells to environmental changes. Mol Biol Cell 11: 4241–4257. doi: 10.1091/mbc.11.12.4241.

26. Jorgensen P, Nishikawa JL, Breitkreutz BJ, Tyers M (2002) Systematic identification of pathways that couple cell growth and division in yeast. Science 297: 395–400. doi: 10.1126/science.1070850.

27. Segel LA (1984) Modeling dynamic phenomena in molecular and cellular biology. Cambridge ; New York: Cambridge University Press.

28. Hughes TR, Roberts CJ, Dai H, Jones AR, Meyer MR, et al. (2000) Widespread aneuploidy revealed by DNA microarray expression profiling. Nat Genet 25: 333–337. doi: 10.1038/77116.

29. Dekel E, Alon U (2005) Optimality and evolutionary tuning of the expression level of a protein. Nature 436: 588–592. doi: 10.1038/nature03842.

30. Murray HD, Schneider DA, Gourse RL (2003) Control of rRNA expression by small molecules is dynamic and nonredundant. Mol Cell 12: 125–134. doi: 10.1016/S1097-2765(03)00266-1.

31. Saldanha AJ, Brauer MJ, Botstein D (2004) Nutritional homeostasis in batch and steady-state culture of yeast. Mol Biol Cell 15: 4089–4104. doi: 10.1091/mbc.E04-04-0306.

This article has supplemental information that is not featured in this version of the text. To view these files, please visit the original version of the article as cited in the beginning of this chapter.

CHAPTER 9

A SCREEN FOR RNA-BINDING PROTEINS IN YEAST INDICATES DUAL FUNCTIONS FOR MANY ENZYMES

TANJA SCHERRER, NITISH MITTAL, SARATH CHANDRA JANGA, AND ANDRÉ P. GERBER

9.1 INTRODUCTION

Immediately when RNA is synthesized by RNA polymerases, RNA binding proteins (RBPs) assemble on the nascent transcript forming ribonucleoprotein (RNP) complexes, which tightly control all of the further steps in a RNA's life. On one hand, RBPs assist the processing and assembly of non-coding (nc) RNAs into RNP complexes, which mediate essential cellular functions such as splicing and translation [1]. On the other hand, RBPs are essential for mRNA maturation, which involves the addition of a 7-methylguanosine cap at the 5′end of mRNA-precursors, the splicing-out of introns, editing, and the addition of a polyadenosine tail at the 3′end of the message. RBPs further guide mRNA export and localization to specific cytoplasmic loci for translation, and ultimately, they control the decay of (m)RNAs [2]. Notably, all these steps are highly connected to each other and linked with other gene regulatory layers to ensure proper expression of every gene in a cell [3].

This chapter was originally published under the Creative Commons Attribution License. Scherrer T, Mittal N, Janga SC, and Gerber AP. A Screen for RNA-Binding Proteins in Yeast Indicates Dual Functions for Many Enzymes. PLoS ONE 5,11 (2010). doi:10.1371/journal.pone.0015499.

The availability of genomic tools now allows the systematic identification of RNA targets for RBPs to obtain a global view of their gene regulatory potential. One of the main approaches include the immunopurification of RNP complexes followed by the analysis of the associated RNAs with DNA microarrays, a method referred to as RNA-immunopurification-microarray (RIP-Chip). Numerous studies applying these genomic tools revealed that many RBPs associate with distinct RNA target sets comprised of a few up to several hundred RNAs, which are often enriched for specific sequence/structural elements that define RBP binding sites. The sets of bound RNAs were often related containing mRNAs coding for functionally or cytotopically related proteins (e.g. [4]–[6]; reviewed in [7]–[9]). These findings lead to a model that proposes important coordinative roles for RBPs in the expression of functionally related groups of messages, referred to as 'RNA regulons' or 'post-transcriptional operons' [7]. Moreover, it underscores that RBPs bind simultaneously and/or sequentially to RNAs generating numerous RNP particles, whose dynamic composition and combinatorial arrangement may be unique for each mRNA expressed in a cell [8]–[10].

RBPs comprise 3 to 11% of the proteomes in bacteria, archaea and eukaryotes underlining the importance of RNA regulation for cell function [11]. In the budding yeast *Saccharomyces cerevisiae*, more than 500 proteins are predicted to function as RBPs [6], [10]. An extensive bioinformatic survey, considering evolutionary conservation, identified almost 100 protein motifs linked to RNA regulation; about half of them have been classified as "enzymatic" domains mostly present in RNA modification enzymes and nucleases. Another 40 motifs or so have been classified as "non-catalytic" RNA-binding domains, which are often part of multi-subunit RNP complexes [11]. Notably, RBPs often contain an array of RNA-binding motifs (RBMs), which further increases the specificity and affinity towards the RNA.

The vast number of protein motifs linked to RNA regulation and the ancient origin of RNA regulation, which is possibly the most evolutionary conserved component of a cell's physiology, proposes that many proteins implicated in other cellular processes could have retained RNA-binding capacity. For instance, several metabolic enzymes in mammals have been shown to bind to and regulate mRNA expression (reviewed in [12]–[15]). Perhaps best characterized are the iron regulatory proteins (IRP), cytoplasmic aconitases that regulate the translation or stability of several messages depending on cellular iron levels [16]. Moreover, a recent comprehensive RIP-Chip study analyzing the RNA targets for more than 40 different RBPs and some other proteins in yeast showed that two metabolic enzymes, for which homologs in mammals have been reported to bind RNA, were reproducibly associated with cellular RNAs, indicating that RNA regulation by these proteins may be evolutionarily conserved [6]. These observations have raised speculations about the existence of yet largely overlooked post-transcriptional regulatory networks between intermediary metabolism and RNA regulation [12]. It furthermore highlights the need of systematic discovery tools to identify novel RBPs as the "universe" of RBPs in eukaryotes could be well underestimated. Possibly many more proteins could have retained or acquired the capacity to bind RNA enabling post-transcriptional gene regulation at yet uncharacterized levels and processes.

In this study, we set out to screen for novel and unconventional RBPs. We therefore used protein microarrays containing 70% of the yeast's proteome and probed them with different sorts of RNA. We selected almost 200 proteins that reproducibly interacted with RNA, most of them not previously annotated to act as RNA-binding proteins such as metabolic enzymes. We further determined in vivo associated RNAs for 13 potential RBPs by RIP-Chip. Most of the RBPs bound to distinct subsets of mRNA, some of them code for functionally related proteins and thus, possibly comprise "RNA regulons". Since this screen is not saturated we expect that many more RBPs—including proteins with dual functions—exist in eukaryotic organisms, forming a dense and robust post-transcriptional scaffold that effectively coordinates gene expression to ensure the integrity and stability of a cells fate.

9.2 RESULTS

9.2.1 DETECTION OF SPECIFIC RNA-PROTEIN INTERACTIONS WITH PROTEIN MICROARRAYS

We used functional protein microarrays to screen for proteins that interact with RNA (Figure 1). Protein microarrays have been previously used to identify proteins that interact with small viral RNAs [17], but to our knowledge, there has been no screen to detect proteins interacting with cellular RNAs. To establish the experimental procedure, we first probed protein microarrays with a short 36 nucleotide (nt) long RNA termed E2B-min, which is a fragment of the Ash1 mRNA known to specifically interact with She2p [18]. She2 is a RBP that facilitates the localization of Ash1 mRNA and other messages to the bud-tip during cell division [19]. Among the 4,088 proteins present on the array, the strongest signal of fluorescent-ly labeled E2Bmin RNA was seen with She2p (24.2 standard deviations [SD] above the mean of signal intensities from two independent experiments; Z-scores are given in Dataset S1). No signals were obtained with an array where proteins were heat-denatured before probing with RNA, indicating that RNA interactions must derive from active proteins. Besides She2, six GTPases (Arf1, Arf3, Arl2, Ypt1, Ypt7, Ypt32; $p<10^{-9}$), a tRNA guanylyltransferase (Thg1), and a single-stranded DNA-binding protein (Rim1) also strongly and reproducibly interacted with E2Bmin (SD>3.5 in replicates). Whether these E2Bmin binders may be implicated in the regulation of Ash1 mRNA in vivo remains to be elucidated. At least, these experiments show that specific RNA-protein interactions can be detected with our experimental set-up.

9.2.2 MANY ENZYMES MAY INTERACT WITH CELLULAR RNAS

To screen for proteins that interact either with total RNA or mRNAs, we basically used the same experimental set-up as applied for the E2Bmin experiments. We probed the protein microarrays with Cy3 labeled 'total RNA', which was isolated from yeast cells grown in different carbon sources, and with Cy5 labeled mRNAs isolated from total RNA via oligo-dT columns (see Materials and Methods). Because data was less reproducible compared to the replicate arrays probed with E2Bmin RNA described above, we assigned each element on the array a percentile rank based on background subtracted signals, and calculated median percentile ranks across the five replicates [20] (raw data is provided in Dataset S2). Thereby, the highly ranked proteins represent those with highest signals on the array (e.g. She2p probed with E2Bmin RNA is ranked = 1 in the above

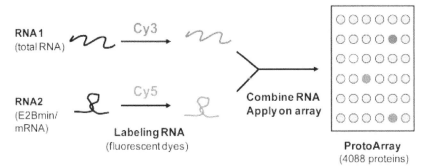

FIGURE 1: Identification of RNA-binding proteins with protein microarrays. Protein microarrays (Protoarrays) contained 4,088 different yeast proteins (~70% of the proteome) individually spotted in duplicates onto a modified glass slide. The arrays were probed with a mixture of fluorescently labeled RNAs. After washing, the arrays were scanned and analyzed for proteins that bound either labeled RNAs.

described experiments). The analysis of ranks instead of Z-scores has been previously applied to analyze chromatin immunoprecipitation-chip data and performs well when magnitude and scale of the actual signals varies between replicates [20]. If there are features that are consistently highly ranked across multiple replicates, the distribution of the median percentile ranks of all features will form a bimodal curve; and the median percentile rank at the trough of this bimodal distribution can be selected as a conservative cut-off to define targets [20]. A histogram of the median ranks across the five replicate protein arrays showed a bimodal distribution, which we assumed to represent non-binders and binders, the latter ones to be consistently highly ranked across replicate experiments (Figure 2). We have therefore chosen the trough of the distribution as a conservative cutoff to define proteins that reproducibly interacted with either total RNA or mRNAs; selecting 67 total RNA and 173 mRNA binders, respectively (a list of the total 180 proteins selected from this analysis is provided in Table S1). 90% of total RNA binders were also found in the pool of mRNA binders, but most of the mRNA binders did not strongly bind total RNA (113; 65%). These proteins may preferentially interact with mRNAs, which are underrepresented in the total RNA fraction. However, we wish to note that this selection procedure was designed to go for a robust list of potential RNA binders. It may thus not provide a comprehensive list of all RNA-binders, and further inspection of the data may reveal additional RNA-binders.

We categorized the selected 180 proteins that either interacted with total RNA or mRNA based on Gene Ontologies (GO) retrieved from the Princeton GO server. 132 out of the 180 proteins (73%) had at least one known function annotated with GO. 28 proteins were annotated with the GO term 'RNA-binding', which is therefore over-represented among the group of all 180 selected proteins (p<10^{-3}, Figure 3; a detailed list of GO terms is provided in Table S2). Further manual inspection of the 180 proteins revealed 18 additional proteins with RNA related functions – adding-up to 46 proteins that act in RNA metabolism (25% of all selected proteins; 35% of proteins with assigned functions; marked in blue in Table S1). In contrast, DNA binding proteins including transcription factors (TFs; 13 proteins, 7%) were not over-represented suggesting that our assay discriminates between DNA and RNA-binders. Moreover, only four of the 180 proteins (Bcy1p, Deg1p, Pfk26p, Yer087p) were among 208 proteins selected in a similar screen applying protein microarrays to identify single- or double-stranded DNA binding proteins [21]. In conclusion, our list of selected proteins bears a substantial fraction of previously known RNA-binders or proteins with RNA-related functions, indicating that our assay likely selected proteins that have RNA-binding properties. However, as outlined above, our stringent cut-off is not expected to identify all of the RNA-binders. Moreover, there are many reasons why diverse known RBPs, which are present on the array did not give reproducible signals across replicates. This includes inactivation of proteins on the slide surface, mis-folding or RNA cross-hybridization in solution, and finally, many annotated RBPs act in protein complexes (e.g. ribosomal proteins) and thus may not specifically interact with RNAs on their own.

Regarding the assigned functions among our list of selected RNA-binders, we were intrigued that many of them have catalytic functions, including oxidoreductases, hydrolases, lyases and transferases (total 94 proteins; 52%, p<0.003) (Figure 3, Table S2). Whereas 17 of these enzymes have been previously linked to RNA related processes, the remaining ones act in unrelated processes such as fatty acid metabolism (p<0.007) or lipid oxidation (p<0.008). Moreover, 25 of these enzymes can be mapped to the yeast metabolic network [22], which are therefore significantly over-

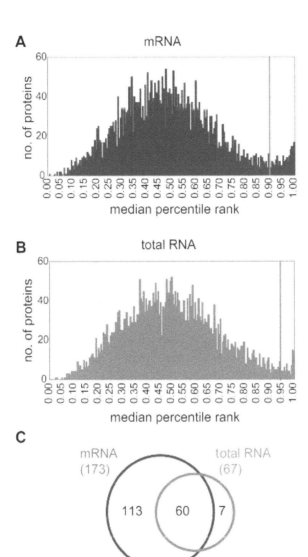

FIGURE 2: Selection of mRNA and total RNA-binding proteins. (A) Distribution of ranked median signal intensities resulting from protein microarrays probed with mRNAs. The trough at 0.9 was taken as cut-off and all proteins with greater ranks were selected as mRNA binders. (B) Distribution of ranked median signal intensities resulting from arrays probed with total RNA. The trough at 0.95 was taken as cut-off and all proteins with greater ranks were considered as total RNA binders. (C) Venn Diagram representing overlap between proteins binding to total RNA and mRNAs.

represented compared to all of the metabolic enzymes in this network present on the protein microarray (397 proteins, p<0.016). In agreement with this bias for enzymes, most of the herein identified potential RBPs are cytoplasmic (141 proteins, $p<10^{-6}$), membrane-associated (p<0.003), and some of them located to peroxisomes ($p<2\times10^{-5}$)(Figure 3). These results indicate that many cytoplasmic enzymes could interact with RNA. In principle, this could provide opportunities to directly connect intermediary metabolism with posttranscriptional gene regulation.

We further analyzed our experimentally defined set of 180 RBPs for the occurrence of protein domains annotated by the Pfam database [23]. 4,049 proteins in *S. cerevisiae* were annotated with 6,119 domains in Pfam, and we analyzed whether some of these domains were over-represented among the 150 proteins (out of 180) that contained at least one Pfam domain (Pfam domains are annotated in Table S1). As expected, most prevalent were known RNA-binding domains such as the K homology (KH_1) domain, the RNA recognition motif (RRM_1) and a subtype of the zinc finger motif, zf-CCHC, which were all significantly enriched (Table 1). Interestingly, several domains were enriched that have not been previously related to RNA function ($p<10^{-3}$, hypergeometric) and occur in proteins devoid of other known RNA-binding motifs. This includes the ubiquitin motif or the weakly conserved repeat module PC_rep, which are found in several proteins involved in protein degradation control [24]. It also includes the WW motif and the TPR_1 (tetratricopeptide repeat), which mediate protein-protein interactions and the assembly of multiprotein complexes [25], and several enzymatic domains contained in metabolic enzymes. Whether any of these domains directly or indirectly mediate RNA-binding has yet to be investigated, but their significant overrepresentation makes them prime candidates for further analysis.

9.2.3 POTENTIAL RBPS COME FROM DIFFERENT EXPRESSION REGIMES

We next asked how the expression of our selected RNA-binders varies across different growth conditions to see whether our selection is biased to any kind of expression characteristics. We therefore compiled a large

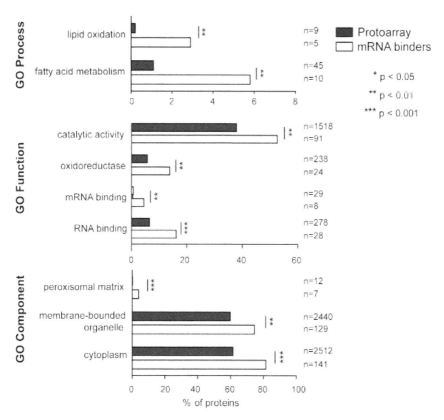

FIGURE 3: Significantly shared GO terms among mRNA binders. The 173 m RNA binders were searched for significantly enriched GO terms as compared to all the 4,088 proteins present on the protein microarray. Bar diagrams indicate relative amount of genes of the respective GO term among all proteins on the array or among the selected mRNA binders, respectively.

collection of microarray data available for a wide range of experimental conditions for *S. cerevisiae* from the M3D database [26] (experimental conditions are indicated in the Table S3). Because this data is available in Robust MultiArray (RMA) normalized format [27], it enables direct comparison of expression levels (see Materials and Methods). Expression profiles could be obtained for 164 of the 180 RBPs identified in this study, and we performed K-means clustering with 10 groups to identify subsets of genes that exhibited similar expression patterns. This analysis revealed that the genes followed very heterogeneous expression patterns; genes clustered into different expression regimes namely ubiquitously highly expressed, ubiquitously poorly expressed and specific to conditions (a heatmap cluster of this analysis is shown in the Figure S1).

We further compared the expression levels of the potential RBPs identified in this study with previously annotated RBPs (see Material and Methods). We found no general difference (p<0.64, Wilcoxon test). However, our herein identified RBPs are generally higher expressed than non-RBPs (p<2×10^{-6}, Wilcoxon test); an observation that has been made previously for conventional RBPs as well [28]. We therefore speculate that in particular the highly expressed unconventional RBPs may give good leads for future experiments as they have the potential to control many RNA targets [28].

TABLE 1: Pfam domains enriched in the list of putative RBPs.

Domain	Occurance (RPBs)	Occurance (genome)	Occurance (protoarray)	P-value (Hypergeometric)
KH_1	10	18	16	5.44E-09
zfCCHC	8	23	15	1.06E-06
RRM_1	12	78	41	4.75E-06
ubiquitin	6	15	15	0.00018
PC_rep	4	7	7	0.00048
TPR_1	6	27	19	0.00077
WW	4	8	8	0.00089
adh_short	4	13	9	0.0015
S1	3	6	5	0.0023
TYA	5	81	18	0.0038
LDH_1_C	2	3	2	0.0041
TBP	2	2	2	0.0041
cNMP_binding	2	5	2	0.0041
Acyl_CoA_thio	2	2	2	0.0041
Ldh_1_N	2	3	2	0.0041
PseudoU_synth_1	2	6	2	0.0041

9.2.4 SELECTED NOVEL RBPS ASSOCIATE WITH DISTINCT SETS OF MRNAS

To examine whether some novel potential RBPs in our selected list of RNA-binders associate with RNA in vivo, we purified endogenously expressed tandem-affinity purification (TAP)-tagged proteins from cells grown in rich media, and identified co-purifying RNAs with yeast DNA oligo arrays. From our list of 180 RBPs, we selected 13 proteins that are expressed at different levels, and for which respective mRNA expression patterns are different across a variety of conditions, providing a representative sample of differentially expressed putative RBPs (marked in Figure S1). Besides a transcriptional regulator (Lap3p) and a co-chaperone (Sti1p), we selected eleven proteins with catalytic activities (Dfr1p, Gre3p, Map1p, Mdh1p, Mdh3p, Meu1p, Pfk2p, Phr1p, Pot1p, Pre10p, Ymr1p), some of them acting in intermediary metabolism, reflecting the fact that many candidate RBPs are enzymes. Seven of the proteins are cytoplasmic, two are peroxisomal, and one representative each are from the nuclear, mitochondrial, ribosomal and proteasome compartment.

We performed three independent affinity isolations with each of the 13 potential RBPs and five independent mock isolations with untagged control cells (= mock isolates). To identify RNA that were significantly associated with the proteins we selected those features that were on average at least 3-fold enriched in the affinity isolates compared to the mock controls with a p-value of less than 0.01 (see Materials and Methods). This analysis revealed that all proteins were associated with unique sets, comprised of a few to dozens of different RNAs (Figure 4; raw data from RIP-Chip experiments and a list of selected features is given in the Dataset S3). Notably, the proteins were almost entirely associated with mRNAs, excluding highly expressed ncRNAs such as rRNAs, tRNAs and snoRNAs. This indicates that these candidate RBPs primarily target mRNAs for potential gene expression control. It also substantiates the specificity of our assays as there is no apparent bias for selection of highly expressed ncRNAs. We also found no correlation between the expression level of these proteins [29] and the number of selected targets (Pearson correlation r = 0.04), further substantiating that the observed associations are selective and not merely driven by expression.

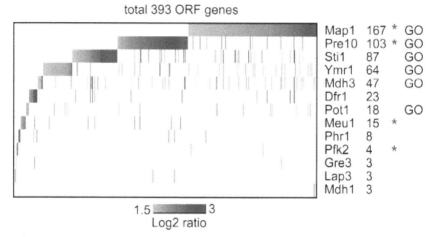

total 393 ORF genes

Map1	167 *	GO
Pre10	103 *	GO
Sti1	87	GO
Ymr1	64	GO
Mdh3	47	GO
Dfr1	23	
Pot1	18	GO
Meu1	15 *	
Phr1	8	
Pfk2	4 *	
Gre3	3	
Lap3	3	
Mdh1	3	

1.5 ▭▬▬ 3
Log2 ratio

FIGURE 4: Selected novel RNA-binding proteins bind to distinct sets of mRNAs. (A) Heat map of mRNAs associated with indicated proteins. The color code (orange-blue) indicates the fold-change (log$_2$ ratio scale) of the respective feature in the affinity isolation compared to mock control microarray data. The number of mRNA targest for each protein is indicated next to the name of the protein. A star (*) denotes association with own mRNA. 'GO' indicates that GO terms are significantly enriched among targets (see Table 2).

Four of the 13 proteins (30%) were associated with their own mRNA (Pfk2, Pre2, Map1, Meu1). Binding to the own mRNA offers the possibility for auto-regulation through the formation of positive or negative feedback loops [30]. This fraction is therefore similar to previous finding with canonical RBPs, where 18 of 46 RBPs (40%) were associated with their own RNA [6]. Remarkably, this fraction is considerably larger compared to TFs, where 10% bound to their own promoter sequences in a global TF-binding site analysis for 106 TFs [30].

Because many RBPs bind to mRNAs coding for functionally related proteins, we searched for common themes among the messages that were associated with the 13 proteins. For six proteins (Map1, Mdh3, Pot1, Pre10, Sti1, and Ymr1) we found significantly enriched GO groups among associated messages, offering the potential to coordinate expression of functionally related groups of messages or 'RNA regulons' (Table 2; a more comprehensive list of GO terms is provided in the Table S4). Noteworthy, proteins associated with only a few messages may be less prone for this analysis as the number of the associated mRNAs might be too small to achieve statistically sound data. Although every protein assembled with unique GO terms (e.g. Sti1p bound mRNAs code preferentially for proteins acting in telomere maintenance and DNA recombination), some of the enriched GO terms appeared with more than one protein. For instance, messages associated with Map1p, Pot1p, and Ymr1p are commonly related to translation. However, the particular messages that added to this term were mostly different and only one message (Rps9b) was shared among the targets for the three proteins. Likewise, three proteins (Map1, Mdh3, Sti1) were preferentially associated with messages coding for proteins annotated with pyrophosphatase activity. Among the many targets for these proteins, only eight mRNA targets are shared, which do not link to pyrophosphatase activity (two pyruvate decarboxylases [Pdc1, Pdc5] were commonly enriched; $p < 2 \times 10^{-4}$). Therefore, it appears that although some GO terms were enriched with more than one of the proteins, it is not because these proteins bound to a common set of messages that connects to one particular GO term, but rather that they were associated with different messages that belong to the same functional class.

TABLE 2: Selected list of GO terms enriched among mRNA targets.

Protein	Category	GO term (p-value)
Map1	Process	translation elongation (10^{-7}), small molecule metabolic process (10^{-5})
	Function	catalytic activity (10^{-11}), pyrophosphatase activity (2×10^{-7})
	Compartment	plasma membrane enriched fraction (10^{-10}), ribosome (4×10^{-6})
Pre10	Function	hydrolase activity (0.007)
Sti1	Process	telomere maintenance via recombination (5×10^{-5})
	Function	helicase activity (10^{-14}), pyrophosphatase activity (10^{-10})
Ymr1	Process	translation (3×10^{-11})
	Function	structural consituent of ribosome (10^{-10})
	Compartment	ribosome (6×10^{-14})
Mdh3	Function	nucleoside-triphosphatase activity (10^{-3}), helicase activity (2×10^{-3})
Pot1	Process	translation (8×10^{-10})
	Function	structural constituent of ribosome (8×10^{-12})

9.2.5 MAP1P NEGATIVELY AFFECTS GENE EXPRESSION OF MRNA TARGETS

To investigate how one of the selected candidate enzymes could affect gene expression of targets, we measured the relative changes of mRNA levels of cells overexpressing MAP1 compared to control cells with DNA microarrays. Map1p is a methionine aminopeptidase (MetAP) that catalyzes the co-translational removal of N-terminal methionine from nascent polypeptides, and it is functionally redundant with Map2p [31], [32]. Notably, Map1p contains two zinc-finger motifs, one CCCC-type and the other of the CCHH-type [33], which occur in DNA-binding proteins and in some RBPs [34] – however these domains were not thought to provide selective RNA-binding but rather to confer interaction of Map1p with the ribosome [35]. Yeast cells bearing a plasmid with *MAP1* under the control of galactose inducible promoter, and control cells containing an empty plasmid, were grown to mid-log phase and expression was induced with 2% galactose for 1.5 hours. Noteworthy, inducible short-time overexpression of RBPs could be beneficial to measure direct effects of proteins on gene expression by minimizing secondary effects that may raise after prolonged alterations of expression levels (Scherrer et al., submitted). We obtained mRNA expression profiles for 6,851 features representing 5,889 yeast genes (raw data is provided in Dataset S4). *MAP1* expression was increased 4.2-fold being the most changed mRNA of all analyzed features. The relative expressions levels of Map1p target mRNAs were very slightly (mean fold change = 0.925) but significantly decreased compared to all non-targets ($p<10^{-5}$, Mann-Whitney U test) (Figure 5). Of note, only 44 genes changed at least 1.5 fold with p<0.05 (one sample t-test); and seven Map1p targets were overrepresented among the 36 down-regulated messages ($p = 7 \times 10^{-5}$, Fisher's exact test). The same analysis with microarray data obtained from cells overexpressing GIS2 (another ZnF protein among the selected RNA-binders) did not reveal reduced expression of Map1p targets (TS and APG, unpublished results), indicating that the observed shift in the distribution of Map1p targets was not a general effect due to protein overexpression. In conclusion, these results suggest that Map1p could negatively affect mRNA expression of selected mRNA targets.

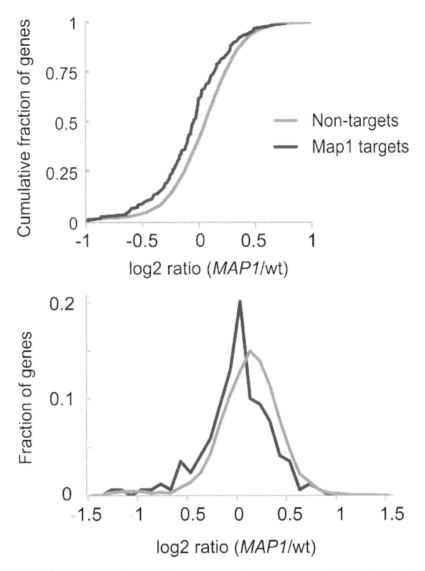

FIGURE 5. Gene expression profiling of yeast cells overexpressing MAP1. Distribution of average Cy5/Cy3 fluorescence ratios from three microarray hybridizations comparing RNA levels of MAP1 over-expressing yeast cells with control cells. In the upper panel, the fraction of transcripts indicated on the y-axis refers to the cumulative fraction of sequences on the microarray; $\log_2$ ratios are plotted on the x-axis. The lower panel shows a histogram depicting the fraction of transcripts (y-axis) that are clustred within bins of 0.1 $\log_2$ ratios (x-axis). One line delineates the distribution of Map1p RNA targets defined from affinity purifications. The other represents non-targets.

9.3 DISCUSSION

Protein microarrays have been applied to detect protein-protein, protein-lipid, protein-DNA and protein-viral RNA interactions [17], [21], [36], [37]. Here, we describe the use of protein microarrays for the detection of protein-RNA interactions. We identified dozens of potentially "novel" RBPs that either interacted with mRNA or total RNA on protein microarrays. Strikingly, among these were many enzymes with well-established cellular functions. For some of them, we have shown significant association with functionally related messages, possibly allowing coordination of the expression of 'RNA regulons' as seen for bona fide RBPs. This was further initially demonstrated for Map1p, for which we observed subtle coordinated down-regulation of target mRNAs upon *MAP1* overexpression, indicating that Map1p preferentially negatively affects gene expression of target messages. This study therefore expands our understanding of post-transcriptional gene networks suggesting regulatory functions to a variety of proteins not connected to gene expression regulation so far.

Our observations that many proteins with enzymatic activities bind to RNAs and potentially participate in RNA regulation are reminiscent to previous observations stating RNA-binding functions for several mammalian enzymes (reviewed in [12], [13]). These mammalian enzymes revealed a striking common denominator—they catalyze reactions that often involve mono- or dinucleotides as substrates or co-factors [14]. Similarly, we found that a large fraction (41 proteins) of the "novel" RBPs with assigned catalytic activities (total 95 proteins) require nucleotide related co-factors/substrates (Table S1): 14 proteins require ATP/AMP as substrate (e.g. kinases), 13 need nicotinamide adenine dinucleotides (NAD) or its 5-phosphate derivate (NADP) as a cofactor, seven employ Coenzyme-A (CoA) found in many enzymes acting in the sterol/fatty acid metabolism, and nine use others such as GTP/GMP or S-adenosyl-methionine. In this regard, the protein binding site for NAD or NADP has been postulated to have occasionally evolved to a binding surface for polyribonucleotides in some mammalian enzymes (e.g. thymidylate synthase (TS) and dihydrofolate reductase (DHFR), as well as glyceraldehyde-3-phosphate, isocitrate, and lactate dehydrogenases) [14]. We tested four (Gre3, Dfr1,

Mdh1, Mdh3) NAD binding proteins for association with cellular RNA with RIP-Chip and found that all of them were reproducibly associated with mRNAs, proposing that their NAD binding sites could also have evolved to conduct some RNA regulatory functions. Interestingly, the strong prevalence for nucleotide binding sites among the putatively novel RBPs is also in analogy to recent observations suggesting the existence of transcription regulators that are metabolic enzymes [38]. This raises the possibility that both TFs as well as RBPs might function as direct sensors of the metabolic state of the cell suggesting novel circuits for gene regulation. In this scenario, the binding of metabolic cofactors in the reduced or oxidized form (e.g. NAD/NADH+) could differentially regulate the activity of responding RBPs, either through impacting RNA-binding or modulating interaction with other RNP components. Therefore, careful evaluation of the redox state and of the substrate availability will be of further need to decipher the molecular roles of enzymatic RBPs. In addition, modulation of RNA-binding function may result from direct competition between RNA and substrates/cofactors as seen with mammalian IRPs, TS and DHFR [15], [16]. In that case, RNA binding can only occur when substrates are limiting and/or enzymes are in excess and thus, this could possibly contribute to some of the weaker associations seen between RNA and some of the enzyme-related RBPs in our RIP-Chip experiments.

Our screen also proposes RNA binding properties for enzymes that act independently of nucleotides or other cofactors such as peptidases and phosphatases (e.g. Map1, Ymr1). Moreover, our analysis for the enrichment of Pfam domains among our selection of RNA-binders revealed several unexpected domains to be associated with proteins identified in our screen, namely protein-protein interaction domains such as the tetratricopeptide repeat superfamily, which includes the PC_rep and TPR1_domains [23]. We confirmed association of a substantial set of mRNAs with one of the representatives of this family, Sti1p, which contains four TPR1 domains. Although we do not know whether the measured interactions occur directly, it is feasible that some TPR domains could have acquired (or lost) RNA binding functions during evolution: The TPR motif consists of three to 16 tandem-repeats of 35 amino acids that fold into a helix-turn-helix structure and hence, the motif is thought to be closely related to pentatricopeptide repeats (PPR) [39]. The PPR domains rapidly expanded in plants (100–500 genes) where proteins bearing these domains have well established functions in RNA binding, making it reasonable to speculate that some closely related TPR motifs might also have RNA-binding properties.

Nevertheless, the RNA-binding site may also be distinct from the enzymatic site. Diverse examples for shuffling of enzymatic domains next to RNA-binding domains are known such as adenosine deaminases acting on RNA or RNA helicases [40]. Some proteins also retained enzymatic functions in metabolism such as Rib2p in yeast [41]. We analyzed the RNA regulatory potential for Map1p (a methionine aminopeptidase), for which the catalytic domain (peptidase) may be well separated from the RNA-binding sites. The protein contains two Zn-finger domains, which are essential for the normal processing function of MetAP in vivo [33] and were thought to provide interaction with the ribosome [35]. However, Zn-finger domains have been widely seen to mediate protein-DNA or protein-RNA interactions [34] and hence, they may act as RNA-binding motifs in Map1p as well. Howsoever, based on your results it appears that Map1p is a dual function enzyme that can negatively affect the expression of some mRNAs targets, including messages coding for proteins that act in translation – in particular translational elongation – and which are therefore in the same process as Map1p.

Several mammalian metabolic enzymes are thought to control the translation or stability of their own mRNAs [13]. For instance, TS binds with high affinity to its own 5'-UTR near the initiator AUG codon and represses translation [15]. Thereby, mRNA binding sites in TS overlap with the binding sites for its substrates, methylenetetrahydrofolate and dUMP and therefore, mRNA and substrate are in direct competition. Likewise, DHFR, a second enzyme in the thymidylate synthesis pathway also binds to its own mRNA, which can be competed by the substrate (folate) antagonist methotrexate [13], [15]. Four proteins, for which mRNA targets were identified with RIP-Chip bound to their own mRNAs, offering the possibility for auto-regulation (Figure 4). Among these was also Pfk2p, which is the β-subunit of the hetero-octameric phosphofructokinase (PFK) involved in glycolysis. Noteworthy, the specific associations of Pfk2p with its own message are independent of the PFK complex, as neither our protein array nor the RIP-Chip analysis revealed significant associations of RNAs with the other subunit of this complex, termed Pfk1p (TS and APG, unpublished results). Since glycolysis is crucial for cell physiology, the activities of enzymes acting in this pathway must be tightly controlled, which is mainly thought being accomplished by transcription and/or the regulation of protein synthesis or degradation [42]. The binding of Pfk2p to its own message could provide an additional layer of expression regulation by controlling the translation, localization or the stability of the message. Such post-transcriptional feedback regulation could add a sensitive mechanism to adapt PFK levels to changing environmental conditions. We wish to note that self-controlling functions among RBPs generally appear to occur more often than among transcription factors, as about 30–40% of RBPs are associated with their own messages compared to 10% of transcription factors that bind to their own promoters [30]. We speculate that such auto-regulation might be beneficial for RBPs in some specific circumstances to control their expression in a temporal and spatial context with respect to other RBPs, and as a means to fine-tune their levels in the cell for appropriate combinatorial interplay.

In conclusion, various instances of enzymes that also act in RNA-metabolism have been previously reported. Our findings put these specific examples into a more general context indicating that RNA regulation by enzymes may be far more common than previously anticipated. A good fraction of (metabolic) enzymes may therefore have a "moonlighting" role in regulating RNA metabolism, which could allow establishing various direct connections between metabolic status and post-transcriptional gene regulation [12]. Future studies on the regulation of mRNA targets by both enzyme-related and conventional RBPs in yeast and other species will help to further shape the RNA-protein interaction network and its regulatory potential and plasticity, and to further establish novel connections between different layers of cellular control.

9.4 MATERIALS AND METHODS

9.4.1 PLASMIDS, STRAINS AND MEDIA

TAP-tagged strains [29] and the isogenic wild-type strain BY4741 (MATa *his3Δ1 leu2Δ0 met15Δ0 ura3Δ0*), as well as plasmid pBG1805-Map1 [44] were obtained from Open Biosystems. Yeast cells were grown in yeast-peptone-dextrose medium (YPD; 1% yeast extract, 2% bacto-tryptone, 2% glucose) or in synthetic complete medium (SC) [45]. YPGal and YPG are identical to YPD except that they contain 2% galactose or 3% glycerol, respectively, instead of glucose; and SG and SR are identical to SC but contain 2% galactose or 2% raffinose, respectively, instead of glucose. SC-Ura corresponds to SC lacking uracil.

9.4.2 RNA PREPARATION AND LABELING

20 pmol of forward and reverse complementary oligonucleotides encoding the E2Bmin sequence [18] and the T7 RNA polymerase promotor were incubated for one minute at 95°C in 20 µl of water and annealed by cooling down the reaction slowly to room temperature (RT). E2Bmin RNA was synthesized by transcription of annealed DNA templates with T7 RNA polymerase (Promega) for two hours at 37°C. The reactions were treated with DNase I (Roche), and RNA was extracted with phenol/chloroform and precipitated with ethanol. The integrity of the RNA fragment was controlled on a 15% polyacrylamide gel containing 8 M urea. Total RNA was isolated from yeast cells by hot phenol extraction [46]. Total RNA was isolated from cells grown either in YPD, SCGal, SD, YPGal, YPG and combined at the ratio (w/w) 2:2:1:1:1. Messenger RNA was isolated from pooled total RNA with the Oligotex mRNA Mini Kit (Qiagen) according to the manufacturer's protocol. Concentration of RNA was generally assessed by UV-spectrometry with a Nanodrop device (Witeg).

RNA was fluorescently labeled with either Cy3 or Cy5 using the MIC-ROMAX ASAP RNA labeling Kit (Perkin Elmer Cat# MPS544) according to the manufacturer's protocol. Labeled RNA was purified with the RNeasy Micro kit (Qiagen) to remove unincorporated dyes and immediately used for array analysis.

9.4.3 PROTEIN MICROARRAYS AND DATA ANALYSIS

We used commercially available protein microarrays containing duplicate probes of 4,088 yeast proteins and additional control proteins spotted on a modified glass slide (ProtoArray™ Yeast Proteome Microarray mg v.1.0; Invitrogen Cat# PA012106; http://www.invitrogen.com). The frozen arrays were thawed at 4°C for 15 min and blocked for 2 hours at 4°C in phosphate-buffered-saline pH 7.4 (PBS; Invitrogen) supplemented with 1% nuclease/protease-free BSA (Equitech-Bio), 1 mM DTT, 50 µg/ml *E. coli* tRNA (Roche), and 50 µg/ml heparin. The arrays were dried by centrifugation at 300 g for 1 min at 4°C and immediately probed with fluorescently labeled RNAs. Therefore, Cy3 labeled total RNA (5–10 µg) were combined with either Cy5 labeled mRNAs (= poly(A)+ RNA; 2 µg) or E2Bmin RNA (1.5 µg) and mixed in 60 µl of RNA-binding buffer (RBB, 20 mM Tris-HCl pH 7.9, 75 mM NaCl, 2 mM $MgCl_2$, 5% glycerol, 0.05% Triton-X100, 1% BSA, 1 mM DTT, 0.2 mg/ml *E. coli* tRNA, 0.02 mg/ml heparin) supplemented with 6 U of RNaseOUT (Invitrogen, Cat# 10777-019) and applied on the protein microarray, which was covered with a lifterslip (22×60 mm; Erie Scientific). The arrays were put into a sealed hybridization chamber to prevent drying-out, and incubated for 90 min at room temperature in the dark. The slides were washed twice for 10 min at 4°C with 25 ml of RBB buffer supplemented with 10 U/ml RNaseOUT, and twice with 1× RBB buffer lacking tRNA. The arrays were dried by centrifugation at 300 g for 5 min and immediately scanned with an Axon Scanner 4200 (Molecular Devices). Data was collected with GenePix Pro 5.1 (Molecular Devices) and imported into Acuity 4.0, which averages data for duplicated spots (Molecular Devices). For data analysis, we removed features representing non-yeast control proteins (e.g. GST) and spots with irregular shapes (FLAG> = 0). Protein microarray raw data have been deposited at ArrayExpress via http://www.ebi.ac.uk/miamexpress/ (accession number: E-MEXP-2897; see below).

To select proteins that bind E2Bmin RNA, we retrieved median signal intensities of background subtracted signals for the red channel (Cy5) probed with E2Bmin RNA. Proteins, for which the signal intensities were at least 3.5 standard deviations (Z score>3.5) above the median of all averaged signals from replicate arrays, were considered as RNA binders (raw data is given in the Dataset S1). To select proteins that interact with total RNA or mRNAs, we retrieved background subtracted median signal intensities of both channels from five replicate arrays, and calculated percentile ranks from 0 to 1 for each channel and array (raw data is given in Dataset S2). The distribution of the median percentile ranks across array replicates was plotted as a histogram and the trough of the bimodal distribution was taken as a conservative cut-off to select proteins that consistently interacted with RNAs (0.90 for mRNA, and at 0.95 for total RNA)(Figure 2).

9.4.4 RNA AFFINITY ISOLATIONS

Affinity purification of TAP-tagged proteins was carried out as described previously [4], [6], except that yeast cells were broken mechanically with glass beads in a Tissue Lyser (Qiagen) for 12 min at 300 Hz and 4°C. RNAs from the extract (input) and from the affinity isolates were purified with the RNeasy Mini or Micro Kit (Qiagen), respectively.

9.4.5 MAP1 OVEREXPRESSION

100 ml of BY4741 cells bearing plasmid pBG1805-Map1 or the empty plasmid pBG1805 (= control) were grown in SR-Ura media at 30°C to an OD_{600} of 0.45–0.5 and expression was induced with 2% galactose for 1.5 hours. Cells were generally harvested by centrifugation and washed twice with 800 µl of ice-cold sterile water. RNA was isolated by hot-phenol extraction for microarray analysis as described above [46].

9.4.6 DNA MICROARRAYS AND DATA ANALYSIS

70-mer oligo arrays representing features for all annotated nuclear yeast genes (including all ORFs and ncRNAs, introns and some intergenic regions), the mitochondrial genome and various control spots were produced at the Center for Integrative Genomics, University of Lausanne. Arrays were processed and hybridized with fluorescently labeled cDNAs as described previously [47]. For RIP-Chip experiments, 5 µg of total RNA isolated from the extract (input) and up to 50% (~500 ng) of the affinity purified RNA were reverse transcribed in the presence of 5-(3-aminoallyl)-dUTP and natural dNTPs with a mixture of random nonamer and dT(20)V primers, and cDNAs were covalently linked to Cy3 and Cy5 NHS-monoesters (GE HealthSciences Cat# RPN5661), respectively, and competitively hybridized on yeast oligo arrays at 42°C for 14 hours in formamide-based hybridization buffer. Gene expression changes upon MAP1 overexpression were obtained by comparative microarray analysis of Cy3 labeled cDNAs derived from cells expressing the empty vector (pBG1805) and of Cy5 labeled cDNAs from MAP1 (pBG1805-Map1) expressing cells. Microarrays were scanned with an Axon Scanner 4200A (Molecular Devices) and analyzed with GenePix Pro 5.1 (Molecular Devices). Arrays were deposited and computer normalized at the Stanford Microarray Database [48]. All DNA microarray data are available at the Stanford Microarray Database (SMD) or at the Gene Expression Omnibus at www.ncbi.nlm.nih.gov/geo (accession nos. GSE21850 and GSE21864).

Log2 median ratios from three independent RBP affinity isolations and five mock control isolations were retrieved from SMD and exported into Microsoft Excel after filtering for signal over background >1.8 in the channel measuring total RNA derived from the extract. We used the web interface for Cyber-T (http://cybert.microarray.ics.uci.edu/) to employ statistical analyses based on regularized t-tests that use a Bayesian estimate of the variance among gene measurements within an experiment [49]. Features, for which data was obtained in more than 60% of the arrays and that were on average 3-fold enriched with a p-value of less than 0.01 in protein affinity isolates compared to mock controls were considered as potential RNA targets (Dataset S3). For MAP1 overexpression profiling, log2 median ratios from three biological replicates were filtered for regression correlation <0.6 and signal over background >2.0 in both channels (Dataset S4).

9.4.7 MICROARRAY DATA FILES

Protein microarray raw data are available at the ArrayExpress database at http://www.ebi.ac.uk/microarray-as/ae/ (acession no. E-MEXP-2897). DNA microarray raw data are available at the Stanford Microarray Database (SMD) or at the Gene Expression Omnibus at www.ncbi.nlm.nih. gov/geo (accession nos. GSE21850 and GSE21864). Microarray data is compliant with MIAME protocol.

9.4.8 DATABASES AND BIOINFORMATICS

Significantly shared GO terms among the selected proteins from the Protoarray screen were identified with the Generic Gene Ontology (GO) Term Finder at the Lewis-Sigler Institute at Princeton University (release 27-Jan-2009; http://go.princeton.edu/cgi-bin/GOTermFinder, [50]) based on annotations in the *Saccharomyces cerevisiae* Genome Database (SGD). Commonly enriched GO terms among mRNAs associated with selected proteins were retrieved with the GO Term Finder that uses a hypergeometric distribution with Multiple Hypothesis Correction to calculate p-values (SGD; www.yeastgenome.org). Thereby, we used 6,336 features representing ORF probes for which microarray data was obtained as the background gene set, and only terms with p<0.01 (Bonferroni corrected) were considered. Domain annotations for all S. cerevisiae proteins were retrieved from the Pfam database (Pfam 24.0) at http:// pfam.sanger.ac.uk/[23]. Significance for enrichment of Pfam domains among RBPs was calculated based on domain content on the Protoarray by using hypergeometric distribution available from the R package for statistical computing.

9.4.9 EXPRESSION ANALYSIS OF SELECTED RNA-BINDERS ACROSS CONDITIONS

247 microarray datasets (Affymetrix data) in the form of Robust Multi Array (RMA) normalized profiles were retrieved from the M3D database [26] (conditions are indicated in Table S3). K-means clustering was performed across conditions with the Euclidean distance metric and added into 10 groups. To compare the expression level of novel RBPs against previously documented RBPs [6] and non-RBPs, the latter defined as those which do not encode for documented or novel RBPs, we calculated the median expression level of a gene across the conditions in the M3D dataset and compared the populations using Wilcoxon rank-sum test or Mann-Whitney U test available in the R statistical package [28].

REFERENCES

1. Matera AG, Terns RM, Terns MP (2007) Non-coding RNAs: lessons from the small nuclear and small nucleolar RNAs. Nat Rev Mol Cell Biol 8: 209–220. doi: 10.1038/nrm2124.
2. Moore MJ (2005) From birth to death: the complex lives of eukaryotic mRNAs. Science 309: 1514–1518. doi: 10.1126/science.1111443.
3. Maniatis T, Reed R (2002) An extensive network of coupling among gene expression machines. Nature 416: 499–506. doi: 10.1038/416499a.
4. Gerber AP, Herschlag D, Brown PO (2004) Extensive association of functionally and cytotopically related mRNAs with Puf family RNA-binding proteins in yeast. PLoS Biol 2: E79. doi: 10.1371/journal.pbio.0020079.
5. Gerber AP, Luschnig S, Krasnow MA, Brown PO, Herschlag D (2006) Genome-wide identification of mRNAs associated with the translational regulator PUM-ILIO in Drosophila melanogaster. Proc Natl Acad Sci U S A 103: 4487–4492. doi: 10.1073/pnas.0509260103.
6. Hogan DJ, Riordan DP, Gerber AP, Herschlag D, Brown PO (2008) Diverse RNA-binding proteins interact with functionally related sets of RNAs, suggesting an extensive regulatory system. PLoS Biol 6: e255. doi: 10.1371/journal.pbio.0060255.
7. Keene JD (2007) RNA regulons: coordination of post-transcriptional events. Nat Rev Genet 8: 533–543. doi: 10.1038/nrg2111.
8. Halbeisen RE, Galgano A, Scherrer T, Gerber AP (2008) Post-transcriptional gene regulation: from genome-wide studies to principles. Cell Mol Life Sci 65: 798–813. doi: 10.1007/s00018-007-7447-6.

9. Morris AR, Mukherjee N, Keene JD (2010) Systematic analysis of posttranscriptional gene expression. Wiley Interdiscip Rev Syst Biol Med 2: 162–180. doi: 10.1002/wsbm.54.

10. Glisovic T, Bachorik JL, Yong J, Dreyfuss G (2008) RNA-binding proteins and post-transcriptional gene regulation. FEBS Lett 582: 1977–1986. doi: 10.1016/j.febslet.2008.03.004.

11. Anantharaman V, Koonin EV, Aravind L (2002) Comparative genomics and evolution of proteins involved in RNA metabolism. Nucleic Acids Res 30: 1427–1464. doi: 10.1093/nar/30.7.1427.

12. Hentze MW, Preiss T (2010) The REM phase of gene regulation. Trends Biochem Sci 35: 423–426. doi: 10.1016/j.tibs.2010.05.009.

13. Ciesla J (2006) Metabolic enzymes that bind RNA: yet another level of cellular regulatory network? Acta Biochim Pol 53: 11–32.

14. Hentze MW (1994) Enzymes as RNA-binding proteins: a role for (di)nucleotide-binding domains? Trends Biochem Sci 19: 101–103. doi: 10.1016/0968-0004(94)90198-8.

15. Tai N, Schmitz JC, Liu J, Lin X, Bailly M, et al. (2004) Translational autoregulation of thymidylate synthase and dihydrofolate reductase. Front Biosci 9: 2521–2526. doi: 10.2741/1413.

16. Hentze MW, Muckenthaler MU, Andrews NC (2004) Balancing acts: molecular control of mammalian iron metabolism. Cell 117: 285–297. doi: 10.1016/S0092-8674(04)00343-5.

17. Zhu J, Gopinath K, Murali A, Yi G, Hayward SD, et al. (2007) RNA-binding proteins that inhibit RNA virus infection. Proc Natl Acad Sci U S A 104: 3129–3134. doi: 10.1073/pnas.0611617104.

18. Jambhekar A, McDermott K, Sorber K, Shepard KA, Vale RD, et al. (2005) Unbiased selection of localization elements reveals cis-acting determinants of mRNA bud localization in Saccharomyces cerevisiae. Proc Natl Acad Sci U S A 102: 18005–18010. doi: 10.1073/pnas.0509229102.

19. Shepard KA, Gerber AP, Jambhekar A, Takizawa PA, Brown PO, et al. (2003) Widespread cytoplasmic mRNA transport in yeast: identification of 22 bud-localized transcripts using DNA microarray analysis. Proc Natl Acad Sci U S A 100: 11429–11434. doi: 10.1073/pnas.2033246100.

20. Buck MJ, Lieb JD (2004) ChIP-chip: considerations for the design, analysis, and application of genome-wide chromatin immunoprecipitation experiments. Genomics 83: 349–360. doi: 10.1016/j.ygeno.2003.11.004.

21. Hall DA, Zhu H, Zhu X, Royce T, Gerstein M, et al. (2004) Regulation of gene expression by a metabolic enzyme. Science 306: 482–484. doi: 10.1126/science.1096773.

22. Lu C, Zhang Z, Leach L, Kearsey MJ, Luo ZW (2007) Impacts of yeast metabolic network structure on enzyme evolution. Genome Biol 8: 407. doi: 10.1186/gb-2007-8-8-407.

23. Finn RD, Mistry J, Tate J, Coggill P, Heger A, et al. (2010) The Pfam protein families database. Nucleic Acids Res 38: D211–222. doi: 10.1093/nar/gkp985.

24. Lupas A, Baumeister W, Hofmann K (1997) A repetitive sequence in subunits of the 26S proteasome and 20S cyclosome (anaphase-promoting complex). Trends Biochem Sci 22: 195–196. doi: 10.1016/S0968-0004(97)01058-X.

25. D'Andrea LD, Regan L (2003) TPR proteins: the versatile helix. Trends Biochem Sci 28: 655–662. doi: 10.1016/j.tibs.2003.10.007.

26. Faith JJ, Driscoll ME, Fusaro VA, Cosgrove EJ, Hayete B, et al. (2008) Many Microbe Microarrays Database: uniformly normalized Affymetrix compendia with structured experimental metadata. Nucleic Acids Res 36: D866–870. doi: 10.1093/nar/gkm815.

27. Irizarry RA, Bolstad BM, Collin F, Cope LM, Hobbs B, et al. (2003) Summaries of Affymetrix GeneChip probe level data. Nucleic Acids Res 31: e15. doi: 10.1093/nar/gng015.

28. Mittal N, Roy N, Babu MM, Janga SC (2009) Dissecting the expression dynamics of RNA-binding proteins in posttranscriptional regulatory networks. Proc Natl Acad Sci U S A 106: 20300–20305. doi: 10.1073/pnas.0906940106.

29. Ghaemmaghami S, Huh WK, Bower K, Howson RW, Belle A, et al. (2003) Global analysis of protein expression in yeast. Nature 425: 737–741. doi: 10.1038/nature02046.

30. Kanitz A, Gerber AP (2010) Circuitry of mRNA regulation. Wiley Interdiscip Rev Syst Biol Med 2: 245–251. doi: 10.1002/wsbm.55.

31. Chang YH, Teichert U, Smith JA (1992) Molecular cloning, sequencing, deletion, and overexpression of a methionine aminopeptidase gene from Saccharomyces cerevisiae. J Biol Chem 267: 8007–8011.

32. Li X, Chang YH (1995) Amino-terminal protein processing in Saccharomyces cerevisiae is an essential function that requires two distinct methionine aminopeptidases. Proc Natl Acad Sci U S A 92: 12357–12361. doi: 10.1073/pnas.92.26.12357.

33. Vetro JA, Chang YH (2002) Yeast methionine aminopeptidase type 1 is ribosome-associated and requires its N-terminal zinc finger domain for normal function in vivo. J Cell Biochem 85: 678–688. doi: 10.1002/jcb.10161.

34. Brown RS (2005) Zinc finger proteins: getting a grip on RNA. Curr Opin Struct Biol 15: 94–98. doi: 10.1016/j.sbi.2005.01.006.

35. Zuo S, Guo Q, Ling C, Chang YH (1995) Evidence that two zinc fingers in the methionine aminopeptidase from Saccharomyces cerevisiae are important for normal growth. Mol Gen Genet 246: 247–253. doi: 10.1007/BF00294688.

36. Zhu H, Bilgin M, Bangham R, Hall D, Casamayor A, et al. (2001) Global analysis of protein activities using proteome chips. Science 293: 2101–2105. doi: 10.1126/science.1062191

37. Hu S, Xie Z, Onishi A, Yu X, Jiang L, et al. (2009) Profiling the human protein-DNA interactome reveals ERK2 as a transcriptional repressor of interferon signaling. Cell 139: 610–622. doi: 10.1016/j.cell.2009.08.037.

38. Shi Y (2004) Metabolic enzymes and coenzymes in transcription—a direct link between metabolism and transcription? Trends Genet 20: 445–452. doi: 10.1016/j.tig.2004.07.004.

39. Small ID, Peeters N (2000) The PPR motif - a TPR-related motif prevalent in plant organellar proteins. Trends Biochem Sci 25: 46–47. doi: 10.1016/S0968-0004(99)01520-0.

40. Lunde BM, Moore C, Varani G (2007) RNA-binding proteins: modular design for efficient function. Nat Rev Mol Cell Biol 8: 479–490. doi: 10.1038/nrm2178.

41. Behm-Ansmant I, Grosjean H, Massenet S, Motorin Y, Branlant C (2004) Pseu-douridylation at position 32 of mitochondrial and cytoplasmic tRNAs requires two distinct enzymes in Saccharomyces cerevisiae. J Biol Chem 279: 52998–53006. doi: 10.1074/jbc.M409581200.

42. Daran-Lapujade P, Rossell S, van Gulik WM, Luttik MA, de Groot MJ, et al. (2007) The fluxes through glycolytic enzymes in Saccharomyces cerevisiae are predomi-nantly regulated at posttranscriptional levels. Proc Natl Acad Sci U S A 104: 15753–15758. doi: 10.1073/pnas.0707476104.

43. Tsvetanova NG, Klass DM, Salzman J, Brown PO (2010) Proteome-Wide Search Reveals Unexpected RNA-Binding Proteins in Saccharomyces cerevisiae. PLoS One 5: e12671. doi: 10.1371/journal.pone.0012671.

44. Gelperin DM, White MA, Wilkinson ML, Kon Y, Kung LA, et al. (2005) Biochem-ical and genetic analysis of the yeast proteome with a movable ORF collection. Genes Dev 19: 2816–2826. doi: 10.1101/gad.1362105.

45. Sherman F (2002) Getting started with yeast. Methods Enzymol 350: 3–41.

46. Kohrer K, Domdey H (1991) Preparation of high molecular weight RNA. Methods Enzymol 194: 398–405.

47. Halbeisen RE, Scherrer T, Gerber AP (2009) Affinity purification of ribosomes to access the translatome. Methods 48: 306–310. doi: 10.1016/j.ymeth.2009.04.003.

48. Demeter J, Beauheim C, Gollub J, Hernandez-Boussard T, Jin H, et al. (2007) The Stanford Microarray Database: implementation of new analysis tools and open source release of software. Nucleic Acids Res 35: D766–770. doi: 10.1093/nar/gkl1019.

49. Baldi P, Long AD (2001) A Bayesian framework for the analysis of microarray ex-pression data: regularized t -test and statistical inferences of gene changes. Bioinfor-matics 17: 509–519. doi: 10.1093/bioinformatics/17.6.509.

50. Boyle EI, Weng S, Gollub J, Jin H, Botstein D, et al. (2004) GO::TermFinder—open source software for accessing Gene Ontology information and finding significantly enriched Gene Ontology terms associated with a list of genes. Bioinformatics 20: 3710–3715. doi: 10.1093/bioinformatics/bth456.

This article has supplemental information that is not featured in this ver-sion of the text. To view these files, please visit the original version of the article as cited in the beginning of this chapter.

CHAPTER 10

GENETIC LANDSCAPE OF OPEN CHROMATIN IN YEAST

KIBAICK LEE, SANG CHEOL KIM, INKYUNG JUNG, KWONEEL KIM, JUNGMIN SEO, HEUN-SIK LEE, GIREESH K. BOGU, DONGSUP KIM, SANGHYUK LEE, BYUNGWOOK LEE, AND JUNG KYOON CHOI

10.1 INTRODUCTION

The genetic basis of gene expression has been studied in various organisms [1]–[5]. For example, two different strains of *Saccharomyces cerevisiae* (BY and RM) were crossed to produce a number of different genetic recombinants, and their expression levels and genotypes were analyzed [1], [6]. We previously utilized this system to separate the *cis*- and *trans*-components of variation in gene expression [7]. Tirosh et al. [8] profiled nucleosome patterns in the inter-specific hybrids of two yeast species to dissect *cis*- and *trans*-effects on nucleosome positioning. Recently, variations in the binding patterns of transcription factors (TFs) have begun to be studied [9]–[11].

This chapter was originally published under the Creative Commons Attribution License. Lee K, Kim SC, Jung I, Kim K, Seo J, Lee H-S, Bogu GK, Kim D, Lee S, Lee B, and Choi JK. Genetic Landscape of Open Chromatin in Yeast. PLoS Genetics 9,2 (2013). doi:10.1371/journal.pgen.1003229.

Chromatin structure controls the access of a wide spectrum of DNA binding proteins involved in not only transcription but also DNA repair, recombination, and replication. Therefore, open chromatin areas can indicate DNA regions accessible to such regulators and thus have been used to identify regulatory regions or elements in the genome. In addition to the well-known DNaseI hypersensitivity assay, the FAIRE technique has been used to capture open chromatin sites in the genome with the aid of massively parallel sequencing (FAIRE-seq) [12]–[14]. In a recent study, the FAIRE DNA was analyzed by genotyping arrays to identify functional regulatory polymorphisms [15]. FAIRE-seq, however, is capable of providing a quantitative measure of chromatin accessibility along with sequence polymorphisms so that the direct effects of DNA sequences on chromatin accessibility can be examined. For example, it has been shown that SNPs located within open chromatin can influence chromatin accessibility, thus demonstrating that chromatin structure can be a heritable feature [11].

As chromatin is a genetically regulated material, a genetic association approach could be used to understand the genetic architecture of chromatin regulation by examining open chromatin in multiple genetically different individuals. A recent study [16] used this approach for chromatin accessibility across 70 human individuals. Because of the large size of the human genome, open chromatin sites were analyzed only in association with local genetic markers to identify *cis*-associations. Transcription factor binding was shown to be one of the main mechanisms by which DNA polymorphisms affect chromatin structure.

In this work, we took advantage of the compact size and comprehensive annotation of the yeast genome to dissect the entire genetic architecture of chromatin regulation, including both *cis*- and *trans*-associations, to better interpret the functional association of *trans*-acting factors. To this end, we generated open chromatin maps of 100 yeast samples, including the parental strains (BY and RM, two replicates of each) and their descendants [6] by means of the FAIRE-seq technique.

10.2 RESULTS

10.2.1 GENERAL CHARACTERIZATION OF OPEN CHROMATIN REGIONS

Open chromatin peaks were first identified for each sample. We then obtained a total of 7,527 OCRs by combining the peak signals of the 96 genetically different yeast strains. For each OCR, the density of the corresponding peak in each strain was calculated and normalized across the strains. The normalized peak density measures showed high reproducibility (R = 0.95~0.99) between the replicates from different FAIRE batches and sequencing libraries (Figure S1). More than half of the OCRs were located at promoters, and 18.6% and 16.4% of the peaks fell near transcription termination sites and within ORFs, respectively (Figure S2). The OCRs mostly coincided with nucleosome-free regions at promoters or transcription termination sites (Figure S3). Approximately 57% of yeast genes contained an OCR at their promoter, and 40% of replication origins overlapped with 14.3% of the OCRs (Figure S2). The average size of the OCRs in BY and RM was 159 bp, while the average size of the OCRs combined across all the strains was 236 bp (Figure S4).

10.2.2 COMPARISON OF CIS- AND TRANS-VARIATION

We sought to estimate the direct influence of underlying DNA sequences on chromatin configuration by quantitatively comparing sequence-dependent (*cis*) variation and sequence-independent (*trans*) variation in chromatin accessibility. Cis-variation indicates variation in chromatin accessibility among individuals in which the DNA sequences of the given open-chromatin locus are different, while *trans*-variation indicates variation

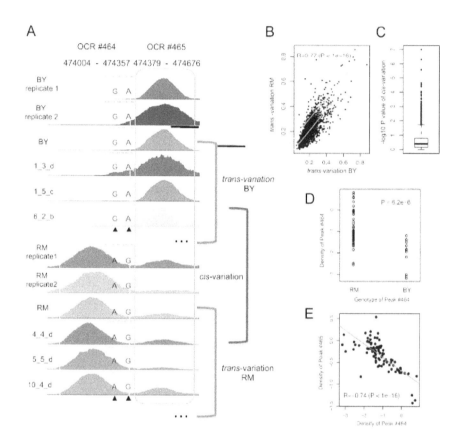

FIGURE 1: Measurement of *trans*- and *cis*-variation. (A) Sequence effects on chromatin regulation. The two peaks (OCR #464 and OCR #465) are shown for strains with the BY genotype and RM genotype, as determined based on the two SNPs found within OCR #464. (B) The two *trans*-variation measures were obtained as illustrated in Figure 1A and compared with each other. (C) The significance of *cis*-variation was measured by the t-test for the 1,738 OCRs. (D) Peak density of OCR #464 as a function of its genotype. (E) Anti-correlation between the peak density of OCR #464 and that of OCR #465 across all yeast strains.

in chromatin accessibility among individuals with an identical geno-type at the given locus. To measure *cis*-variation as the magnitude of chromatin variation caused primarily by *cis*-acting elements residing directly beneath open chromatin, we sought to determine the genotype of each OCR based on the SNP profiles generated from our sequence data. This enabled the classification of OCRs into either BY or RM groups according to each strain's inheritance of the locus (Figure 1A). The *cis*-variation of each OCR was defined as the variance of peak density among the strains with the same genotype at that OCR. The two *cis*-variation measures (each from the BY and RM group) were highly consistent (Figure 1B). Approximately 23% (1,738 OCRs) had more than ten individuals in each group. We assessed the statistical significance of *trans*-variation by considering the within-group variance (*cis*-variation): 11.8% (P<0.05) or 4.8% (P<0.01) of the 1,738 OCRs were called significant (Figure 1C).

10.2.3 CIS- AND TRANS-ASSOCIATIONS IN QTL MAPPING

QTL mapping was performed by interrogating the 7,527 OCRs against the genetic markers selected and processed based on the previous genotype data [6] (see Materials and Methods). A total of 11,048 associations were identified at a false discovery rate (FDR) of 0.01 by our chromatin QTL mapping. Approximately 7.9% of the associations involved *cis*-acting loci within 100 kb (12.66% within 1 Mb), whereas the majority of chromatin traits were linked to *trans*-regulatory loci. The OCRs associated in *trans* tended to display a higher *trans*-variation (P<2×10^{-16}), while those associated in *cis* had a higher *cis*-variation (P = 1.1×10^{-4}), indicating consistency between sequence-based genotyping and microarray-based genotyping. We employed the gene expression data for the 96 strains [6] and carried out expression QTL mapping by repeating the procedures used for the chromatin QTL mapping (see Materials and Methods). At an FDR of 0.01, 12,317 associations between genotypes and expression levels were identified.

10.2.4 CHARACTERIZATION OF CIS-ASSOCIATIONS

We identified a total of 2,234 OCRs in which there was a TF-binding motif that contained a polymorphism and found that these OCRs were twice as likely to be associated in *cis* than other OCRs (P = 4.6×10^{-7}). However, there was no difference with respect to *trans*-association. This implies that the effect of DNA sequence variation on chromatin structure is often manifested through underlying TF-binding motifs independently of *trans*-acting regulators.

To determine whether *cis*-associations can also be explained by differential nucleosome formation, we searched for *cis*-QTL SNPs in the well-known poly A/T tract nucleosome depletion signature. We extracted the reference genome sequences surrounding the SNP locations within the OCRs from our FAIRE-seq data and then looked for the presence of a poly A/T tract. Even with a very loose threshold (five consecutive A/Ts), we could only identify five such instances. This is contradictory to the major role of the AT-rich sequences in the divergence of nucleosome positioning between different species [8]. We propose that poly A/T tracts residing in open chromatin may be under strong selective pressure and thus resistant to sequence changes because of their importance in regulatory function.

Because the *cis*-associations between DNA sequences and chromatin accessibility are likely to be mediated by TF binding, a sequence polymorphism that affects chromatin accessibility in *cis* should also affect gene expression in the neighborhood. Indeed, a sizeable fraction (45%) of the chromatin-associated SNPs were associated with the expression of nearby genes. By contrast, only 15% of the expression-associated SNPs turned out to influence the accessibility of nearby chromatin, indicating that there are mechanisms by which sequence polymorphisms can affect the expression of nearby genes without affecting chromatin accessibility.

Reciprocal regulation of two chromatin loci by DNA sequences could be observed in OCR #464 and OCR #465. These two OCRs were associated with multiple *cis*-markers encompassing 100 kb upstream to 15 kb downstream of the loci. Sequence analysis detected two underlying SNPs that were associated with the peak density of OCR #464 (Figure 1A and 1D). Interestingly, the density of the adjacent peak (OCR #465) was negatively correlated with that of OCR #464 across the strains (Figure 1E),

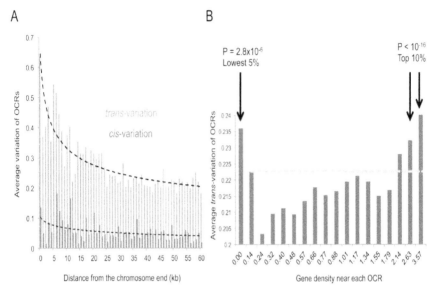

FIGURE 2: The magnitude of *trans*- and *cis*- variation and the number of genes within 50 kb upstream and downstream of the peak boundaries. (A) The magnitude of *trans*- and *cis*-variation as a function of the distance from chromosome ends. The average variation of OCRs within 2 kb windows was plotted for 1 kb bins. The *trans*-variations within 10 kb of the chromosome ends were significantly higher than those farther away ($P<6.6\times10^{-25}$). For *cis*-variation, the P value was 5×10^{-4} when the t-test was used. (B) The number of genes falling within 50 kb upstream and 50 kb downstream of the peak boundaries was obtained for each OCR. This number was divided by the size of the peak for the normalized gene density. Gene-rich OCRs (top 10%) and gene-poor OCRs (lowest 5%) were compared with the other OCRs by the t-test.

demonstrating a reciprocal regulation of the two chromatin loci. In line with our sequencing-based genotypes, all the *cis*-markers indicated that the RM genotype increases the peak density of OCR #464.

10.2.5 CHARACTERIZATION OF TRANS-ASSOCIATIONS

The sum of *trans*-variation in the *trans*-associated OCRs was divided by the sum of *trans*-variation across all the OCRs, revealing that 45.2% of the total *trans*-variation across the OCRs could be explained by genetic factors. To examine how much of the *trans*-variation of each OCR is explained by *trans*-acting genetic factors, we computed the explanatory power of the linear regression (R^2) for each OCR and its associated *trans*-loci. The average R^2 of the *trans*-associated OCRs was 33%. Enrichment of high *trans*-variation OCRs was observed in the vicinity of telomeres (Figure 2A and green marks in Figure S5). This pattern was not observed for *cis*-variation (Figure 2A and Figure S5). High *trans*-variation OCRs also coincided with gene-rich regions (Figure 2B and blue ticks in Figure S5) and gene-poor regions (Figure 2B and light-blue ticks in Figure S5).

Approximately 50% of chromatin QTLs were gene expression QTLs and vice versa, indicating that the *trans*-associations we identified are technically robust and biologically meaningful. However, only 17.6% of these dual QTLs were associated with chromatin and expression traits at the same locus. In other words, many of the dual QTLs were responsible for chromatin traits and gene expression traits that are distantly located (e.g., in different chromosomes). It is possible that regulatory SNPs affect chromatin accessibility for DNA regulation other than *trans*cription (e.g., DNA repair, recombination, etc.), which in turn leads to secondary gene expression changes, and that regulatory loci affect the expression of downstream regulators in *trans*, which in turn causes secondary changes in the accessibility of the target chromatin regions.

We examined the number of *trans*-linkages for each OCR. Most OCRs were responsive to a small number of regulatory loci. Only a few (6.8%) had more than five linkages with the average number being three times lower than for gene expression traits (2.1 versus 5.9) (Figure 3A). This implies that chromatin traits are rather specifically governed by a handful of *trans*-regulators, whereas gene expression processes are responsive to more regulatory inputs. An opposite trend was observed for regulatory loci (Figure 3B). There were regulatory loci responsible for an extremely large number of chromatin traits, with a few cases in which >200 OCRs were linked to a single promiscuous chromatin QTL (Figure 3B). The horizontal dots observed in the chromatin association map (Figure 3C) illustrate 'extensive' regulation by chromatin regulatory loci (Figure 3D), as opposed to the 'intensive' regulation of gene expression traits (Figure 3E).

To investigate the multi-target chromatin regulatory loci, or hotspot QTLs, we first selected those with >65 *trans*-associated OCRs. We annotated each locus by searching for known DNA or chromatin regulators flanking the marker within 10 kb [17] and merged the adjacent markers

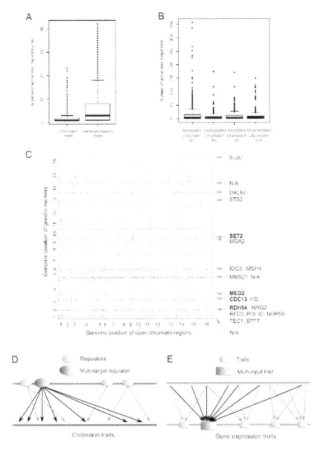

FIGURE 3: Characterization of *trans*-associations. (A) The number of *trans*-regulatory loci associated with each chromatin trait (left) and gene expression trait (right). (B) The number of target traits of each *trans*-regulatory locus was examined for chromatin QTLs and expression QTLs. Annotated QTLs were defined as having at least one known regulator in the vicinity. (C) In this chromatin association map, each dot indicates a linkage between a genetic marker (QTL; y axis) and a trait (OCR; x axis); red or blue indicates that the BY or RM genotype positively regulates the OCR, respectively. The annotation of the 17 QTL hotspots is shown on the right side. The names of the regulators associated with the same genetic marker are separated by a semicolon and those associated with closely located markers by a dot. N/A denotes an unannotated QTL. (D–E) Different regulation architectures of chromatin traits (D) and gene expression traits (E). On the regulator side, most chromatin regulatory loci are responsible for a few traits; however, certain regulatory loci can have upwards of 100 targets. On the target side, individual chromatin traits are usually targeted by less than five loci. The average number of associated loci is three times higher for gene expression traits than for chromatin traits, an indication that the *trans*cription process is responsive to more regulatory inputs or stimuli.

covering the same regulator. A total of 32 initial hotspot loci were merged into 17 hotspots, 14 of which flanked at least one known regulator (master regulators listed in Figure 3C). The annotated (regulator-containing) loci tended to influence more chromatin traits than the unannotated loci (P = 5×10^{-4}) (Figure 3B). By contrast, no enrichment of known regulators near multi-target expression regulatory loci was observed (Figure 3B).

Among the master regulators (Figure 3C) were three TFs with se-quence-specific DNA binding activity: *DAL82, TEC1*, and *NRG2*. Posi-tion weight matrices were available for the DNA-binding motif of Dal82p and Tec1p. Remarkably, 62% of the 71 DAL82-associated OCRs con-tained the Dal82p-binding motif. However, no Tec1p-binding motif en-richment was observed in the associated OCRs. The influence of Tec1p might be exerted not through direct binding but via interaction with other factors under normal growth conditions. Data for Nrg2p binding sites are not available. *SET2* and *MED2* are involved in the transcription of many genes in a non-sequence-specific manner. Set2p is a histone methyltrans-ferase that plays a role in general transcription elongation, and Med2p is a subunit of the mediator complex that forms the RNA polymerase II ho-loenzyme. Their target OCRs were identified in gene-rich regions (Figure 4A and Figure S6).

Rdh54p is a Swi2/Snf2-like factor that plays a role in recombinational repair of DNA double-strand breaks (DSBs) during mitosis and meiosis by interacting with Rad51p and Rad54p [18]–[20]. DSBs occurring at recom-bination hotspots in yeast are found near open chromatin [21]. We employed a measure of "recombination hotness" that was globally obtained based on DSB distribution [22]. The *RDH54* OCRs showed the highest recombination hotness among the master regulators (Figure 4B), with a P value of 9×10^{-25} (Figure S7), and tended to fall near the recombination hotspots (Figure 4C). Cdc13p is a multi-functional telomere-binding protein that participates in telomere replication and maintenance especially by mediating telomer-ase access to telomeric chromatin [23]–[25]. Among the hotspot loci, the *CDC13* locus had the largest number of associated OCRs in close proximity to telomeres (seven OCRs within 1 kb from telomeres). The enrichment of *CDC13*-associated OCRs near telomeres is shown in Figure 4D. Telomeres are associated with recombination coldspots [22]. Indeed, the recombination hotness of the *CDC13* OCRs was very low (Figure 4B).

10.3 DISCUSSION

In this work, we sought to dissect the genetic architecture of chromatin regulation. The multi-target regulatory structure reflects the wide-ranging nature of certain chromatin regulators. Surprisingly, however, many chromatin QTLs were found to govern only a few target traits. It is conceivable that the chromatin structures at particular loci are not susceptible to genetic perturbations or that the technical limitations of our method for detecting subtle changes in chromatin traits may prevent the identification of weakly associated targets. In this case, there may be numerous potential regulatory targets that have not passed our statistical threshold.

On the other hand, the chromatin traits that were responsive to certain genetic perturbations had only a few regulatory inputs, in contrast to the high responsiveness of gene expression traits to multiple regulatory signals. Therefore, chromatin states alone may not be sufficient to explain the precise level of transcription. Once upstream regulators set the stage by priming the chromatin structure, various downstream regulatory inputs may add additional layers of complexity to gene expression control. This is also reflected in the lack of common targets between chromatin QTLs and expression QTLs. Only 18% of the dual QTLs (i.e., SNPs that are both chromatin QTLs and expression QTLs) were associated with chromatin accessibility and gene expression at the same locus simultaneously. However, the identification of many dual QTLs was encouraging itself because it suggests that the detected QTLs are likely to contain functional regulators. We successfully annotated chromatin QTLs, particularly those responsible for a large number of target chromatin traits. The identification of functionally relevant *trans*-regulators from expression QTL mapping has been reported to be difficult [26].

Sequence-specific TF binding appears to be very important in *cis*-associations. We observed an enrichment of *cis*-associations for TF-motif-containing OCRs and common QTLs linking chromatin accessibility and nearby gene expression. This is consistent with the finding that human SNPs associated with chromatin in *cis* are frequently found in TF-binding sites [16]. Moreover, consistency in allele frequencies were observed between the sequence reads for open chromatin and those for TF binding.

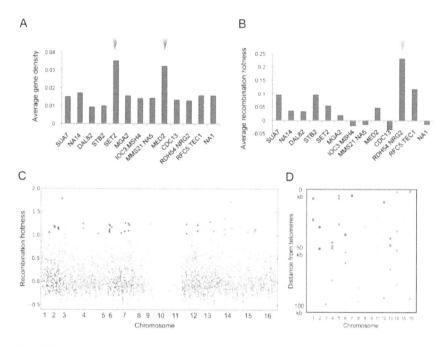

FIGURE 4: Functional analysis of the OCRs of multi-target regulators. (A–B) The average (A) gene density and (B) recombination hotness score ($\log_2$ ratio) for the OCRs associated with the multi-target regulators listed in Figure 3C. Unannotated QTLs were denoted as NA concatenated with the chromosome number (e.g., NA14 is on chromosome XIV). (C) Each spot corresponds to a genomic locus having a score for recombination hotness. Loci with a hotness score >1 located near the RDH54-associated OCRs are highlighted. (D) The dots indicate the OCRs of the multi-target regulators. The *CDC13* OCRs are colored according to the chromosome they belong to. The *CDC13* OCRs within 50 kb of telomeres are highlighted.

In contrast to the previous study [16] in which only *cis*-regulation was thoroughly examined, here we took advantage of the compact size and comprehensive annotation of the yeast genome to dissect the architecture of *trans*-regulatory mechanisms as well. In conclusion, our work provides insight into the genetic basis of chromatin regulation and its relationship with *trans*cription control. Genetic variation in open chromatin in the human genome can underlie disease phenotypes, and thus, the current study has medical implications. For example, previous studies [13], [15] identified regulatory polymorphisms in open chromatin that were previously linked through genome-wide association studies with diabetes and HDL cholesterol levels.

10.4 MATERIALS AND METHODS

10.4.1 IDENTIFICATION OF OCRS AND ESTIMATION OF PEAK DENSITY

We obtained the BY-RM cross strains from the original authors [1], [6]. FAIRE experiments were performed based on the published protocol [12]. We selected 94 yeast segregants and subjected them and the BY and RM strains to 100-bp sequencing on Illumina HiSeq2000. To identify the FAIRE-seq read peaks, we ran F-Seq [27] as previously suggested for FAIRE-seq data analysis [13]. Small-sized peaks (<15 bp) were extended in both directions such that all the peaks were at least15 bp long. To identify all possible OCRs, we combined the extended peaks of the 96 yeast strains (exclusive of the replicates) and merged overlapping peaks into a single peak using BEDTools [28], resulting in 7,527 unique OCRs. The number of FAIRE-seq reads that mapped uniquely to each OCR was counted in each yeast strain. The read count of each OCR was normalized by taking into account the size of the peak and the total number of tags produced from each FAIRE library as $\log_2(\#tags_{OCR}/size_{OCR})/(\#tags_{genome}/size_{genome})$. After the log2 *trans*formation, the negative values were set to zero (ceiling). This normalization scheme was used in our previous work [29]. We further normalized the final matrix of the 7,527 OCRs and 96 strains by scaling the 96 sample vectors to zero mean and unit variance. To assess reproducibility, the FAIRE-seq reads of the parental replicates were mapped to the predefined OCRs and the same normalization scheme was repeated for the four independent samples.

10.4.2 GENOTYPING OF THE OCRS AND ESTIMATION OF TRANS- AND CIS-VARIATION

SNPs were detected from the FAIRE-seq reads using the Illumina's CASAVA suite. SNP calls with fewer than five reads were discarded. For heterogeneous calls, only the major polymorphism with a certain frequency (>80%) was taken. The genotype of each OCR was determined based on its SNP profile. The OCR in the given strain was considered to have inherited the BY (or RM) allele if its genotype perfectly matched with the genotype of the OCR in the BY (or RM) strain. For genotyping at a less stringent threshold, the OCRs whose SNP profile matched with either the BY or RM profile for >50% of the SNPs were also classified as BY or RM. To compute *trans*-variation, the standard deviations of the normalized peak density measures within the BY and RM groups was measured. We identified a total of 1,738 OCRs for which at least ten individuals inherited either a BY or RM allele; we then re-grouped the yeast strains according to the genotype of the given OCR. To assess the statistical significance of *cis*-variation, we used the two-sample t test to measure the difference in the means of the BY and RM groups.

10.4.3 CHROMATIN QTL MAPPING AND EXPRESSION QTL MAPPING

The genetic markers from the original study [6] were used for QTL mapping. As suggested by Lee et al. [17], adjacent markers with no more than two genotypic mismatches across the 96 strains were merged into one average genotype profile, resulting in 1,533 markers. As suggested previously [17], we identified the genes located within 10 kb upstream or downstream of the genomic region covered by the merged genetic marker. To identify potential regulators, we used Gene Ontology to identify 495 genes involved in "DNA binding", and 508 genes known to be involved in *trans*cription and chromatin regulation, resulting in a total of 752 unique genes. For QTL mapping, we measured associations by means of the correlation coefficient or the linear regression between the genotypes represented as a categorical variable (0: RM, 0.5: missing, 1: BY) and the chromatin traits represented as the normalized peak-density measure. False discovery rates (FDRs) were computed based on the permutation test, as follows. The matrix of peak density was shuffled by resampling the sample vectors (yeast strains) to generate B randomized matrices, .b = 1,...B. The P value was determined by comparing the observed association with the expected associations $\hat{r}^b$ from the permuted data as:

$$P = \frac{[1 + \sum_{b=1}^{B} I(|\hat{r}^b| \geq |\hat{r}|)]}{B + 1}$$

where I is an interpretation function. B=1000 was used. The P values were adjusted for multiple testing to yield FDRs, as suggested by Benjamini and Hochberg [30]. An FDR of 0.01 was used. A distance of 100 kb between the marker and the trait was used to differentiate *cis*- and *trans*-associations. We employed the gene expression data for the 96 strains [6] and performed expression QTL mapping by repeating the same procedures.

10.4.4 ANALYSIS OF CHROMATIN QTL MAPPING RESULTS

A total of 11,048 marker-trait associations involving 3,522 OCRs were identified at an FDR of 0.01 when the correlation coefficient was used. To evaluate the consistency between the FDR-based non-parametric approach and the parametric method, we obtained a P value for each marker-trait pair based on the linear regression. At parametric P values$<10^{-3}$ and $<10-5$, 91.1% and 81.1% of the identified associations were called significant, respectively. Adjacent genetic markers (<10 kb) associated with a common trait in QTL mapping were combined. Trans-loci were examined to determine whether the corresponding genetic marker covered at least one of the 752 regulators. According to this criterion, all *trans*-loci were classified into annotated loci or unannotated loci. We defined hotspot chromatin loci as having more than 65 genetic linkages. Adjacent genetic markers covering the same regulator were manually merged.

10.4.5 ADDITIONAL DATA ANALYSIS

To calculate the density of the genes surrounding each OCR, the number of genes located within 50 kb upstream and 50 kb downstream of the OCR peak boundaries was determined. This number was divided by the size of the peak for normalization. The microarray data for the recombination hotspots of the yeast genome [22] were downloaded from http://derisilab. ucsf.edu/hotspots/. The cy5/cy3 ratios from seven ORF arrays were averaged and log2 *trans*formed. The positions of replication origins were downloaded from http://cerevisiae.oridb.org. For TF motif analysis, we used position weight matrices [31] based on in vivo binding assays by chromatin immunoprecipitation for 203 yeast TFs [32] and another set of position weight matrices based on systematic in vitro assays of 112 yeast TFs [33]. TF motifs occurring in OCRs were identified by means of the HOMER package [34] using the two position weight matrix sets.

10.4.6 DATA AVAILABILITY

The FAIRE-seq data for the 96 yeast strains are available at the GEO database with accession number GSE33466.

The following link has been created to allow review of the record GSE33466: http://www.ncbi.nlm.nih.gov/geo/query/acc.cgi?token=zvyzn qwickewmto&acc=GSE33466

REFERENCES

1. Brem RB, Yvert G, Clinton R, Kruglyak L (2002) Genetic dissection of *trans*criptional regulation in budding yeast. Science 296: 752–755. doi: 10.1126/science.1069516.
2. Schadt EE, Monks SA, Drake TA, Lusis AJ, Che N, et al. (2003) Genetics of gene expression surveyed in maize, mouse and man. Nature 422: 297–302. doi: 10.1038/nature01434.
3. Morley M, Molony CM, Weber T, Devlin JL, Ewens KG, et al. (2004) Genetic analysis of genome-wide variation in human gene expression. Nature 430: 743–747. doi: 10.1038/nature02797.

4. Brem RB, Storey JD, Whittle J, Kruglyak L (2005) Genetic interactions between polymorphisms that affect gene expression in yeast. Nature 436: 701–703. doi: 10.1038/nature03865.

5. Cheung VG, Spielman RS, Ewens KG, Weber TM, Morley M, et al. (2005) Mapping determinants of human gene expression by regional and genome-wide association. Nature 437: 1365–1369. doi: 10.1038/nature04244.

6. Brem RB, Kruglyak L (2005) The landscape of genetic complexity across 5,700 gene expression traits in yeast. Proc Natl Acad Sci 102: 1572–1577. doi: 10.1073/pnas.0408709102.

7. Choi JK, Kim YJ (2008) Epigenetic regulation and the variability of gene expression. Nat Genet 40: 141–147. doi: 10.1038/ng.2007.58.

8. Tirosh I, Sigal N, Barkai N (2010) Divergence of nucleosome positioning between two closely related yeast species: genetic basis and functional consequences. Mol Syst Biol 6: 365. doi: 10.1038/msb.2010.20.

9. Zheng W, Zhao H, Mancera E, Steinmetz LM, Snyder M (2010) Genetic analysis of variation in *trans*cription factor binding in yeast. Nature 464: 1187–1191. doi: 10.1038/nature08934.

10. Kasowski M, Grubert F, Heffelfinger C, Hariharan M, Asabere A, et al. (2010) Variation in *trans*cription factor binding among humans. Science 328: 232–235. doi: 10.1126/science.1183621.

11. McDaniell R, Lee B-K, Song L, Liu Z, Boyle AP, et al. (2010) Heritable individual-specific and allele-specific chromatin signatures in humans. Science 328: 235–239. doi: 10.1126/science.1184655.

12. Giresi PG, Kim J, McDaniell RM, Iyer VR, Lieb JD (2007) FAIRE (Formaldehyde-Assisted Isolation of Regulatory Elements) isolates active regulatory elements from human chromatin. Genome Res 17: 877–885. doi: 10.1101/gr.5533506.

13. Gaulton KJ, Nammo T, Pasquali L, Simon JM, Giresi PG, et al. (2010) A map of open chromatin in human pancreatic islets. Nat Genet 42: 255–259. doi: 10.1038/ng.530.

14. Song L, Zhang Z, Grasfeder LL, Boyle AP, Giresi PG, et al. (2011) Open chromatin defined by DNaseI and FAIRE identifies regulatory elements that shape cell-type identity. Genome Res 21: 1757–1767. doi: 10.1101/gr.121541.111.

15. Smith AJP, Howard P, Shah S, Eriksson P, Stender S, et al. (2012) Use of allele-specific FAIRE to determine functional regulatory polymorphism using large-scale genotyping arrays. PLoS Genet 8. e1002908 doi:10.1371/journal.pgen.1002908.

16. Degner JF, Pai AA, Pique-Regi R, Veyrieras J-B, Gaffney DJ, et al. (2012) DNase I sensitivity QTLs are a major determinant of human expression variation. Nature 482: 390–394. doi: 10.1038/nature10808.

17. Lee S-I, Pe'er D, Dudley AM, Church GM, Koller D (2006) Identifying regulatory mechanisms using individual variation reveals key role for chromatin modification. Proc Natl Acad Sci 103: 14062–14067. doi: 10.1073/pnas.0601852103.

18. Klein HL (1997) RDH54, a RAD54 homologue in *Saccharomyces cerevisiae*,is required for mitotic diploid-specific recombination and repair and for meiosis. Genetics 147: 1533–1543.

19. Petukhova G, Sung P, Klein H (2000) Promotion of Rad51-dependent D-loop formation by yeast recombination factor Rdh54/Tid1. Genes & Dev 14: 2206–2215. doi: 10.1101/gad.826100.

20. Shah PP, Zheng X, Epshtein A, Carey JN, Bishop DK, et al. (2010) Swi2/Snf2-related *trans*locases prevent accumulation of toxic Rad51 complexes during mitotic growth. Mol Cell 39: 862–872. doi: 10.1016/j.molcel.2010.08.028.

21. Wu TC, Lichten M (1994) Meiosis-induced double-strand break sites determined by yeast chromatin structure. Science 263: 515–518. doi: 10.1126/science.8290959.

22. Gerton JL, DeRisi J, Shroff R, Lichten M, Brown PO, et al. (2000) Global mapping of meiotic recombination hotspots and coldspots in the yeast *Saccharomyces cerevisiae*. Proc Natl Acad Sci 97: 11383–11390. doi: 10.1073/pnas.97.21.11383.

23. Nugent CI, Hughes TR, Lue NF, Lundblad V (1996) Cdc13p: a single-strand telomeric DNA-binding protein with a dual role in yeast telomere maintenance. Science 274: 249–252. doi: 10.1126/science.274.5285.249.

24. Evans SK, Lundblad V (1999) Est1 and Cdc13 as comediators of telomerase access. Science 286: 117–120. doi: 10.1126/science.286.5437.117.

25. Lustig AJ (2001) Cdc13 subcomplexes regulate multiple telomere functions. Nat Struct Biol 8: 297–299. doi: 10.1038/86157.

26. Yvert G, Brem RB, Whittle J, Akey JM, Foss E, et al. (2003) Trans-acting regulatory variation in *Saccharomyces cerevisiae* and the role of *trans*cription factors. Nat Genet 35: 57–64. doi: 10.1038/ng1222.

27. Boyle AP, Guinney J, Crawford GE, Furey TS (2008) F-Seq: a feature density estimator for high-throughput sequence tags. Bioinformatics 24: 2537–2538. doi: 10.1093/bioinformatics/btn480.

28. Quinlan AR, Hall IM (2010) BEDTools: a flexible suite of utilities for comparing genomic features. Bioinformatics 26: 841–842. doi: 10.1093/bioinformatics/btq033.

29. Choi JK (2010) Contrasting chromatin organization of CpG islands and exons in the human genome. Genome Biol 11: R70. doi: 10.1186/gb-2010-11-7-r70.

30. Benjamini Y, Hochberg Y (1995) Controlling the false discovery rate: a practical and powerful approach to multiple testing. J Roy Statist Soc Ser B 57: 289–300.

31. MacIsaac KD, Wang T, Gordon DB, Gifford DK, Stormo GD, et al. (2006) An improved map of conserved regulatory sites for *Saccharomyces cerevisiae*. BMC Bioinformatics 7: 113. doi: 10.1186/1471-2105-7-113.

32. Harbison CT, Gordon DB, Lee TI, Rinaldi NJ, MacIsaac KD, et al. (2004) Transcriptional regulatory code of a eukaryotic genome. Nature 431: 99–104. doi: 10.1038/nature02800.

33. Badis G, Chan ET, Bakel Hv, Pena-Castillo L, Tillo D, et al. (2008) A Library of Yeast Transcription Factor Motifs Reveals a Widespread Function for Rsc3 in Targeting Nucleosome Exclusion at Promoters. Mol Cell 32: 878–887. doi: 10.1016/j.molcel.2008.11.020.

34. Heinz S, Benner C, Spann N, Bertolino E, Lin YC, et al. (2010) Simple Combinations of Lineage-Determining Transcription Factors Prime *cis*-Regulatory Elements Required for Macrophage and B Cell Identities. Mol Cell 38: 576–589. doi: 10.1016/j.molcel.2010.05.004.

This article has supplemental information that is not featured in this version of the text. To view these files, please visit the original version of the article as cited in the beginning of this chapter.

PART IV

COMPARATIVE GENOMICS

CHAPTER 11

HIGH QUALITY *DE NOVO* SEQUENCING AND ASSEMBLY OF THE *SACCHAROMYCES ARBORICOLUS* GENOME

GIANNI LITI, ALEX N. NGUYEN BA, MARTIN BLYTHE, CAROLIN A. MbLLER, ANDERS BERGSTRLIM, FRANCISCO A. CUBILLOS, FELIX DAFHNIS-CALAS, SHIMA KHOSHRAFTAR, SUNIR MALLA, NEEL MEHTA, CHEUK C. SIOW, JONAS WARRINGER, ALAN M. MOSES, EDWARD J. LOUIS, AND CONRAD A. NIEDUSZYNSKI

11.1 BACKGROUND

The budding yeast, *Saccharomyces cerevisiae*, is a leading system in genomics due to the small genome size (12 Mb) and the availability of powerful genetic techniques. Genome sequencing of multiple hemiascomycete yeasts and multiple individuals from several species have allowed the application of a range of powerful comparative approaches. Comparative genomics have revealed evolutionary mechanisms that shape genomes and provided a formidable tool for assigning function to DNA sequence [1,2].

This chapter was originally published under the Creative Commons Attribution License. Liti G, Ba ANN, Blythe M, Müller CA, Bergström A, Cubillos FA, Dafhnis-Calas F, Khoshraftar S, Malla S, Mehta N, Siow CC, Warringer J,Moses AM, Louis EJ, and Nieduszynski CA. High Quality de novo Sequencing and Assembly of the Saccharomyces arboricolus Genome. BMC Genomics **14**,69 (2013). doi:10.1186/1471-2164-14-69.

The closely related sensu stricto *Saccharomyces* species (*S. cerevisiae, S. paradoxus, S. mikatae, S. kudriavzevii, S. arboricolus* and *S. bayanus*) provide a clade with multiple genetically tractable species [3]. The genome sequence of several sensu stricto species [4,5] revealed a level of nucleotide divergence comparable to that between humans and birds yet a level of structural variation comparable to that between humans and chimps [6]. Comparisons of genome structures have provided insight into mechanisms of genome evolution and speciation. For example, the presence of a limited number of genomic rearrangements that are not consistent with the phylogeny, provide strong evidence against the chromosomal speciation model [7].

Sequence comparisons between the sensu stricto species have allowed improved genome annotation [4,5]. Sequence conservation allowed the identification of additional small open reading frames and the refinement of translation start and stop positions. Lack of sequence conservation resulted in the elimination of spurious open reading frames. Combining experimental data for protein binding sites with sequence conservation allowed the identification of functional DNA sequences [8,9]. The power of these and other comparative genomic approaches [10] rely upon the number of species sequenced, the evolutionary divergence of the selected species and the quality of the assembled genome sequence.

Recently the yeast *Saccharomyces arboricolus* was isolated from the bark of the Fagaceae tree in China [11]. The *S. arboricolus* karyotype is consistent with the other sensu stricto species in terms of chromosome number and size. Sequence information (limited to a portion of the rDNA) unambiguously grouped this species within the sensu stricto complex. *S. arboricolus* can form viable hybrids with the other sensu stricto species but resulting gametes are not viable [12]. Together these data demonstrate that *S. arboricolus* is a novel sensu stricto species.

Here, we report high-quality sequence and assembly of the *S. arboricolus* genome (type strain H-6T; CBS 10644TT) by combining two deep sequencing platforms. We report chromosome size scaffolds, genome annotation and synteny analysis. Genome wide phylogenetic analysis places *S. arboricolus* between *S. bayanus* and *S. kudriavzevii* in the sensu stricto phylogenetic tree. Finally, we considered the phenotypic profile of *S.*

arboricolus under multiple environmental conditions in the light of its gene content and phylogeny.

11.2 RESULTS

11.2.1 GENOME SEQUENCE AND ASSEMBLY

We generated a high quality genome assembly for *S. arboricolus* using a combination of high-throughput sequencing platforms and strategies (Table 1). First, we generated single-end reads using the Roche 454 pyrosequencing platform. This gave long reads that facilitated assembly. Second, we used Roche 454 paired-end reads, with ~8 kb insert size, to join contigs into chromosome size scaffolds (combined Roche 454 sequence coverage ~49X). We anticipated that the large insert size of the paired-end library would be sufficient to span any repeat elements (e.g. full length single Ty elements). Finally, we used 50 bp reads from SOLiD (Life Technologies) sequencing (~100X sequence coverage) to correct homopolymer errors present in the Roche 454 sequence. This combination strategy resulted in high quality sequence with chromosome-sized scaffolds.

TABLE 1: Deep sequencing metrics

Library	Reads	Mapped reads	Mean mapped read length (bp)
Roche 454 Fragment	734,353	726,488	360
Roche 454 8 kb Paired	1,711,390	1,520,755	200
Life Technologies SOLiD	31,316,590	21,753,029	50

De novo assembly of the Roche 454 (fragment and pair-end) reads was performed using the Newbler algorithm (see Methods). This resulted in 290 contigs (≥500 bp; N50 117,280 bp) that were joined using the Roche 454 paired-end reads to give 35 scaffolds. There are 17 scaffolds that are comprised of a single contig (2024 - 5644 bp) and one scaffold comprising of two contigs (9948 bp; Additional file 1: Table S1). The remaining 17 scaffolds account for >99% of the assembly and are between 72 and 1246 kb long. The total base count of the assembly, 11.6 Mb, is comparable to the physical genome size predicted from the karyotype [11] and is similar to the completed *S. cerevisiae* genome sequence (12.1 Mb) and genome sequence of other sensu stricto yeasts (11.6 - 11.9 Mb) [3,13].

Pyrosequencing suffers from an inherent difficulty in determining the number of incorporated nucleotides in homopolymer regions, due to the non-linear signal from the incorporation of >5 identical nucleotides [14]. Comparing our *S. arboricolus* genome sequence to *S. cerevisiae*, identified >700 open reading frames (ORFs) with putative frame-shifts (Figure 1A). These putative frame-shifts are predominantly in homopolymer runs and are therefore likely due to errors in the pyrosequencing (Figure 1). Indeed the *S. paradoxus* assembly [15], which was based on Sanger sequence reads (that do not suffer from homopolymer errors), shows less than half the number of frame-shifts than *S. arboricolus* (Figure 1A). We further analyzed the homopolymeric runs that cause frame-shifts and found that they tended to be longer and more A-biased than the corresponding frame-shifts in *S. paradoxus* (Figure 1B and C). To overcome this problem we used SOLiD sequencing that relies on a different chemistry and is not sub-ject to the same error. We used the Roche 454 *de novo* assembly to map the SOLiD reads, identify errors and then correct the assembly. This mapping, errors calling and correction process was then repeated a further 4 times. In total we corrected 121 single base substitutions and 1682 indels. This resulted in a dramatic reduction in the number of putative frame-shifts to levels comparable to that seen with conventional Sanger sequencing (as represented by the *S. paradoxus* genome, Figure 1A). The corrected assembly also improved the distribution of frame-shifts such that the over-abundance of long homopolymeric runs and the A-bias were greatly re-duced (Figure 1B and C).

The coverage of mapped SOLiD reads gives a measure of sequence copy number and can be used to reveal repeat regions that have collapsed during assembly. Overall we find a scarcity of high-coverage regions (Ad-ditional file 2: Figure S1), implying that there has been very little collapse of repeat regions during assembly. Short (<100 bp) regions of elevated

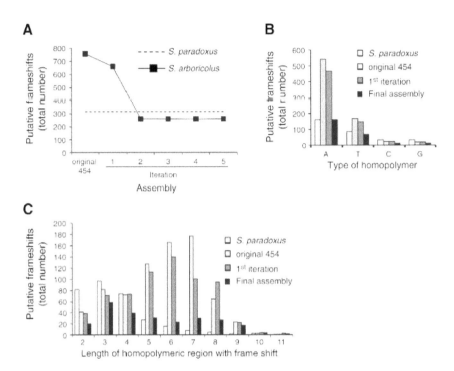

FIGURE 1: Solving the homopolymer problem. (A) Graph showing the number of frameshifts identified in each *S. arboricolus* assembly (filled symbols) compared to the *S. paradoxus* assembly (dashed trace). (B) Bar chart showing the number of frameshifts caused by homopolymers identified in each *S. arboricolus* assembly (filled bars) for each base compared to the *S. paradoxus* assembly (unfilled bars). (C) Bar chart showing the number of frameshifts caused by homopolymers of different lengths in each *S. arboricolus* assembly (filled bars) compared to the *S. paradoxus* assembly (unfilled bars).

copy number frequently correspond to highly repetitive tandem repeats and/or homopolymer tracks. Longer regions of elevated copy number are predominantly subtelomeric with the noteworthy exception of the rDNA repeats on chromosome XII.

The resulting *S. arboricolus* genome assembly comprises of whole chromosome scaffolds with only 186 gaps. These gaps have an average size of 1206 bp, the smallest two are just 1 bp, and the largest is 5846 bp. These regions consist of complex or repetitive sequences resulting in poor mapping of the SOLiD data. By comparison to recently improved assemblies of other sensu stricto yeasts [3], our *S. arboricolus* genome sequence has a higher proportion of the sequence (>99% compared with 96-98%) in a smaller number of scaffolds (35 compared with 147-226). Therefore, after the 'gold standard' *S. cerevisiae*, our *S. arboricolus* genome sequence represents the next most complete assembly.

11.2.2 GENOME STRUCTURE AND ANNOTATION

We compared our *S. arboricolus* genome assembly to the *S. cerevisiae* reference genome using LASTZ. We found that the 17 long scaffolds are each syntenic with a single *S. cerevisiae* chromosome or the mitochondrial genome with the exception of one predicted reciprocal translocation (Figure 2A) between the right arms of chromosome IV and XIII. The breakpoints are intergenic regions between ORFs *MRPL1* and *TMA64* on chromosome IV and *YKU80* and *SPG4* on chromosome XIII (Figure 2B). Interestingly, the breakpoint on chromosome XIII is adjacent to a tRNA gene, a feature previously reported to be associated with reciprocal translocations [7]. We used diagnostic PCR to experimentally confirm this reciprocal translocation (Figure 2C). The reciprocal translocation is unique to *S. arboricolus*, it is not present in *S. bayanus* or other sensu stricto assemblies [3] and therefore occurred after the *S. arboricolus* radiation.

The gene content of sensu stricto budding yeast species is thought to be similar [18,19], therefore we used comparative gene-annotation methods based on the well-annotated *S. cerevisiae* proteome to identify and annotate the ORFs in the *S. arboricolus* genome. Using exonerate [20], we aligned each *S. cerevisiae* protein to the *S. arboricolus* genome (see Methods). We assigned the top matching *S. arboricolus* ORF (based on the exonerate score) as a putative ortholog to each *S. cerevisiae* protein. We then compared the neighbouring genes of each *S. cerevisiae* gene with the neighbours of the putative orthologous *S. arboricolus* ORF to define a first set of 4798 orthologous gene pairs where the gene order has been conserved, which we refer to as "syntenic orthologs". Because this method uses the best sequence match, missing assignments of syntenic orthologous ORFs may occur when the match with greatest sequence similarity is not the syntenic ortholog. To overcome this problem, we again used exonerate but emphasized the position of the predicted ORF, allowing the score to be slightly below the best scoring match (see Methods). An additional 519 *S. cerevisiae* genes had a high-scoring, but not top match with the expected syntenic gene pair. We considered these to be syntenic orthologs as sequence similarity together with gene order conservation is thought to be a more reliable indicator of orthology than sequence similarity alone [21]. To identify genes that may be found in *S. arboricolus*, but not in *S. cerevisiae*, we used Genemark [22], which is a *de novo* gene prediction method, and does not rely on sequence similarity (see Methods) and identified 106 genes that were not predicted using exonerate. These Genemark predictions contain novel genes and ORFs that were missed by exonerate as only the best hit from exonerate was considered in our gene prediction.

We explored the possibility that our annotation of the *S. arboricolus* genome contained novel genes. As was observed with the *S. bayanus* genome [19], the vast majority (96%) of the genes in *S. arboricolus* have conserved gene order with *S. cerevisiae*. The remaining "non-syntenic" genes include 104 that have similarity to another *S. cerevisiae* gene but are not syntenic (by our definition) and the 106 the genes predicted *de novo* (within the 16 assembled chromosomes). Analysis of the non-syntenic genes allowed the detection of at least two small local rearrangements relative to *S. cerevisiae* due to inversion of a large portion of DNA. The first one occurs on chromosome VI between ORFs *FAR7* and *YFR017C* (Figure 3A) and the second one on chromosome XIV between *YNL034W* and *COG6* (data not shown). To determine whether these were specific to the *S. arboricolus* genome, we compared these regions to the other sensu stricto genomes, and found that the *S. cerevisiae* gene order is likely to be the derived state, as *S. bayanus* and *S. kudriavzevii* shows the same gene order as *S. arboricolus* (Figure 4). Other synteny breaks occur predominantly in the subtelomeric regions: there is a significant enrichment of non-syntenic and novel genes predicted in the first and last 10% of the chromosomes (P-value = 5×10^{-37}, Figure 3B) [23]. We also considered the genes predicted in *S. arboricolus* that were not syntenic with *S. cerevisiae*. Of these 210 genes, 44 had no BLAST hits within the *S. cerevisiae* genome (e-value cutoff 1e-10). Interestingly, 3 of these 44 genes are likely to be *S. cerevisiae* specific gene losses, rather than new genes arising in *S. arboricolus*, as they are found in *S. bayanus* (Figure 3C). Two of the non-telomeric *S. cerevisiae* gene losses are SIR1 genes as previously

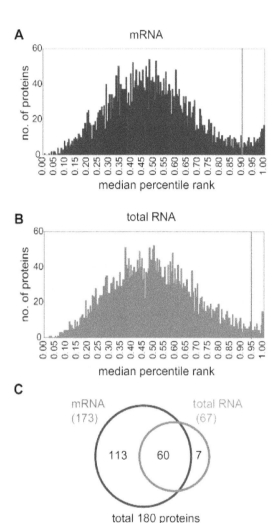

FIGURE 2: Identification of a single reciprocal translocation. (A) Dot plot representation of DNA sequence identity between the *S. cerevisiae* and *S. arboricolus* genomes. A single reciprocal translocation is apparent between chromosomes IV and XIII. (B) Cartoon representation of the location of the reciprocal translocation including the flanking features and the primer locations (not to scale) used to confirm the translocation. (C) PCR-based confirmation of the reciprocal translocation. Various primer and template combinations (as indicated) were used to amplify products corresponding to either the *S. cerevisiae* or the *S. arboricolus* gene order. In each case the resulting PCR products support the reciprocal translocation identified by the genome assembly.

reported [24]. Of the remaining 41 genes, 20 have no blast hit within Uniref90 (e-value cutoff 1e-10) and we considered the possibility that these were truly novel genes. After manual inspection based on presence of stop codons within the predicted peptide, protein sequence lengths, Pfam analysis andadditional blast searches, we concluded that 4 of these genes are likely to represent novel genes in *S. arboricolus* (Additional file 3).

We also searched for tRNA coding sequences within the 16 chromosomes using tRNAscan-SE [25] and annotated whether they were syntenic using a similar strategy to that described above (see Methods). In total, 257 tRNAs were found, 252 of which are syntenic with *S. cerevisiae* tRNAs. Next we used BLAST to search for the presence of repetitive elements in the genome such as subtelomeric genes and Ty elements. We detected the most distal subtelomeric element, Y', in the genome sequence. This element is therefore present in all the sensu stricto species except *S. bayanus*[26,27]. We also detected Ty2 element sequences using as a query the region that does not share similarity with the Ty1 element (1.7-kb *ClaI* Ty2- specific sequences [26]).

11.2.3 PHYLOGENETIC ANALYSIS

We tested five possible placements of *S. arboricolus* within the sensu stricto complex (Additional file 4: Figure S2), by sampling 100 sets of 50 random proteins for which we have data for all 6 sensu stricto species, as well as *S. castellii* as outgroup. These protein sequences were concatenated, and we computed the likelihood of the five phylogenetic trees using PAML. All 100 trees supported the grouping of *S. arboricolus* as diverging after the common ancestor with *S. bayanus* and before *S. kudriavzevii*, and all but 1 of these trees obtained bootstrap scores >0.9. Originally *S. arboricolus* was placed, based on a limited amount of ribosomal DNA sequence, between *S. mikatae* and *S. kudriavzevii*[11], however our genome-scale phylogenetic analysis has much greater power and unambiguously supports the new tree structure (Figure 5).

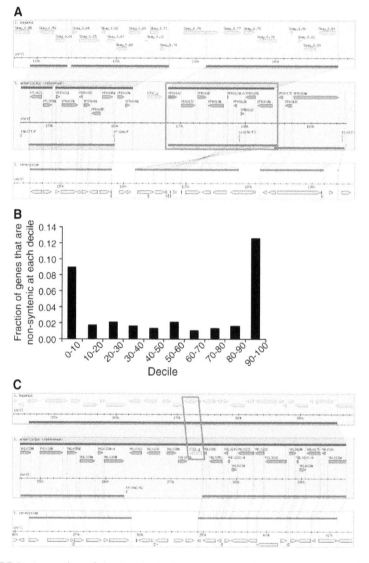

FIGURE 3: Annotation of the *S. arboricolus* genome. (A) A Gbrowse_syn visualization of the inversion on chromosome VI within *S. cerevisiae* with respect to *S. arboricolus* and *S. bayanus*. (B) Novel and non-syntenic genes are predominantly telomeric. Each chromosome was divided in deciles and the fraction of genes predicted at each decile that we define as novel and non-syntenic is shown. (C) Gene complement differences from *S. cerevisiae* can be explained by *S. cerevisiae* specific gene loss. A Gbrowse_syn visualization, showing a gene (labeled as 2721_g) present in both *S. bayanus* and *S. arboricolus*, but not in *S. cerevisiae*.

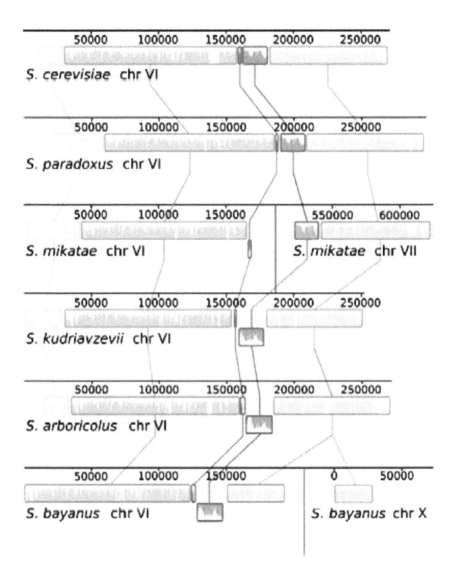

FIGURE 4: Structure of *Saccharomyces* sensu stricto chromosome VI. Chromosomal blocks of high sequence similarity are given the same colour and are connected by vertical lines. The average conservation level of the sequence is displayed within each block. Blocks placed below the horizontal center of each chromosome are showing sequence similarity in the inverse direction. In *S. mikatae* and *S. bayanus*, parts of chromosome VI have been translocated to chromosome VII and chromosome X, respectively [7].

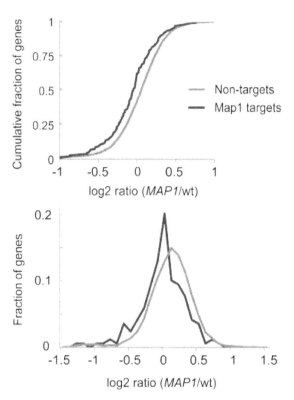

FIGURE 5: Phylogenetic analyses of the *Saccharomyces* sensu stricto group. A phylogenetic tree of the sensu stricto *Saccharomyces* species places *S. arboricolus* between *S. bayanus* and *S. kudriavzevii*. Tree topology was obtained using random concatenations of protein sequences with S. castellii as the outgroup. Branch lengths represent the maximal likelihood estimate of the evolutionary distance in DNA substitutions per codon using all genes that we identified as 'syntenic orthologs'.

We next set out to estimate the evolutionary distances between the species in the sensu stricto clade. To do so, we used the phylogeny as determined above, and computed codon-based maximum likelihood estimates (see Methods) of evolutionary distance based on alignments of 3899 genes for which we had 1 to 1 orthologs in all of the sensu stricto species, and that were syntenic between *S. arbicolus* and *S. cerevisiae*. We computed the median branch lengths (in substitutions per codon) for these, and they are shown in Figure 5.

11.2.4 S. ARBORICOLUS *WEB AND STRAIN RESOURCES*

To make the *S. arboricolus* genome sequence available and to facilitate analysis, we have made available a number of web resources (http://www.moseslab.csb.utoronto.ca/sarb webcite). These include genome sequence, annotation and datasets for genes and proteins. The gene and protein sets are annotated based on the *S. cerevisiae* ortholog systematic name. Novel putative ORFs identified by Genemark are also reported. A BLAST server and a genome browser (Gbrowse [28] and GBrowse_syn [29]) are available. The *S. arboricolus* genome browser offers the opportunity to view and compare the genome structure of *S. cerevisiae*, *S. arboricolus* and *S. bayanus*. Using *S. arboricolus* as the central reference species, gene order conservation or chromosomal rearrangements between the three species can easily be observed. Genes in *S. arboricolus* are coloured differently based on their annotation (e.g. syntenic orthologs, Genemark predictions, etc.). Finally, to facilitate experimental analysis of *S. arboricolus* the *HO* gene was disrupted (in the type strain H-6) and stable haploids were generated.

11.2.5 S. ARBORICOLUS *PHENOTYPIC LANDSCAPE*

Taking the phylogeny and gene content of *S. arboricolus* (described above) into account, we revisited recently generated data on its phenotypic diversity. Although included for completeness in our publication on the phenotypic landscape of *Saccharomyces* sensu stricto species [30], the *S. arboricolus* phenotypes were not specifically analyzed or commented on. The sequenced strain (CBS 10644^{TT}) and two additional, genetically distinct lineages isolated from similar habitats in Southern China were subjected to high resolution phenotyping of proliferative capability across >120 environments selected to represent variations in common yeast habitats, such as carbon and nitrogen source variations, tolerance to metabolites and toxins produced by plants and bacteria, and variations in vitamins and mineral availability (Additional file 5: Table S2). The fitness components lag, rate (population doubling time) and efficiency of reproduction (population density change) were extracted from high density growth curves and normalized to those of the *S. cerevisiae* reference strain, providing >360 precise measures of organism-environment interactions (Figure 6A). In the absence of stress, *S. arboricolus* proliferated slightly slower than *S. cerevisiae* and one strain (CBS 10644TT) also showed reduced efficiency (Figure 6A). However, these *S. arboricolus* growth aberrations in conditions with no stress were marginal compared to the dramatic proliferation deviations observed in a vast range of stress-inducing niche environments (Figure 6B). Remarkably, almost all of these aberrations constituted grave defects, many corresponding to more than 10-fold reductions in mitotic performance. Thus, *S. arboricolus* showed drastically reduced tolerance to fruit organic acids such as citric, tartaric and oxalic acid and to high temperatures and very poor utilization of adenine, serine and threonine as nitrogen sources. Notably, *S. arboricolus* failed to proliferate during conditions of elevated Li^+ and Cu^{2+}, traits likely explained by absence of the amplifications of the lithium exporter (*ENA1*) and the copper metallo-thionein (*CUP1*) that determine these traits in *S. cerevisiae*[30] (Figure 6C). The many niche specific proliferation deficiencies of *S. arboricolus* may explain its limited geographical and ecological distribution compared to *S. cerevisiae*.

FIGURE 6: Phenomics analyses of *S. arboricolus*. (A) Reproductive lag, rate (population doubling time) and efficiency (change in population density) of *S. arboricolus* CBS 10644TT, AS 2.3317 and AS 2.3319 were extracted from high density growth curves in no stress conditions. The performance of the *S. cerevisiae* strain BY4741 is shown as reference. (B) Relative reproductive performance of *S. arboricolus* strains CBS 10644[TT], AS 2.3317 and AS 2.3319 in a wide array of environments. The performance of each strain (n=2) was normalized to the *S. cerevisiae* reference strain BY4741 (n=20), or its auxotrophic mother S288C, providing a relative measure (log_2 [BY4741/isolate]). Broken line shows average performance in basal (no stress) conditions. Strong phenotype deviations from *S. cerevisiae* are labeled with the respective condition. (C) Mitotic reproduction of *S. arboricolus* strains during conditions of elevated concentrations of Li^+ (0.3 M) and Cu^{2+} (1 mM) or utilizing galactose and melibiose as sole carbon sources. *S. cerevisiae* S288C is included as a reference. (D) Average reproductive lag, rate and efficiency of *S. arboricolus* (CBS 10644[TT], AS 2.3317 and AS 2.3319) plotted against the corresponding averages for *S. bayanus* (CBS1001, GL274 and GL388) and *S. kudriavzevii* (GL22, GL23, GL391 and GL392) [30]. Grey diagonals indicate 1:1 correlations, numbers indicate Pearson correlation coefficients. (E) Phenotypes distinguishing *S. arboricolus* from *S. bayanus* and *S. kudriavzevii* respectively. Significant phenotype differences were defined at α<0.2 (Students t-test, Bonferroni correction). Error bars = Standard Error of the Mean.

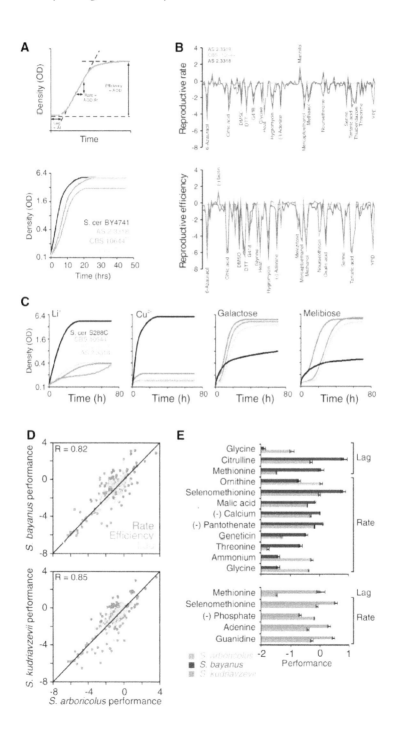

Among the rare examples of superior *S. arboricolus* performance were better utilization of the sugar alcohol mannitol, one of the most abundant energy storage molecules in nature [31] and tolerance to biotin depletion, rarely observed in *S. cerevisiae* due to ancestral loss of the biotin synthesis genes *BIO1* and *BIO6* [32]. Both *BIO1* and *BIO6* are present as conserved cistrons in *S. arboricolus* strain CBS 10644^{TT} (Figure 6B), explaining the biotin auxotrophy. *S. arboricolus* also featured consistently good utilization of the monosaccharide galactose (Figure 6C), a highly variable trait in both *S. cerevisiae*[30] and *S. kudriavzevii*[33] due to frequent loss-of-function mutations emerging in different lineages of these species that impair growth on galactose (e.g. the reference strain S288C). The coding sequences of the GAL pathway genes are also fully conserved in *S. arboricolus* strain CBS 10644^{TT}. Interestingly, *S. arboricolus* strain CBS 10644^{TT} has also retained an intact melibiase encoding *MEL1*, which is lost in most *S. cerevisiae* lineages [30]; all three *S. arboricolus* isolates also utilized the disaccharide melibiose, a less common plant energy storage compound, with a vastly superior rate and efficiency.

Overall, the three *S. arboricolus* isolates showed virtually identical trait profiles (Pearson correlation, r=0.73-0.91). Moreover, using data from Warringer et al. [30], we found *S. arboricolus* traits to closely mimic those of its relatives, *S. bayanus* and *S. kudriavzevii* (Figure 6D). These remarkable trait similarities, encompassing mitotic performance in a wide variety of environments, imply that these three species, despite billions of generations of separation, are adapted to rather similar ecological conditions. The few cases of *S. arboricolus* deviations from *S. bayanus* and *S. kudriavzevii* primarily included nitrogen utilization traits, such as superior *S. arboricolus* utilization of ammonium, glycine and ornithine, but inferior utilization of methionine, serine and citrulline (Figure 6E). Presumably, this reflects differences in nitrogen storage compounds among plant species that dominate the main habitats of these species and hint at ecological factors that may have driven speciation of the ancestral lineages. Further indications as to the nature of these factors may also be found in the reduced tolerance of *S. arboricolus* to malic acid, concentrated in e.g.

apples, and to the toxin geneticin, produced by bacteria of the *Micromono-spora* genus (Figure 6E).

11.3 DISCUSSION

Our approach of using multiple high-throughput sequencing strategies resulted in high quality genome sequence that continuously covers the large majority of the *S. arboricolus* genome. The *de novo* assembly revealed that *S. arboricolus* is largely syntenic to *S. cerevisiae*, similar to the other sensu stricto species. We mapped and validated a single reciprocal translocation that occurred in the *S. arboricolus* lineage and identified a few additional small-scale rearrangements. Our assembly extends into the subtelomeric regions of most chromosome ends. However, these repetitive sequences pose a major challenge to genome assemblies and the subtelomeric structure presented here will benefit from further experimental validation.

Sequence analysis unambiguously revealed the position of *S. arboricolus* in the phylogenetic tree (Figure 5; correcting the previously reported position based on a limited amount of ribosomal sequence [11]). Phylogeny of individual genes revealed a limited number of conflicting tree topology as has been previously reported for other sensu stricto species [34]. We did not observe large segments of the genome (for example, equivalent in size to the average gene) with high similarity with other species as signature of introgression as previously reported in other species and strains [35-37].

We have looked for the presence and absence of middle repetitive elements such as Ty and subtelomeric genes. For some of these elements the presence and absence is consistent with the phylogeny [26]. We detected the subtelomeric element Y' in *S. arboricolus*, indicating that this element entered in the sensu stricto ancestor after the divergence of *S. bayanus*. Much more puzzling is the phylogeny of Ty2, present in *S. cerevisiae*, *S. mikatae* and *S. arboricolus* but absent in *S. paradoxus*, *S. kudriavzevii* and

S. bayanus. Both multiple loss and acquisition can explain the scattered phylogeny but are unlikely events. One possibility is a recent exchange of Ty2 among these species (horizontal transfer) as supported by high sequence similarity. A possible mechanism is the ability of these species to fuse their cytoplasms, without progressing to karyogamy, and allowing the exchange of Ty particles that can self-propagate in the genome.

Phenotype analysis demonstrated a remarkable similarity in trait profiles between *S. arboricolus* and both *S. bayanus* and *S. kudriavzevii* (Figure 6D). These similarities suggest that these three species have adapted to similar environmental niches. The limited number of phenotypic differences between these species may reflect the specific nutrients available within each species habitat. Phenotypic comparisons between *S. cerevisiae* and *S. arboricolus* frequently reflect differences in gene content, including the sensitivity of *S. arboricolus* to elevated Li^+ and Cu^{2+} and the ability of *S. arboricolus* to utilize melibiose.

So far, the Chinese isolates of *S. arboricolus* are the only ones available. Future surveys will reveal whether this species is limited to this region or whether other geographic populations exist. It is interesting to note that two of the *Saccharomyces* species have only been isolated in Asia (*S. mikatae* in Japan and *S. arboricolus* in China) despite extensive surveys in similar environments in other continents [38-40].

11.4 CONCLUSIONS

The *Saccharomyces* sensu stricto complex offers a powerful range of sequence divergences that have allowed the mapping of functional elements [4,5], improved genome annotation and comparisons of genome organization [7,19,41]. Genome sequencing has revealed levels of divergence ranging from 0.1 - 0.6% among *S. cerevisiae* strains, 1.5 - 4.5% between geographic subpopulations of *S. paradoxus*[15], 6% between *S. bayanus* var. uvarum and S. eubayanus[42], and 15% - 30% between *S. cerevisiae* and the other sensu stricto species [4,5]. The *S. arboricolus* genome sequence should enhance the power of comparative genomics by increasing the total sequence divergence and improve the quality of alignments by adding a new branch between *S. bayanus* and *S. kudriavzevii*, the more divergent species.

11.5 METHODS

11.5.1 GENOMIC DNA AND LIBRARY PREPARATION

We extracted DNA from the type strain of *S. arboricolus* H-6^T (CBS 10644^{TT}) isolated in China from the bark of Quercus fabri[11]. For the Roche 454 library construction and sequencing, 5 µg of high molecular weight genomic DNA was used to make standard shotgun DNA library as described in the Roche GS FLX Titanium General Library Preparation Method Manual with the exception of DNA fragmentation, which was done with Covaris S2 sonicator (fragmentation parameters: Duty cycle - 5%, Intensity - 1, cycles/burst - 200, time - 85 s, bath temperature - 5°C). 15 µg of high molecular weight genomic DNA was used to make the 8 Kb paired end library as stated in the Roche GS FLX Titanium 8 Kb Span Paired end library preparation method manual. Exceptions include: DNA extraction was done using QIAquick Gel extraction kit (Qiagen, Cat no. 28760) instead of Electroelution as stated in the manual and fragmentation of circularised DNA was done using Covaris S2 sonicator (Duty cycle −5%, Intensity - 3, cycles/burst - 200, time - 120 s, bath temperature - 5°C). Sequencing of standard shotgun fragment library was carried out on ¾ of a PTP and the 8 Kb paired end library was sequenced on a full PTP using Roche 454 Titanium sequencing chemistry. For SOLiD library construction and sequencing, 500 ng of high molecular weight genomic DNA was used to make a barcoded DNA fragment library as stated in the SOLiD 4 library preparation guide. Enzymes and reagents were used from NEBNext DNA sample prep Master mix set 3 (NEB, Cat no. E6060L). The barcoded DNA fragment library was quantified using Kapa Library Quantification kit (Kapa Biosystems, Cat no. KK4823). 200-300 bp library size selection was carried out using 2% SizeSelect E-Gel (Life Technologies, Cat no. G6610-02). SOLiD EZ Bead System was used according to manufacturer's guide to prepare ePCR and templated bead enrichment. Sequencing was performed on a SOLiD 4 analyser according to the manufacturer's instructions to generate 50 bp reads in colour space.

11.5.2 GENOME ASSEMBLY

We assembled the genome of *S. arboricolus* using the Newbler algorithm (v2.3, Roche) for *de novo* assembly of reads generated by the 454 pryo-sequencing platform. Combinations of read datasets, reads added in assembly iterations, and assembler parameters were explored before selecting the optimal combination according to assembly metrics (number of scaffold sequences and contigs, the average and longest contig length and N50 value). All reads were trimmed against a dataset of adapter and vector sequences in the initial step of the assembly process.

The selected assembly parameters used an expected coverage value of 40X with all other settings remaining at default values. Two assembler iterations were employed; the first iteration included all 734,353 single fragment reads and one set of 583,674 paired-end reads. The second iteration incorporated an additional set of 518,434 paired-end reads. A third set of paired-end reads was excluded from the assembly due to decreased performance with their inclusion.

The resulting genome assembly comprised of 32 scaffold sequences with a total length of 11,465,281 bp. The scaffolds were comprised of 266 contigs ($\geq$500 bp) with an average length of 43,102 bp (538,482 bp max.) and an N50 value of 136,945 bp. The mapped read coverage of the assembly was 49X.

11.5.3 PYROSEQUENCING ERROR CORRECTION

In order to resolve small errors in the assembly arising from pyrosequencing artifacts, such as homopolymer sequence regions [43,44], we acquired deep sequence coverage (~100X) from short reads. We generated a total of 31,316,59 short (50 bp) reads from a SOLiD 4 single fragment library. Subsequent gapped read alignment and variant calling was achieved using Bioscope 1.3.1 (Life Technologies).

An iterative correction process was devised in which errors in the assembled sequence were identified from the SOLiD read alignment data as either a single nucleotide polymorphisms (SNP) for single base errors, or as small InDels (insertion/deletion) for homopolymer pyrosequencing errors. Each iteration of the assembly correction process involved the initial mapping of SOLiD reads against the 454 assembly, followed by SNP calling. Selected putative SNPs were then integrated into the assembly sequence and SOLiD reads were remapped to allow InDels to be called and integrated. This process was repeated until no additional variants were detected. In subsequent iterations additional reads were mapped allowing the identification and correction of a small number of further errors. Both SNPs and InDels were calculated from alignment data using Bioscope 'high stringency' variant parameter settings. Additionally, integrated variants were required to represent a minimum of 60% of the alignment data.

11.5.4 GENE ANNOTATION AND ORTHOLOGY ASSIGNMENTS

S. cerevisiae was used as the reference proteome for the program exonerate, which uses comparative approaches for gene finding based on protein sequence similarity. An initial pass with the protein2dna model and a refine boundary of 2000 was used to find the best orthologous candidate of each *S. cerevisiae* gene. For intronic genes, the max intron size was limited to 1500 bp and the model used was protein2genome.

To annotate genes within the *S. arboricolus* genome, we first identified gene orthologs with conserved synteny. To do so, we analysed the top hit by exonerate for each *S. cerevisiae* gene. When three neighbouring genes within *S. cerevisiae* all identified three neighbouring genes within *S. arboricolus*, we assigned the *S. arboricolus* gene in the middle (flanked by its two neighbours) as a syntenic ortholog. This initial step discovered most of the syntenic genes within *S. arboricolus*. Other genes within *S. cerevisiae* that had not been assigned an ortholog were further analysed with the hypothesis that these may have exonerate hits within the expected positions but were not the most similar sequence within *S. arboricolus*. We looked at the top 10 exonerate hits of the remaining *S. cerevisiae* genes for matches in *S. arboricolus* between the initially assigned syntenic ortholog. When only one hit was found between these syntenic orthologs, we used this hit as a newly discovered syntenic ortholog. This process was repeated until no more syntenic orthologs could be found. Finally, we assigned the top exonerate hit of few remaining *S. cerevisiae* genes that were still not assigned a syntenic ortholog as the non-syntenic ortholog provided that they did not overlap with another gene prediction.

De novo gene prediction on the *S. arboricolus* genome was performed using GeneMark-ES, version 2 [45]. The total number of the predicted genes was 5005 within the 16 assembled chromosomes (5038 in total). Of these, 95 genes had non-overlapping coordinates with the genes predicted by Exonerate within the 16 assembled chromosomes (106 in total when including the 19 scaffolds that did not assemble into the chromosomes).

A significant issue when using a comparative-based method, such as exonerate, for gene prediction is that gene boundaries are often incorrectly predicted if there is a lack of homology at these ends. Initially, a large number of predicted genes did not contain a start or stop codon (637 genes and 1121 genes respectively). We have attempted to rectify these starts and ends by extending or truncating the predicted CDS. First, CDSs were extended if a stop codon could be found within 9 codons from the end of our gene prediction. This corrected 857 cases of missing stop codons and further extension only slightly improved the annotation. Second, for start codons, the methionine can be on either side of the predicted gene start. We therefore extended the predicted gene until a methionine was found, but only when a methionine could be found within 9 codons and without any intervening stop codons. In the cases where a stop codon occurred before a suitable methionine was identified, we truncated the CDS to a downstream methionine if it occurred within 9 codons. This corrected 348 cases of missing start codons. Finally, intron-containing genes were left untouched as missing starts and ends for these genes could be due to a missing exon. We note that for intron-containing genes, we specifically use the protein2genome model in exonerate that explicitly attempts to predict all exons found in *S. cerevisiae* genes. This assumes that the presence of introns is conserved between *S. cerevisiae* and *S. arboricolus*.

We aligned the protein sequence orthologs for the sensu stricto using MAFFT [46] with default settings, either with or without *S. castellii* orthologs as an outgroup. For the coding sequence analysis we inserted the gaps back into the DNA sequences. Phylogenetic analysis was performed using PAML [47], either with the codon model for the DNA sequence analysis or with empirical model for the amino acid analysis. Because we are only concerned with the placement of *S. arboricolus* within the established sensu stricto yeast phylogeny, we compared the likelihood of several putative tree topologies that differ only in the position of *S. arboricolus* (Figure S2).

To annotate tRNA coding sequences, we predicted tRNAs using tRNAscan-SE [25] with default settings on the 16 assembled chromosomes. To determine whether or not these predicted tRNAs are syntenic with respect to *S. cerevisiae* we used an analogous strategy to that described above for gene annotations. tRNA coding sequences were annotated as syntenic orthologs if they were flanked by genes within *S. arboricolus* that were assigned as syntenic orthologs and if a tRNA was also found in *S. cerevisiae* between those genes. In all but one cases, the syntenic tRNAs code for the same amino acids.

11.5.5 CHROMOSOMAL STRUCTURE PLOTS

Chromosome structure plots for the *Saccharomyces* sensu stricto species were constructed using Mauve [48]. Assembled chromosomes for *S. paradoxus* (strain CBS432) were obtained from [15] and for *S. mikatae* (IFO 1815 T), *S. kudriavzevii* (IFO 1802 T) and *S. bayanus* var. uvarum (strain CBS 7001) from [3]. As these chromosome assemblies have been constructed partly by using the *S. cerevisiae* genome to orient and order scaffolds, alignments were also made to the unordered scaffolds using MUMmer [49] to confirm the relative orientation of chromosomal segments inverted between species.

11.5.6 MAPPING OF THE PHENOTYPE LANDSCAPE OF S. ARBORICOLUS

The bulk of the phenotypic data was taken from our recent publication [30] on sensu stricto phenotypes where it was included for completeness but where *S. arboricolus* phenotypes were not specifically analyzed or considered. The data displayed as growth curves in this study correspond to novel confirmatory runs performed to ensure the reliability of specific statements. Three diploid isolates of *Saccharomyces arboricolus* were collected as described previously [11] and long time stored in 20% glycerol at -80C. Isolates were subjected to high throughput phenotyping by micro-cultivation (n=2) in an array of environments (Additional file 5: Table S2) essentially as previously described [50]. For pre-cultivations, strains were inoculated in 350 μL of SD medium (0.14% yeast nitrogen base, 0.5% ammonium sulfate and 1% succinic acid; 2% (w/v) glucose; 0.077% Complete Supplement Mixture (CSM, ForMedium), pH set to 5.8 with NaOH or KOH) and incubated for 48 h at 30°C. For experiments where the removal of a specific media component was studied, the pre-culture was performed in absence of this component in order to completely deplete the component in question. For experiments where alternative nitrogen sources were used, two consecutive pre-cultures were performed, the first containing low amounts of ammonium sulphate (0.05%), the second replacing ammonium with the indicated nitrogen source in amounts corresponding to equivalent moles of N. For all experimental runs, strains were inoculated to an OD of 0.03 - 0.1 in 350 μL of SD medium and cultivated for 72 h in a Bioscreen analyser C (Growth curves Oy, Finland). Optical density was measured using a wide band (450-580 nm) filter. Incubation was at 30.0°C (±0.1°C) with ten minutes preheating time. Plates were subjected to shaking at highest shaking intensity with 60s of shaking every other minute. OD measurements were taken every 20 min. Strains were run in duplicates on separate plates with ten replicates of the universal *S. cerevisiae* reference strain BY4741 or its prototrophic mother S288C, in randomised (once) positions on each plate as a reference. The reproductive rate (population doubling time), lag (population adaptation time) and efficiency (population total change in density) were extracted from high density growth curves and put in relation to the corresponding fitness variables of the reference strain BY4741, or in conditions directly involving alterations of nitrogen content, its prototrophic mother S288C, as described previously [30]. The derived Log2 ratios (Log2 (BY4741/ isolate) or, in case of efficiency, Log2 (isolate/BY4741) were used for subsequent analysis.

11.5.7 ACCESSION NUMBERS

Raw sequencing reads are available from the European Nucleotide Archive (EBI ENA) for the SOLiD reads [EMBL: ERP001702], Roche 454 single fragment reads [EMBL: ERP001703] and Roche 454 paired-end reads [EMBL: ERP001704]. The assembled genome is available from NCBI as *Saccharomyces arboricola* [GenBank: ALIE00000000].

REFERENCES

1. Dujon B: Yeast evolutionary genomics. Nat Rev Genet 2010, 11(7):512-524.
2. Nieduszynski CA, Liti G: From sequence to function: insights from natural variation in budding yeasts. Biochim Biophys Acta 2011, 1810(10):959-966.
3. Scannell DR, Zill OA, Rokas A, Payen C, Dunham MJ, Eisen MB, Rine J, Johnston M, Hittinger CT: The awesome power of yeast evolutionary genetics: new genome sequences and strain resources for the saccharomyces sensu stricto genus. G3 2011, 1:11-25.
4. Cliften P, Sudarsanam P, Desikan A, Fulton L, Fulton B, Majors J, Waterston R, Cohen BA, Johnston M: Finding functional features in *Saccharomyces* genomes by phylogenetic footprinting. Science 2003, 301(5629):71-76.
5. Kellis M, Patterson N, Endrizzi M, Birren B, Lander ES: Sequencing and comparison of yeast species to identify genes and regulatory elements. Nature 2003, 423(6937):241-254.
6. Dujon B: Yeasts illustrate the molecular mechanisms of eukaryotic genome evolution. Trends Genet 2006, 22(7):375-387.
7. Fischer G, James SA, Roberts IN, Oliver SG, Louis EJ: Chromosomal evolution in *Saccharomyces*. Nature 2000, 405(6785):451-454.
8. Harbison CT, Gordon DB, Lee TI, Rinaldi NJ, Macisaac KD, Danford TW, Hannett NM, Tagne JB, Reynolds DB, Yoo J: Transcriptional regulatory code of a eukaryotic genome. Nature 2004, 431(7004):99-104.
9. Nieduszynski CA, Knox Y, Donaldson AD: Genome-wide identification of replication origins in yeast by comparative genomics. Genes Dev 2006, 20(14):1874-1879.
10. Müller CA, Nieduszynski CA: Conservation of replication timing reveals global and local regulation of replication origin activity. Genome Res 2012, 22(10):1953-1962.
11. Wang SA, Bai FY: *Saccharomyces arboricolus* sp. nov., a yeast species from tree bark. Int J Syst Evol Microbiol 2008, 58(Pt 2):510-514.

12. Naumov GI, Naumova ES, Masneuf-Pomarede I: Genetic identification of new bio-logical species *Saccharomyces arboricolus* Wang et Bai. Antonie Van Leeuwenhoek 2010, 98(1):1-7.

13. Goffeau A, Barrell BG, Bussey H, Davis RW, Dujon B, Feldmann H, Galibert F, Ho-heisel JD, Jacq C, Johnston M: Life with 6000 genes. Science 1996, 274(5287):546-563-547.

14. Ronaghi M: Pyrosequencing sheds light on DNA sequencing. Genome Res 2001, 11(1):3-11.

15. Liti G, Carter DM, Moses AM, Warringer J, Parts L, James SA, Davey RP, Roberts IN, Burt A, Koufopanou V: Population genomics of domestic and wild yeasts. Na-ture 2009, 458(7236):337-341.

16. Siow CC, Nieduszynska SR, Muller CA, Nieduszynski CA: OriDB, the DNA repli-cation origin database updated and extended. Nucleic Acids Res 2012, 40(1):D682-D686.

17. Cherry JM, Hong EL, Amundsen C, Balakrishnan R, Binkley G, Chan ET, Christie KR, Costanzo MC, Dwight SS, Engel SR: *Saccharomyces* Genome Database: the genomics resource of budding yeast. Nucleic Acids Res 2012, 40(1):D700-D705.

18. Bon E, Neuveglise C, Casaregola S, Artiguenave F, Wincker P, Aigle M, Durrens P: Genomic exploration of the hemiascomycetous yeasts: 5. *Saccharomyces* bayanus var. uvarum. FEBS Lett 2000, 487(1):37-41.

19. Fischer G, Neuveglise C, Durrens P, Gaillardin C, Dujon B: Evolution of gene order in the genomes of two related yeast species. Genome Res 2001, 11(12):2009-2019.

20. Slater GS, Birney E: Automated generation of heuristics for biological sequence comparison. BMC Bioinformatics 2005, 6:31.

21. Byrne KP, Wolfe KH: The Yeast Gene Order Browser: combining curated homol-ogy and syntenic context reveals gene fate in polyploid species. Genome Res 2005, 15(10):1456-1461.

22. Besemer J, Lomsadze A, Borodovsky M: GeneMarkS: a self-training method for prediction of gene starts in microbial genomes. Implications for finding sequence motifs in regulatory regions. Nucleic Acids Res 2001, 29(12):2607-2618.

23. Brown CA, Murray AW, Verstrepen KJ: Rapid expansion and functional divergence of subtelomeric gene families in yeasts. Curr Biol 2010, 20(10):895-903.

24. Zill OA, Scannell D, Teytelman L, Rine J: Co-evolution of transcriptional silenc-ing proteins and the DNA elements specifying their assembly. PLoS Biol 2010, 8(11):e1000550.

25. Lowe TM, Eddy SR: tRNAscan-SE: a program for improved detection of transfer RNA genes in genomic sequence. Nucleic Acids Res 1997, 25(5):955-964.

26. Liti G, Peruffo A, James SA, Roberts IN, Louis EJ: Inferences of evolutionary rela-tionships from a population survey of LTR-retrotransposons and telomeric-associ-ated sequences in the *Saccharomyces* sensu stricto complex. Yeast 2005, 22(3):177-192.

27. Naumov GI, James SA, Naumova ES, Louis EJ, Roberts IN: Three new species in the *Saccharomyces* sensu stricto complex: *Saccharomyces* cariocanus, *Saccharomy-ces* kudriavzevii and *Saccharomyces* mikatae. Int J Syst Evol Microbiol 2000, 50(Pt 5):1931-1942.

28. Stein LD, Mungall C, Shu S, Caudy M, Mangone M, Day A, Nickerson E, Stajich JE, Harris TW, Arva A: The generic genome browser: a building block for a model organism system database. Genome Res 2002, 12(10):1599-1610.

29. McKay SJ, Vergara IA, Stajich JE: Using the Generic Synteny Browser (GBrowse_ syn). Current protocols in bioinformatics. 2010, 31:9.12.1-9.12.25.

30. Warringer J, Zorgo E, Cubillos FA, Zia A, Gjuvsland A, Simpson JT, Forsmark A, Durbin R, Omholt SW, Louis EJ: Trait variation in yeast is defined by population history. PLoS Genet 2011, 7(6):e1002111.

31. Song SH, Vieille C: Recent advances in the biological production of mannitol. Appl Microbiol Biotechnol 2009, 84(1):55-62.

32. Hall C, Dietrich FS: The reacquisition of biotin prototrophy in *Saccharomyces cerevisiae* involved horizontal gene transfer, gene duplication and gene clustering. Genetics 2007, 177(4):2293-2307.

33. Hittinger CT, Goncalves P, Sampaio JP, Dover J, Johnston M, Rokas A: Remarkably ancient balanced polymorphisms in a multi-locus gene network. Nature 2010, 464(7285):54-58.

34. Rokas A, Williams BL, King N, Carroll SB: Genome-scale approaches to resolving incongruence in molecular phylogenies. Nature 2003, 425(6960):798-804.

35. Argueso JL, Carazzolle MF, Mieczkowski PA, Duarte FM, Netto OV, Missawa SK, Galzerani F, Costa GG, Vidal RO, Noronha MF: Genome structure of a *Saccharomyces cerevisiae* strain widely used in bioethanol production. Genome Res 2009, 19(12):2258-2270.

36. Liti G, Barton DB, Louis EJ: Sequence diversity, reproductive isolation and species concepts in *Saccharomyces*. Genetics 2006, 174(2):839-850.

37. Novo M, Bigey F, Beyne E, Galeote V, Gavory F, Mallet S, Cambon B, Legras JL, Wincker P, Casaregola S: Eukaryote-to-eukaryote gene transfer events revealed by the genome sequence of the wine yeast *Saccharomyces cerevisiae* EC1118. Proc Natl Acad Sci USA 2009, 106(38):16333-16338.

38. Johnson LJ, Koufopanou V, Goddard MR, Hetherington R, Schafer SM, Burt A: Population genetics of the wild yeast *Saccharomyces* paradoxus. Genetics 2004, 166(1):43-52.

39. Kuehne HA, Murphy HA, Francis CA, Sniegowski PD: Allopatric divergence, secondary contact, and genetic isolation in wild yeast populations. Curr biol 2007, 17(5):407-411.

40. Sampaio JP, Goncalves P: Natural populations of *Saccharomyces* kudriavzevii in Portugal are associated with oak bark and are sympatric with *S. cerevisiae* and *S. paradoxus*. Appl Environ Microbiol 2008, 74(7):2144-2152.

41. Langkjaer RB, Nielsen ML, Daugaard PR, Liu W, Piskur J: Yeast chromosomes have been significantly reshaped during their evolutionary history. J Mol Biol 2000, 304(3):271-288.

42. Libkind D, Hittinger CT, Valerio E, Goncalves C, Dover J, Johnston M, Goncalves P, Sampaio JP: Microbe domestication and the identification of the wild genetic stock of lager-brewing yeast. Proc Natl Acad Sci USA 2011, 108(35):14539-14544.

43. Quince C, Lanzen A, Davenport RJ, Turnbaugh PJ: Removing noise from pyrosequenced amplicons. BMC Bioinformatics 2011, 12:38.

44. Quince C, Lanzen A, Curtis TP, Davenport RJ, Hall N, Head IM, Read LF, Sloan WT: Accurate determination of microbial diversity from 454 pyrosequencing data. Nat Methods 2009, 6(9):639-641.
45. Ter-Hovhannisyan V, Lomsadze A, Chernoff YO, Borodovsky M: Gene prediction in novel fungal genomes using an ab initio algorithm with unsupervised training. Genome Res 2008, 18(12):1979-1990.
46. Katoh K, Misawa K, Kuma K, Miyata T: MAFFT: a novel method for rapid multiple sequence alignment based on fast Fourier transform. Nucleic Acids Res 2002, 30(14):3059-3066.
47. Yang Z: PAML: a program package for phylogenetic analysis by maximum likelihood. Comput Appl Biosci 1997, 13(5):555-556.
48. Darling AC, Mau B, Blattner FR, Perna NT: Mauve: multiple alignment of conserved genomic sequence with rearrangements. Genome Res 2004, 14(7):1394-1403.
49. Kurtz S, Phillippy A, Delcher AL, Smoot M, Shumway M, Antonescu C, Salzberg SL: Versatile and open software for comparing large genomes. Genome Biol 2004, 5(2):R12.
50. Warringer J, Anevski D, Liu B, Blomberg A: Chemogenetic fingerprinting by analysis of cellular growth dynamics. BMC Chem Biol 2008, 8:3.
51. Homer N, Merriman B, Nelson SF: BFAST: an alignment tool for large scale genome resequencing. PLoS One 2009, 4(11):e7767.
52. Benson G: Tandem repeats finder: a program to analyze DNA sequences. Nucleic Acids Res 1999, 27(2):573-580.

This article has supplemental information that is not featured in this version of the text. To view these files, please visit the original version of the article as cited in the beginning of this chapter.

CHAPTER 12

COMPARATIVE GENE EXPRESSION BETWEEN TWO YEAST SPECIES

YUANFANG GUAN, MAITREYA J. DUNHAM,
OLGA G. TROYANSKAYA, AND AMY A. CAUDY

12.1 BACKGROUND

The *Ascomycete* yeasts present one of the most promising systems for comparative functional genomics. Fungi have been densely sampled by a number of sequencing projects [1], covering an enormous range of divergence. Genome sequence analyses of the *Saccharomyces* yeasts and related species have been used to establish the history of gene duplication [2-6], conservation at binding sites [7,8], and co-evolution of binding sites with regulators [9]. Thus, a range of evolutionary phenomena can be studied in these species based on their genomic sequence. However, sequence conservation is not always completely predictive of functional conservation. As just one example, we recently reported that only a subset of conserved promoter motifs actually drive periodic gene expression over the cell cycle in two closely related species [10].

Most of the experimental characterization of gene function has been performed in a small number of model fungal systems, which can provide

This chapter was originally published under the Creative Commons Attribution License. Guan Y, Dunham MJ, Troyanskaya OG and Caudy AA. Comparative Gene Expression Between Two Yeast Species. BMC Genomics 14,33 (2013). doi:10.1186/1471-2164-14-33.

an anchor for these broad genome sequencing surveys. These species include *Saccharomyces cerevisiae, Neurospora crassa, Candida albicans,* and *Schizosaccharomyces pombe*, along with several other emerging models such as *Ashbya gossypii*. Comparative studies between these species, which by some estimates cover a billion years of divergence, have been informative [11,12]. Analysis of gene expression changes over growth [13], the cell cycle [13], and stress treatments [14,15] highlighted both similarities and differences in ortholog expression. Unfortunately, the ability to link individual gene expression divergence with the causative molecular factors has been limited because of the vast evolutionary distances involved.

Experimental protocols developed in the model systems are often readily portable to less well-studied sister species, allowing us to choose species well-placed to identify and study functional divergence. Comparisons of gene expression across particular species with interesting characteristics can not only highlight how patterns of gene expression change over evolutionary time, but can also discover genes with particular functions. A comparison among xylose-metabolizing species of yeasts, for example, was able to couple sequence analysis with gene expression profiling to identify important genes via their presence in the genomes of interest and their induction when grown on xylose [16]. Followup studies in *S. cerevisiae* confirmed these associations.

Due to their close proximity to *S. cerevisiae*, studies in the sensu stricto yeasts have also been particularly informative. These species cover a range of conservation, have high quality annotated reference genomes [17], and are becoming even more attractive as the sequences of many strains within each species are forthcoming using new high-throughput sequencing tools (e.g. [18]). Furthermore, their ability to form interspecific hybrids leverages the resources available in *S. cerevisiae* and allows tests of gene function and regulation in shared cellular environments [19-23]. Recent work on expression-based full-genome characterization is reported in [24,25], which used *S. cerevisiae* microarrays to measure the gene expression consequences of heat shock stress and mating induction on three other yeast species. Their data suggest that expression divergence can occur relatively rapidly and is correlated to gene function, though relatively uncorrelated to sequence conservation [26]. Due to the *S. cerevisiae* arrays used, they were unable to examine more divergent species. In order to broaden these studies to more divergent yeasts, species-specific arrays must be used, as has been done, for example, for *Candida glabrata* [27]. Most importantly, due to the limited condition space of just a small number of treatments in these studies, conclusions about evolution of gene function and regulation have been difficult to generalize.

To address these challenges, we previously developed a computational framework [28,29] to identify a set of experiments that could best characterize gene function in a naive species. Based on available expression date in the *S. cerevisiae* literature, we identified and carried out a set of 304 experiments over 46 conditions in the sensu stricto species *S. bayanus* var *uvarum*. By choosing only the most informative experiments from the vast *S. cerevisiae* literature, we were able to survey a large phenotypic space at high accuracy with a modest amount of experimentation.

To compare these expression datasets more carefully, we developed a statistical metric, Local Network Similarity (LNS), to assess correlation patterns of orthologs. This metric is general and robust—it can be used for analysis of individual matched datasets without the need to assume identical response time for the two species, or for integrated analysis of diverse compendia of experimental or genetic perturbations. Using the LNS metric to compare our large *S. bayanus* expression compendium with a collection of published *S. cerevisiae* expression data, we show that gene expression networks are largely conserved between the species, though much less than within-species comparisons constructed by comparing different conditions. Furthermore, we demonstrated strong and statistically significant evidence for correlation between the divergence of expression and open reading frame sequence, which previous studies using more limited datasets failed to detect (see review [26]). Despite this general conservation pattern, we observed that a quarter of orthologs exhibit condition-specific differences in expression, and 4% show strong differences in global co-expression patterns. Genes involved in the same functional groups share similar divergence patterns, indicating that pathways or processes may share characteristics. In sum, our wide-ranging survey of expression profiles and generic metric of expression divergence allowed us to identify both global and local aspects of regulatory evolution and relate these to sequence divergence.

12.2 RESULTS

S. bayanus var *uvarum* (referred to as "*S. bayanus*" from here forward) is a sequenced but relatively unstudied species of budding yeast typically associated with fermentation environments and diverged by approximately 20 million years from *S. cerevisiae*. We have recently computationally chosen 46 biological treatments for gene expression analysis that would maximally cover biological processes in yeast [28]. Using the resulting dataset of over 300 arrays in *S. bayanus* along with 2569 arrays collected from *S. cerevisiae* [30-34], we carried out both global and dataset-specific comparisons of gene expression.

12.2.1 A GENERIC STATISTICAL METRIC TO QUANTIFY EXPRESSION CONSERVATION BETWEEN SPECIES

To measure the divergence in gene expression between the two species, we developed a metric to compare the expression networks surrounding or-

FIGURE 1. Quantification of expression conservation by local network similarity (LNS). Pair-wise Pearson correlation between genes was calculated for individual *S. bayanus* and *S. cerevisiae* datasets, generating a matrix of gene-gene correlations. The data used to create this illustration are the actual diauxic shift, cell cycle synchronization, and alpha factor treatment. The distribution of these correlation values is between −1 and 1, and can be drastically different from one dataset to another. Therefore, Fisher's z-transformation and normalization of these z-values were applied on each correlation matrix, so that the correlations were comparable across datasets. The resulting correlation matrices are normally distributed and centered at 0 with standard deviation equal to 1. For each orthologous pair i and i', their z-transformed, normalized correlation to all other matched orthologs form two vectors, indicating the relative positions of this pair of ortholog in their respective expression network. The correlation of these two vectors was taken as LNS. To calculate the correlation matrix for global LNS, the average values of individual datasets for a specific gene-gene pair were used to form a new global correlation matrix. According to the properties of normal distribution, the values within this matrix are still normally distributed and centered at 0 with standard deviation equal to 1. This global matrix was then used to calculate global LNS for each ortholog. To simulate the case of non-conservation, orthologous pairs were randomized along one axis of the expression correlation matrix. Therefore in calculating background LNS, only the ortholog match was disturbed, but not the expression network structure (in contrast to randomizing along both axes).

Arrange expression data from two species by ortholog pairs

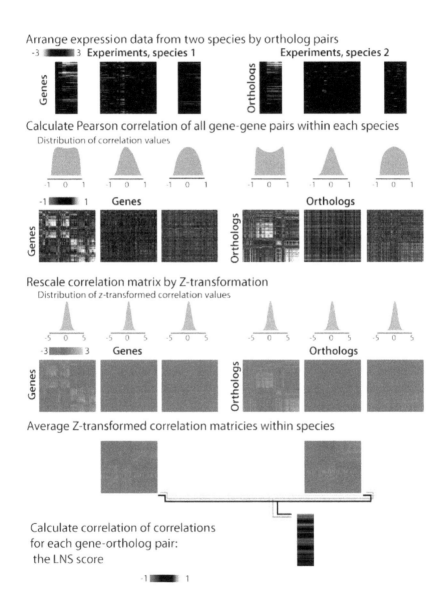

Calculate Pearson correlation of all gene-gene pairs within each species

Rescale correlation matrix by Z-transformation

Average Z-transformed correlation matricies within species

Calculate correlation of correlations
for each gene-ortholog pair:
 the LNS score

thologous genes. Since gene co-expression is strongly linked to functional similarities, differences in expression neighborhoods over a reasonably comprehensive set of perturbations provide a robust proxy of functional relationships. Previously developed methods of analyzing co-expression patterns across species have relied on producing matched datasets, in which comparable timepoints were collected from multiple species exposed to the same conditions [24]. However, exactly matched datasets are difficult to gather, and such an approach relies on the assumption that the precise timing of expression change should be conserved. We sought to develop a method that could detect large-scale patterns of co-expression in addition to those found under specifically matched conditions.

In order to quantify expression divergence at this global level, we developed a metric, Local Network Similarity (LNS), which measures the similarity of expression connections around each member of an ortholog pair (Figure 1). This metric is conceptually inspired by previous analyses that summarize the entire network of co-expression that exists between a gene and the rest of the genes in the genome [35,36]. The distribution of correlation coefficients of different datasets varies greatly, as is demonstrated in Figure 1, which plots actual expression datasets from the two species. LNS is calculated by stabilizing the variance of the distribution of correlation coefficients by first normalizing the within-species gene-gene correlation coefficients using the Fisher transformation and then further normalizing these data to the standard normal distribution. This normalization prevents domination by few changes of large magnitude or loss of signal from changes that occur in only a subset of perturbations (see Methods for details). The expression conservation of a pair of orthologous genes is thus quantified based on the preservation of all the expression connections around the pair of orthologs, i.e., the similarity of the local, first-degree expression networks around the two genes. The distribution of LNS of a completely non-conserved network resembles the normal distribution, making direct hypothesis tests possible.

Notably, and in contrast to previous global comparison approaches, this definition does not rely on alignment of individual datasets but defines a gene's expression pattern in the context of the genome-wide co-expression network. Therefore, the LNS concept could be extended to integrate any number of genome-scale expression datasets—or even other types of genomic data—to quantify conservation between any two species pairs.

12.2.2 GLOBAL EXPRESSION CONSERVATION AND DIVERGENCE IDENTIFIED BY LNS

LNS revealed considerable conservation of correlation structure between the two species' expression networks. We simulated the case of zero conservation by randomizing the orthologous pairs while preserving the network structure (Figure 1). This simulation resulted in LNS scores normally distributed and centered at 0 (Figure 2A). Unlike the randomized network, the LNS scores based on the matched orthologs were distributed from −0.63 to 0.83 with the median of 0.45 (Figure 2A, Additional file 1: Table S1), showing a clear shift towards positive values. This demonstrates that on average, orthologs preserved their expression network. A right shoulder in the LNS distribution suggests that there is a subset of genes with very highly conserved coexpression networks; this population of genes with high LNS persists even when ribosomal genes are removed from the data (data not shown).

In order to test whether this conservation is more or less than what would be expected at this degree of sequence divergence, we would ideally compare a similar gene expression compendium collected from genetically diverse isolates within each species or in species at different genetic distances. Although some data exist exploring differences in gene expression among *S. cerevisiae* strains [37-39], there are not sufficient datasets available for a full test. However, we were able to test the limit case of LNS as compared between different subsets of the *S. cerevisiae* literature. These experiments survey a wide diversity of conditions and have been performed in different strain backgrounds, though mostly focused on a small number of related lab strains. The LNS distribution within the subsampled *S. cerevisiae* datasets was significantly more positive than both the random distribution and the between species comparison (Figure 2A, Additional file 2: Table S2). This result demonstrates the robustness of LNS for identifying patterns of co-expression even across very different

A.

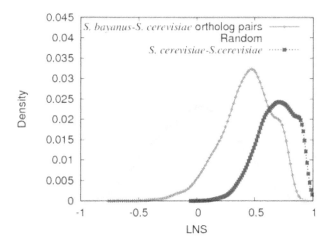

B.

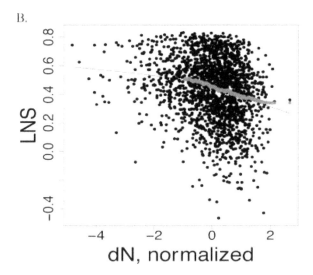

FIGURE 2: Expression conservation relationship with sequence conservation. A. Compared to randomized gene pairs, distribution of LNS scores of all one-to-one ortholog pairs is shifted towards positive values, though this distribution is less positive than comparisons within species. B. LNS is correlated with sequence divergence with a Pearson correlation of -0.235 (p<2.2e-16). Black line is loess smoothed curve line and red line is running average of LNS summed over 100-gene windows. Graphs for other measures of sequence divergence are in Figure S2.

environmental and genetic conditions, and suggests that the level of co-expression difference observed between *S. cerevisiae* and *S. bayanus* is probably beyond what is present within a species.

Cross-species LNS revealed some dramatic changes in correlation patterns in addition to the overall pattern of conservation. We found 183 genes out of the 4701 orthologous pairs (4%) with an LNS lower than 0, suggesting these genes changed globally in their expression network (Additional file 1: Table S1). The orthologs with low LNS represent several underlying biological causes. One obvious category is genes known to carry deficiencies in laboratory strains. For example, the promoter of *CTR3* is disrupted by a Ty2 insertion in many lab strains of *S. cerevisiae*[40], but is intact in *S. bayanus*. Most of the data in the *S. cerevisiae* compendium is derived from lab strains in which *CTR3* expression is driven by the transposon promoter, while the native promoter is present in the *S. bayanus* strains used in our compendium, leading to very different expression patterns and a low LNS score of −0.07166.

Similarly, we noted a number of targets of alpha factor signaling among the lowest LNS scores, and subsequently examined a list of known transcriptional targets of signaling through Gpa1, the alpha component of the heterotrimeric G protein that activates the MAP kinase cascade in yeast [41]. The median LNS of the *GPA1* targets is 0.27, significantly lower than the set of all genes. S288c derived laboratory strains of *S. cerevisiae* carry a S469I *GPA1* mutation that increases signaling through the MAP kinase cascade, but *S. bayanus* and all other members of the sensu stricto group carry the ancestral allele [42]. Therefore, the low overall LNS scores of the *GPA1* target genes may reflect the difference between wild type *GPA1* activity in *S. bayanus* and hyperactive signaling in laboratory strains of *S. cerevisiae*. Background mutations in *S. bayanus* can be identified as well, including the alpha factor protease gene [43], BAR1, which is mutant in the *S. bayanus* strain used in nearly every experiment in our expression compendium, and scored a low 0.26 LNS.

The second category of low LNS scores represents incorrectly annotated ortholog pairs. In our initial analysis, we discovered a number of cases of improperly assigned orthologs (Table 1). For instance, the *S. bayanus* gene *620.38* had been assigned the ortholog *RAS1* during the initial annotation effort [8], and had an LNS of −0.38. However, the protein sequence of the *620.38* ORF has only partial homology to Ras1 and is in fact a close homolog of the vacuolar protein Vps21. In addition, *620.38* is syntenic with VPS21[44]. This example demonstrates that LNS provides a functional criterion for ortholog identification and validation.

TABLE 1: Mis-annotated genes identified by LNS

Gene	Old ortholog	New ortholog	LNS	Comments
620.38	*YOR101W*	*YOR089C*	−0.379	Blast e-value to *VPS21* is 1e-92. Blast to *RAS1* is 11th on list.
576.11	*YGL157W*	*YGL039W*	0.184	Note similar synteny conflict with *674.45*. This gene is second best blast, e-value 1e-160.
674.45	*YGL039W*	*YGL157W*	−0.077	Synteny preserved by change, new gene is best blast hit with e-value 3e-146. Note similar problem with *576.11*.
635.17	*YOR267C*	*YOR233W*	−0.633	Synteny preserved by change. *YOR233W* is best blast hit with e-value 0.
636.13 (cell cycle)	*YPR119W* Best hit with *CLB5* (*YPR120C*).	*YPR120C*	−0.03	

Ortholog pairings were initially taken from the original genome annotation [7]. Several mis-pairs were identified by their low global or condition-specific LNS.

The remaining genes with low LNS are from diverse biological processes and functions. These genes are enriched for orthologs whose *S. cerevisiae* annotations include genes involved in ascospore cell wall formation (GO:0009272 fungal cell wall-type biogenesis, Bonferroni-corrected p-value 8.97×10^{-5} and related terms), and other developmental processes involved in mating and meiotic division, leading to the intriguing hypothesis that gene expression network differences may be related to speciation between these two yeasts. However, genes associated with these biological processes accounted for only 15% of the highly divergent genes. Multiple genes in this set are associated with DNA synthesis and repair, signaling, chromatin organization, metabolism, and transcription, among many other processes, emphasizing that differences are present throughout the cellular network. This list of low LNS genes in known functions should assist the prioritization of experimental work to determine the basis of evolutionary changes in expression. A full quarter of the genes with low LNS scores are of unknown function. Further experiments will be required to determine the mechanisms by which these genes diverge in their expression networks, and the degree to which these differences may contribute to phenotypic differences between the species.

12.2.3 SEQUENCE CONSERVATION PREDICTS GENE EXPRESSION DIVERGENCE

Despite the requirement for experimental followup of individual ortholog pairs, LNS analysis on our large data collection allows us to test several hypotheses regarding the overall role of genome sequence in determining interspecies variation of gene expression. First, we considered the effect of promoter structure by grouping ortholog pairs into TATA-containing in both species, TATA-containing in one of the members, and TATA-less. Though not statistically significant (Additional file 3: Figure S1, $r = -0.02$, $p = 0.08$), TATA-containing orthologs have lower LNS, indicating higher interspecies variation, consistent with previous results [24] using other yeast species and measurements of expression divergence.

Secondly, changes in promoter sequence could potentially cause changes in gene expression, so we extended the evaluation of promoters to examine the relationship between upstream sequence conservation and local network similarity. Upstream sequence conservation is weakly predictive of expression conservation (Additional file 3: Figure S1, $r = 0.047$, $p = 0.00016$, $n = 4701$).

Thirdly, in contrast to the small effects of TATA promoter type or upstream sequence conservation, we found a stronger correlation (Figure 2B, $r = -0.235$, $p < 2.2e-16$) between protein sequence similarity and local network similarity. This relationship was observed when using several different methods of calculating sequence similarity (Additional file 3: Figure S2). This correlation between protein sequence and expression similarity is consistent with the majority of results in mammalian systems

[45-47], *Xenopus*[48] and *Drosophila*[49,50]. This result contrasts with the conclusions of previous studies in yeast that did not detect any relationship between sequence conservation and expression conservation [26]. Our ability to detect this relationship may result from our use of a large, diverse expression compendium and a more generic measurement of expression divergence. Indeed, if we focus solely on a pair of cell cycle datasets and align them by time points, similar to previous works [24,25], we do not detect correlation between sequence conservation and expression conservation (not shown). As a result, although using aligned datasets could help identify orthologs responding with a similar pattern to a particular biological perturbation, calculation of expression correlation of orthologs in a single dataset cannot satisfactorily represent the expression conservation level.

12.2.4 MAJOR CONDITION-SPECIFIC CHANGES IN EXPRESSION BETWEEN S. BAYANUS AND S. CEREVISIAE

Using an expression compendium, the global LNS measures general expression change between species but may not be sensitive to changes in condition-specific expression patterns in response to specific environmental or genetic perturbations. For instance, a gene may be expressed at a basal level in one condition, but be strongly activated relative to other genes in a second condition. For example, the effect of divergence of Ste12 binding sites on alpha factor gene expression response was only detected in an alpha factor dataset and not in a larger expression compendium [25], likely because there is minimal alpha factor signaling under conditions where alpha factor is absent. Since we anticipate that genes highly conserved in expression response in one condition could diverge significantly in another condition, so we investigated expression conservation using a dataset-specific approach. In order to more sensitively examine condition-specific gene expression conservation, we calculated the LNS for individual datasets or small, related groups of datasets similar in size and perturbation and examined the behavior of different functional groups (Table 2). This measure, which we call condition-specific LNS (Additional file 4: Table S3), is still independent of the timing of perturbation response in the two species, but provides a condition-specific measure of expression divergence, enabling us to more sensitively detect divergence that is specific to a particular condition.

TABLE 2: Paired datasets of *S. bayanus* and *S. cerevisiae* for condition-specific LNS calculation

S. bayanus dataset	*S. cerevisiae* dataset	Number of diverged genes (lower than average of randomization, cutoff)	P -value	Median LNS
MMS	[51]	1498 (−0.02515)	1.67E-153	0.2348
Heat shock	[52]	1032 (−0.00467)	4.25E-271	0.2259
Sorbitol	[52]	1330 (0.00561)	3.35E-163	0.1678
Uracil	[53]	1590 (0.001)	5.75443E-82	0.1031
Rapamycin	[54]	1896 (−0.0101)	1.63545E-33	0.07464
Alpha factor	[55]	1438 (−0.006)	5.5186E-102	0.0848
Sporulation	[56]	922 (0.0038)	3.7792E-288	0.08923
Cell cycle	[57]	1124 (0.0047)	1.7278E-226	0.09546
Diauxic shift	[58]	947 (0.00656)	1.1322E-257	0.3741

We permuted gene-gene correlations to simulate the LNS distribution of a randomized network. The numbers of genes below the average LNS for these random networks were counted. We also calculated the p-value by randomizing the matches between orthologs to simulate non-conservation. The average LNS for each matched dataset pair was calculated.

We found that, overall, orthologs exhibited some level of conservation in expression pattern regardless of the treatment, as expressed in the bias towards positive condition-specific LNS across matched datasets (Figure 3 and Table 2). However, individual experimental treatments and genetic perturbations demonstrated different patterns of expression conservation. For example, in the heat shock and diauxic shift treatments, genes on average showed high expression conservation (Figure 3 and Table 2). Other perturbations, for example, alpha factor treatment, uracil limitation, and rapamycin treatment, exhibited a relatively higher number of genes with divergent expression, although the majority of genes were still well-conserved in these datasets. The differences among the LNS distributions are most likely due to both the properties of the experimental treatment and the quality and size of the two datasets compared for each treatment. In order to normalize these effects and attempt to quantify the number of genes with divergent expression across matched datasets, we randomized the ortholog match for each dataset pair so as to simulate the situation of no conservation. In other words, to generate the randomized set, for all ortholog pairs, one member of the pair was matched with a random ortholog in the other species (with the random ortholog coming from some other orthologous pair). In general, around a quarter of the orthologs had a condition-specific LNS lower than the average LNS of randomized datasets. This indicates even very conserved biological responses elicit different gene expression consequences in the two species.

Genes of different functional categories showed differential levels of expression divergence under specific conditions. In general, genes of ribonucleoprotein complex biogenesis and assembly (a term which contains primarily genes involved in ribosome structure and assembly) showed highly conserved expression patterns regardless of the nature of the expression perturbation. Other categories of genes showed more specific patterns of conservation. For example, cell cycle genes were most conserved in the cell cycle dataset and alpha factor treatment. In datasets such as rapamycin or uracil treatment, these genes did not show specific conservation in their coexpression network. This result indicates that conclusions on gene expression conservation can be generic (e.g., ribosomal-related genes) or true under only very specifically defined conditions.

Condition-specific LNS identified mis-annotated genes in *S. bayanus* that were overlooked by the global LNS analysis. For example, *S. bayanus* gene *636.13* was matched to *CLB2* (*YPR119W*) in the initial annotation efforts by [8]. However, it has a low cell cycle LNS (-0.03) and this lack of expression conservation is evident by its shift in peak expression during the cell cycle (Additional file 3: Figure S3). 636.13 instead corresponds to *CLB5* based on both synteny and Blast alignment [44]. This change in gene expression was observable in the condition-specific LNS analysis of cell cycle synchronized cells because this alternate phase of expression changed the correlations with other periodically regulated genes. In the global dataset that consisted of primarily asynchronous cells, these phase specific correlations were not present.

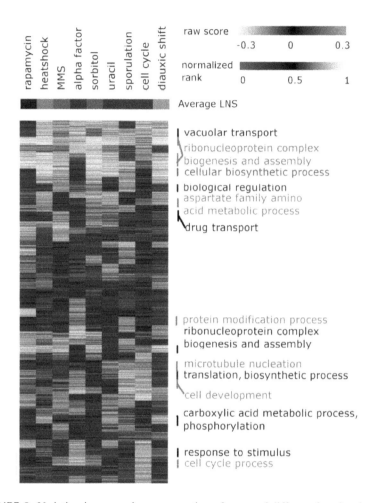

FIGURE 3: Variation in expression conservation of genes of different functional groups under different perturbations. The LNS for each gene for matched datasets of the two species was calculated. LNS scores were first k-means clustered and then arranged hierarchically by the centers of these clusters, and the scores presented by a heat map. GO biological process enrichment for each cluster was determined. Enriched terms with a Bonferonni-corrected p value lower than 0.01 are labeled. The expression patterns of ribosome-related genes are conserved upon most perturbations. However, the expression of many of the genes of other functions is only conserved under specific conditions. The range of LNS scores varies for different datasets. Datasets with large magnitude expression changes tend to have a greater LNS range; dataset size and quality also influence the range of LNS.

We take two examples here to illustrate major changes in the expression response to environmental change between the two species. First, we quantized the expression data from the diauxic shift in *S. bayanus* and *S. cerevisiae*[58] based on their diauxic shift condition-specific LNS (Figure 4A-B). We observed that although the majority of the genes preserved their expression response upon diauxic shift, the lower quartile (by condition-specific LNS) of the orthologs displayed largely anti-correlated expression. Accordingly, 941 orthologs displayed a negative condition-specific LNS in diauxic shift. *S. bayanus* and *S. cerevisiae* have several observed differences in fermentation behavior [60-62], some of which could be associated with the changes in gene expression that we observe.

A similarly large fraction of divergent genes was observed for the paired cell cycle data. We identified 591 (in *S. bayanus*) and 626 (in *S. cerevisiae*) cell cycle-regulated genes whose one-to-one orthologs have data in the other species, making approximately the same proportion of genes cell cycle-regulated in both species. In total this represents 1106 unique genes, with 111 pairs of orthologous genes periodic in both *S. bayanus* and *S. cerevisiae* (p < 0.002, two tailed t-test). We also assessed the conservation of cell cycle specific gene expression using the 800 genes previously identified by a different dataset [55] as periodically expressed in synchrony with the cell cycle in *S. cerevisiae*, and observed that 226 of these orthologs were periodically expressed in synchronous cultures of *S. bayanus* (p<0.001). A large fraction of the periodically expressed genes were only cell cycle regulated in one of the species (Figure 4C-D): of the 1106 cell cycle regulated genes identified in either species, 258 had a cell cycle-specific LNS lower than 0, indicating significant change in their behavior over the cell cycle.

During the cell cycle, either phase changes or a presence of periodic expression in only one species could contribute to low LNS. We have recently correlated some of these changes in periodic gene expression with differential motif presence and nucleosome occupancy in these genes' promoters [10]. Some of the differences in timing could result from changes in the regulation of the cell cycle, but coherent cycling of protein levels could also be achieved even when gene expression is divergent. For example, if one member of a protein complex was periodically expressed in one species, another member of the same complex could be periodically expressed in the other species. This scenario would result in divergence of expression pattern even though the protein complex was periodically regulated through the availability of the cycling component [45,63,64]. Indeed we observe that although most cell cycle protein complexes retain cell cycle-regulated genes in both species, the identity of dynamic members differs between species (Additional file 3: Figure S4). These observations of expression divergence are not limited to the specific examples described above: all datasets have a fraction of divergently expressed genes despite the general trend of expression conservation observed over all data, underlining the importance of a dataset-specific measurement of expression conservation.

12.3 DISCUSSION

In this study, we employed a scalable measurement of expression conservation between species, Local Network Similarity, or LNS, to study the conservation of gene expression networks using large microarray data compendia from *S. bayanus* and *S. cerevisiae*. This unsupervised metric allowed us to quantify expression divergence between orthologs for datasets that are different in time-course sampling, or for species that have differential response kinetics to environmental perturbations. This distance metric scales the measurement of expression conservation between -1 and 1, with the null-hypothesis distribution centered at zero. Future research directions include extension of this metric to greater evolutionary distances and diverse data types beyond gene expression. We expect that the normalization inherent in the LNS metric will make it particularly robust for RNA-seq data in the face of the larger noise component that has been observed for genes with low expression levels [65-68].

Using the newly developed LNS metric, we found that patterns of expression-level divergence vary among different biological processes and functional groups. Certain central processes such as ribosome biogenesis are highly conserved on both the sequence and expression level. However, other functional groups involved in seemingly conserved behavior (e.g. cell cycle, diauxic shift) in fact include a significant fraction of orthologs whose expression diverges. This indicates that specific expression patterning

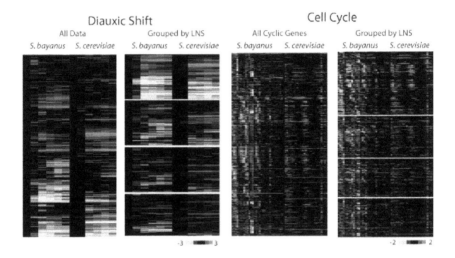

FIGURE 4: Condition-specific LNS sorting of expression data into conserved and non-conserved patterns. A. Expression data from the diauxic shift in both *S. bayanus* and *S. cerevisiae* were separately zero-transformed, and then aligned so that each horizontal line of data contains the expression of an orthologous pair of genes. The paired expression data were hierarchically clustered by uncentered Pearson correlation [59]. The data are presented by a heat map of log2 expression values. *S. cerevisiae* diauxic shift data were from [58]. B. The clustered data were partitioned by LNS score into quartiles, preserving the order of the genes in the original cluster. Genes with different expression between the species have low LNS scores. C. Cell cycle data from MATa cells synchronized by alpha factor arrest were mean centered. *S. cerevisiae* cell cycle data was obtained from [57]. For each species, the phase of the cell cycle was determined by Fourier transformation, and the top genes mapped to the phase were determined as cell cycle-regulated. Orthologs with either member of the ortholog pair determined as cell cycle-regulated are presented. The orthologs were arranged by the time of peak expression in *S. bayanus*.

of some genes is not critical to an organism's response to environmental change. On the other hand, the overall conserved expression patterns between the two species might represent genes with key functional roles in responding to specific environmental changes.

One limitation of the current study is that we cannot determine to what degree sequence divergence explains overall expression network divergence between *S. bayanus* and *S. cerevisiae*. The selective forces acting on gene expression are as yet unclear and deriving a null model for gene expression evolution is a topic of active research (reviewed in [69-71]). Our observation of a bimodal distribution of LNS scores suggests that some genes could be evolving under selection for conserved expression patterns, while others may evolve more neutrally and thus show greater variance [72-74]. However, those genes that appear to be evolving neutrally could be simply not specifically perturbed in the conditions used for the available expression data. Addressing this challenge experimentally requires further collection of diverse expression datasets in genetically divergent strains within these species, as well as in other species across a range of evolutionary distances. Using such datasets, LNS can be used to delineate further how expression differences change with varying levels of sequence change.

12.4 CONCLUSIONS

This study focuses on the response and expression profiling of *S. bayanus* under different perturbations. Complementary studies of regulatory networks, such as interrogation of transcription factor occupancy [75] and nucleosome positioning [10,76], will be useful to more fully characterize changes between species. Furthermore, while here we provide a prototype application of our network-based divergence measure (LNS) to gene expression, this approach should be extendable to other types of genome-wide data and can encompass diverse types of quantification of co-expression and/or network similarity. Extending our comparative approach to other groups of related species, such as *Candida* yeasts, *Drosophila* species, and mammals, could extend the observations made here. Since our experimental and analytical frameworks are agnostic to species and platform, they should be transferable to other systems. Such studies can be combined with sequence analysis to yield a more complete understanding of the mode of phenotype evolution and its relationship with sequence changes.

12.5 METHODS

12.5.1 PRE-PROCESSING OF S. CEREVISIAE AND S. BAYANUS DATA

We assembled 303 arrays covering 46 datasets in *S. bayanus* [GEO: GSE16544] and 2569 arrays covering 125 datasets in *S. cerevisiae*. To allow reasonable comparison between datasets, we performed the following normalization steps. For each dataset, genes that are represented in less than 50% of the arrays were removed from this dataset, and missing values were estimated using KNNimpute with K = 10, Euclidean distance [77]. Finally, biological replicates are averaged, resulting in datasets with each gene followed by a vector representing its expression values in a series of arrays.

12.5.2 CALCULATION OF PAIR-WISE CORRELATION

For each dataset i, between gene pair expression vectors j and k, we calculated the correlation coefficient of their expression pattern:

$$\rho(j_i, k_i) = \frac{\text{cov}(j_i, k_i)}{\sigma_j, \sigma_{k_i}} \tag{1}$$

To allow comparison between datasets, we Fisher's z-transformed these correlation values [78]:

$$z(j_i, k_i) = \frac{1}{2} \ln \frac{1 + \rho(j_i, k_i)}{1 - \rho(j_i, k_i)}$$

$$(2)$$

For each dataset, these z values were then normalized to $Z' \sim N(0,1)$. We define the connection weights z(j,k) between any two genes j and k in a species as the average of the normalized z values over all datasets. This forms a pair-wise connection weight matrix for each species.

12.5.3 CALCULATION OF LOCAL NETWORK SIMILARITY (LNS) AS A MEASUREMENT OF EXPRESSION DIVERGENCE

Connection weights between a specific gene j to all others form a vector W(j) = (z(1,j), z(2,j)...z(N,j)), where N is the total number of matched orthologs. LNSj,j' is defined as the correlation between the matched vectors (by orthology) of the two species, quantifying the conservation of the overall expression pattern of gene j (with its ortholog being j') (Figure 1):

$$LNS_{j,j'} = \frac{\text{cov}(W(j), W(j'))}{\sigma_{W(j)} \sigma_{W(j')}}$$

$$(3)$$

To assess the conservation level of orthologs upon specific biological perturbation (condition-specific LNS), we manually selected 10 datasets in *S. bayanus* that have matched time-course data in *S. cerevisiae*. For each data pair, we define the connection weight as the standard normalized value of formula (2), followed by the calculation of condition-specific LNS according to formula (3) (Figure 3).

12.5.4 CORRELATION BETWEEN SEQUENCE DIVERGENCE AND LNS

Measures of sequence divergence were used as in [26], including dN, dN/dS, Ka, and Ka/Ks as previously calculated [7,79]. A normal distribution was obtained by $\log_2$ transforming the data, mean subtracting it, and normalizing by the standard deviation. Correlation was calculated using the Pearson correlation.

12.5.5 CLUSTERING OF CONDITION-SPECIFIC LNS AND FUNCTIONAL ENRICHMENT ANALYSIS

We determined the number of clusters in the condition-specific LNS matrix according to [80], resulting in 48 clusters. The enrichment of genes in each cluster was calculated through the program GOTermFinder [81].

12.5.6 IDENTIFICATION OF CELL CYCLE REGULATED EXPRESSION

We acquired *S. cerevisiae* cell cycle data from [57]. The following steps of identification of cell cycle regulated genes were applied to each species. For each gene in the cell cycle data, the expression values were centered so that the average over the time course equals to 0. A Fourier transformation was applied to the dataset of individual species so as to identify the period of cell cycle [55]. In *S. bayanus* the top 613 genes mapping to the phase were chosen as cell cycle regulated; and 785 genes in *S. cerevisiae* were chosen. This corresponds to 601 and 644 genes in *S. bayanus* and *S. cerevisiae* having one-to-one orthologs in the other species respectively. Among them, 591 (*S. bayanus*) and 626 (*S. cerevisiae*) have expression data in the other species, with 111 pairs overlapping.

REFERENCES

1. Galagan JE, Henn MR, Ma LJ, Cuomo CA, Birren B: Genomics of the fungal kingdom: insights into eukaryotic biology. Genome Res 2005, 15(12):1620-1631.
2. Kellis M, Birren BW, Lander ES: Proof and evolutionary analysis of ancient genome duplication in the yeast *Saccharomyces cerevisiae*. Nature 2004, 428(6983):617-624.
3. Dietrich FS, Voegeli S, Brachat S, Lerch A, Gates K, Steiner S, Mohr C, Pohlmann R, Luedi P, Choi S, et al.: The *Ashbya gossypii* genome as a tool for mapping the ancient *Saccharomyces cerevisiae* genome. Science 2004, 304(5668):304-307.
4. Wapinski I, Pfeffer A, Friedman N, Regev A: Natural history and evolutionary principles of gene duplication in fungi. Nature 2007, 449(7158):54-61.
5. Scannell DR, Frank AC, Conant GC, Byrne KP, Woolfit M, Wolfe KH: Independent sorting-out of thousands of duplicated gene pairs in two yeast species descended from a whole-genome duplication. Proc Natl Acad Sci USA 2007, 104(20):8397-8402.
6. Dujon B, Sherman D, Fischer G, Durrens P, Casaregola S, Lafontaine I, De Montigny J, Marck C, Neuveglise C, Talla E, et al.: Genome evolution in yeasts. Nature 2004, 430(6995):35-44.
7. Kellis M, Patterson N, Endrizzi M, Birren B, Lander ES: Sequencing and comparison of yeast species to identify genes and regulatory elements. Nature 2003, 423(6937):241-254.
8. Cliften P, Sudarsanam P, Desikan A, Fulton L, Fulton B, Majors J, Waterston R, Cohen BA, Johnston M: Finding functional features in Saccharomyces genomes by phylogenetic footprinting. Science 2003, 301(5629):71-76.
9. Gasch AP, Moses AM, Chiang DY, Fraser HB, Berardini M, Eisen MB: Conservation and evolution of cis-regulatory systems in ascomycete fungi. PLoS Biol 2004, 2(12):e398.
10. Guan Y, Yao V, Tsui K, Gebbia M, Dunham MJ, Nislow C, Troyanskaya OG: Nucleosome-coupled expression differences in closely-related species. BMC Genomics 2011, 12:466.
11. Tsong AE, Miller MG, Raisner RM, Johnson AD: Evolution of a combinatorial transcriptional circuit: a case study in yeasts. Cell 2003, 115(4):389-399.
12. Tsong AE, Tuch BB, Li II, Johnson AD. Evolution of alternative transcriptional circuits with identical logic. Nature 2006, 443(7110):415-420.
13. Ihmels J, Bergmann S, Berman J, Barkai N: Comparative gene expression analysis by differential clustering approach: application to the *Candida albicans* transcription program. PLoS Genet 2005, 1(3):e39.
14. Gasch AP: Comparative genomics of the environmental stress response in ascomycete fungi. Yeast 2007, 24(11):961-976.

15. Rustici G, van Bakel H, Lackner DH, Holstege FC, Wijmenga C, Bahler J, Brazma A: Global transcriptional responses of fission and budding yeast to changes in copper and iron levels: a comparative study. Genome Biol 2007, 8(5):R73.

16. Wohlbach DJ, Kuo A, Sato TK, Potts KM, Salamov AA, Labutti KM, Sun H, Clum A, Pangilinan JL, Lindquist EA, et al.: Comparative genomics of xylose-fermenting fungi for enhanced biofuel production. Proc Natl Acad Sci 2011, 108(32):13212-13217.

17. Scannell DR, Zill OA, Rokas A, Payen C, Dunham MJ, Eisen MB, Rine J, Johnston M, Hittinger CT: The Awesome Power of Yeast Evolutionary Genetics: New Genome Sequences and Strain Resources for the Saccharomyces sensu stricto Genus. G3 (Bethesda, Md) 2011, 1(1):11-25.

18. Liti G, Carter DM, Moses AM, Warringer J, Parts L, James SA, Davey RP, Roberts IN, Burt A, Koufopanou V, et al.: Population genomics of domestic and wild yeasts. Nature 2009, 458(7236):337-341.

19. Bullard JH, Mostovoy Y, Dudoit S, Brem RB: Polygenic and directional regulatory evolution across pathways in Saccharomyces. Proc Natl Acad Sci 2010, 107(11):5058-5063.

20. Horinouchi T, Yoshikawa K, Kawaide R: Genome-wide expression analysis of Saccharomyces pastorianus orthologous genes using oligonucleotide microarrays. J Biosci Bioeng 2010, 110(5):602-607.

21. Martin OC, DeSevo CG, Guo BZ, Koshland DE, Dunham MJ, Zheng Y: Telomere behavior in a hybrid yeast. Cell Res 2009, 19(7):910.

22. Tirosh I, Reikhav S, Levy AA, Barkai N: A yeast hybrid provides insight into the evolution of gene expression regulation. Science (New York, NY) 2009, 324(5927):659-662.

23. Zill OA, Scannell D, Teytelman L, Rine J: Co-evolution of transcriptional silencing proteins and the DNA elements specifying their assembly. PLoS Biol 2010, 8(11):e1000550.

24. Tirosh I, Weinberger A, Carmi M, Barkai N: A genetic signature of interspecies variations in gene expression. Nat Genet 2006, 38(7):830-834.

25. Tirosh I, Weinberger A, Bezalel D, Kaganovich M, Barkai N: On the relation between promoter divergence and gene expression evolution. Mol Syst Biol 2008, 4:159.

26. Tirosh I, Barkai N: Evolution of gene sequence and gene expression are not correlated in yeast. Trends Genet 2008, 24(3):109-113.

27. Lelandais G, Tanty V, Geneix C, Etchebest C, Jacq C, Devaux F: Genome adaptation to chemical stress: clues from comparative transcriptomics in *Saccharomyces cerevisiae* and *Candida glabrata*. Genome Biol 2008, 9(11):R164.

28. Guan Y, Dunham M, Caudy A, Troyanskaya O: Systematic planning of genome-scale experiments in poorly studied species. PLoS Comput Biol 2010, 6(3):e1000698.

29. Flintoft L: Learning to prioritize. Nat Rev Genet 2010, 11(5):315.

30. Parkinson H, Kapushesky M, Shojatalab M, Abeygunawardena N, Coulson R, Farne A, Holloway E, Kolesnykov N, Lilja P, Lukk M, et al.: ArrayExpress--a public database of microarray experiments and gene expression profiles. Nucleic Acids Res 2007, 35(Database issue):D747-D750.

31. Ball CA, Jin H, Sherlock G, Weng S, Matese JC, Andrada R, Binkley G, Dolinski K, Dwight SS, Harris MA, et al.: Saccharomyces Genome Database provides tools

to survey gene expression and functional analysis data. Nucleic Acids Res 2001, 29(1):80-81.

32. Edgar R, Domrachev M, Lash AE: Gene Expression Omnibus: NCBI gene expression and hybridization array data repository. Nucleic Acids Res 2002, 30(1):207-210.

33. Le Crom S, Devaux F, Jacq C, Marc P: yMGV: helping biologists with yeast microarray data mining. Nucleic Acids Res 2002, 30(1):76-79.

34. Marinelli RJ, Montgomery K, Liu CL, Shah NH, Prapong W, Nitzberg M, Zachariah ZK, Sherlock GJ, Natkunam Y, West RB, et al.: The Stanford Tissue Microarray Database. Nucleic Acids Res 2008, 36(Database issue):D871-D877.

35. Dutilh BE, Huynen MA, Snel B: A global definition of expression context is conserved between orthologs, but does not correlate with sequence conservation. BMC Genomics 2006, 7:10.

36. Tirosh I, Barkai N: Comparative analysis indicates regulatory neofunctionalization of yeast duplicates. Genome Biol 2007, 8(4):R50.

37. Brem RB, Yvert G, Clinton R, Kruglyak L: Genetic dissection of transcriptional regulation in budding yeast. Science (New York, NY) 2002, 296(5568):752-755.

38. Fay JC, McCullough HL, Sniegowski PD, Eisen MB: Population genetic variation in gene expression is associated with phenotypic variation in Saccharomyces cerevisiae. Genome Biol 2004, 5(4):R26.

39. Smith EN, Kruglyak L: Gene-environment interaction in yeast gene expression. PLoS Biol 2008, 6(4):e83.

40. Knight SA, Labbe S, Kwon LF, Kosman DJ, Thiele DJ: A widespread transposable element masks expression of a yeast copper transport gene. Genes Dev 1996, 10(15):1917-1929.

41. Lang GI, Murray AW, Botstein D: The cost of gene expression underlies a fitness trade-off in yeast. Proc Natl Acad Sci USA 2009, 106(14):5755-5760.

42. Yvert G, Brem RB, Whittle J, Akey JM, Foss E, Smith EN, Mackelprang R, Kruglyak L: Trans-acting regulatory variation in Saccharomyces cerevisiae and the role of transcription factors. Nat Genet 2003, 35(1):57-64.

43. Zill OA, Rine J: Interspecies variation reveals a conserved repressor of alpha-specific genes in Saccharomyces yeasts. Genes Dev 2008, 22(12):1704-1716.

44. Gordon JL, Byrne KP, Wolfe KH: Additions, losses, and rearrangements on the evolutionary route from a reconstructed ancestor to the modern Saccharomyces cerevisiae genome. PLoS Genet 2009, 5(5):e1000485.

45. Jordan IK, Marino Ramirez L, Koonin EV: Evolutionary significance of gene expression divergence. Gene 2005, 345(1):119-126.

46. Khaitovich P, Hellmann I, Enard W, Nowick K, Leinweber M, Franz H, Weiss G, Lachmann M, Paabo S: Parallel patterns of evolution in the genomes and transcriptomes of humans and chimpanzees. Science 2005, 309(5742):1850-1854.

47. Liao BY, Zhang J: Evolutionary conservation of expression profiles between human and mouse orthologous genes. Mol Biol Evol 2006, 23(3):530-540.

48. Sartor MA, Zorn AM, Schwanekamp JA, Halbleib D, Karyala S, Howell ML, Dean GE, Medvedovic M, Tomlinson CR: A new method to remove hybridization bias for interspecies comparison of global gene expression profiles uncovers an association between mRNA sequence divergence and differential gene expression in Xenopus. Nucleic Acids Res 2006, 34(1):185-200.

49. Nuzhdin SV, Wayne ML, Harmon KL, McIntyre LM: Common pattern of evolution of gene expression level and protein sequence in Drosophila. Mol Biol Evol 2004, 21(7):1308-1317.

50. Lemos B, Bettencourt BR, Meiklejohn CD, Hartl DL: Evolution of proteins and gene expression levels are coupled in Drosophila and are independently associated with mRNA abundance, protein length, and number of protein-protein interactions. Mol Biol Evol 2005, 22(5):1345-1354.

51. Shalem O, Dahan O, Levo M, Martinez MR, Furman I, Segal E, Pilpel Y: Transient transcriptional responses to stress are generated by opposing effects of mRNA production and degradation. Mol Syst Biol 2008, 4:223.

52. Gasch AP, Spellman PT, Kao CM, Carmel-Harel O, Eisen MB, Storz G, Botstein D, Brown PO: Genomic expression programs in the response of yeast cells to environmental changes. Mol Biol Cell 2000, 11(12):4241-4257.

53. Saldanha AJ, Brauer MJ, Botstein D: Nutritional homeostasis in batch and steady-state culture of yeast. Mol Biol Cell 2004, 15(9):4089-4104.

54. Urban J, Soulard A, Huber A, Lippman S, Mukhopadhyay D, Deloche O, Wanke V, Anrather D, Ammerer G, Riezman H, et al.: Sch9 is a major target of TORC1 in *Saccharomyces cerevisiae*. Mol Cell 2007, 26(5):663-674.

55. Spellman PT, Sherlock G, Zhang MQ, Iyer VR, Anders K, Eisen MB, Brown PO, Botstein D, Futcher B: Comprehensive identification of cell cycle-regulated genes of the yeast *Saccharomyces cerevisiae* by microarray hybridization. Mol Biol Cell 1998, 9(12):3273-3297.

56. Primig M, Williams RM, Winzeler EA, Tevzadze GG, Conway AR, Hwang SY, Davis RW, Esposito RE: The core meiotic transcriptome in budding yeasts. Nat Genet 2000, 26(4):415-423.

57. Pramila T, Wu W, Miles S, Noble WS, Breeden LL: The Forkhead transcription factor Hcm1 regulates chromosome segregation genes and fills the S-phase gap in the transcriptional circuitry of the cell cycle. Genes Dev 2006, 20(16):2266-2278.

58. Brauer MJ, Saldanha AJ, Dolinski K, Botstein D: Homeostatic adjustment and metabolic remodeling in glucose-limited yeast cultures. Mol Biol Cell 2005, 16(5):2503-2517.

59. Eisen MB, Spellman PT, Brown PO, Botstein D: Cluster analysis and display of genome-wide expression patterns. Proc Natl Acad Sci USA 1998, 95(25):14863-14868.

60. Arroyo-López FN, Salvadó Z, Tronchoni J, Guillamón JM, Barrio E, Querol A: Susceptibility and resistance to ethanol in Saccharomyces strains isolated from wild and fermentative environments. Yeast (Chichester, England) 2010, 27(12):1005-1015.

61. Masneuf-Pomarède I, Bely M, Marullo P, Lonvaud-Funel A, Dubourdieu D: Reassessment of phenotypic traits for Saccharomyces bayanus var. uvarum wine yeast strains. Int J Food Microbiol 2010, 139(1–2):79-86.

62. Salvadó Z, Arroyo-López FN, Guillamón JM, Salazar G, Querol A, Barrio E: Temperature adaptation markedly determines evolution within the genus Saccharomyces. Appl Environ Microbiol 2011, 77(7):2292-2302.

63. de Lichtenberg U, Jensen LJ, Brunak S, Bork P: Dynamic complex formation during the yeast cell cycle. Science 2005, 307(5710):724-727.

64. Jensen LJ, Jensen TS, de Lichtenberg U, Brunak S, Bork P: Co-evolution of transcriptional and post-translational cell-cycle regulation. Nature 2006, 443(7111):594-597.

65. Tarazona S, Garcia-Alcalde F, Dopazo J, Ferrer A, Conesa A: Differential expression in RNA-seq: A matter of depth. Genome Res 2011, 21(12):2213-2223.

66. McIntyre LM, Loplano KK, Morse AM, Amin V, Oberg AL, Young LJ, Nuzhdin SV: RNA-seq: technical variability and sampling. BMC Genomics 2011, 12(1):293.

67. Liu S, Lin L, Jiang P, Wang D, Xing Y: A comparison of RNA-Seq and high-density exon array for detecting differential gene expression between closely related species. Nucleic Acids Res 2011, 39(2):578-588.

68. Bloom JS, Khan Z, Kruglyak L, Singh M, Caudy AA: Measuring differential gene expression by short read sequencing: quantitative comparison to 2-channel gene expression microarrays. BMC Genomics 2009, 10:221.

69. Gilad Y, Oshlack A, Rifkin SA: Natural selection on gene expression. Trends Genet 2006, 22(8):456-461.

70. Zheng W, Gianoulis TA, Karczewski KJ, Zhao H, Snyder M: Regulatory variation within and between species. Annu Rev Genomics Hum Genet 2011, 12:327-346.

71. Romero IG, Ruvinsky I, Gilad Y: Comparative studies of gene expression and the evolution of gene regulation. Nat Rev Genet 2012, 13(7):505-516.

72. Whitehead A, Crawford DL: Neutral and adaptive variation in gene expression. Proc Natl Acad Sci USA 2006, 103(14):5425-5430.

73. Rifkin SA, Kim J, White KP: Evolution of gene expression in the Drosophila melanogaster subgroup. Nat Genet 2003, 33(2):138-144.

74. Khaitovich P, Pääbo S, Weiss G: Toward a neutral evolutionary model of gene expression. Genetics 2005, 170(2):929-939.

75. Borneman AR, Gianoulis TA, Zhang ZD, Yu H, Rozowsky J, Seringhaus MR, Wang LY, Gerstein M, Snyder M: Divergence of transcription factor binding sites across related yeast species. Science 2007, 317(5839):815-819.

76. Tsankov AM, Thompson DA, Socha A, Regev A, Rando OJ: The role of nucleosome positioning in the evolution of gene regulation. PLoS Biol 2010, 8(7):e1000414.

77. Troyanskaya O, Cantor M, Sherlock G, Brown P, Hastie T, Tibshirani R, Botstein D, Altman RB: Missing value estimation methods for DNA microarrays. Bioinformatics (Oxford, England) 2001, 17(6):520-525.

78. Fisher RA: Frequency distribution of the values of the correlation coefficients in samples from an indefinitely large population. Biometrika 1915, 10(4):507-521.

79. Wall DP, Hirsh AE, Fraser HB, Kumm J, Giaever G, Eisen MB, Feldman MW: Functional genomic analysis of the rates of protein evolution. Proc Natl Acad Sci USA 2005, 102(15):5483-5488.

80. Mardia KV, Kent T, Bibby JM: Multivariate Analysis. sixth edition. London, New York, Toronto, Sydney, San Francisco: Academic Press; 1997.

81. Boyle EI, Weng S, Gollub J, Jin H, Botstein D, Cherry JM, Sherlock G: GO:TermFinder–open source software for accessing Gene Ontology information and finding significantly enriched Gene Ontology terms associated with a list of genes. Bioinformatics 2004, 20(18):3710-3715.

This article has supplemental information that is not featured in this version of the text. To view these files, please visit the original version of the article as cited in the beginning of this chapter.

CHAPTER 13

WHOLE-GENOME COMPARISON REVEALS NOVEL GENETIC ELEMENTS THAT CHARACTERIZE THE GENOME OF INDUSTRIAL STRAINS OF *SACCHAROMYCES CEREVISIAE*

ANTHONY R. BORNEMAN, BRIAN A. DESANY, DAVID RICHES, JASON P. AFFOURTIT, ANGUS H. FORGAN, ISAK S. PRETORIUS, MICHAEL EGHOLM, AND PAUL J. CHAMBERS

13.1 INTRODUCTION

During its long history of association with human activity, the genomic makeup of the yeast *S. cerevisiae* is thought to have been shaped through the action of multiple independent rounds of wild yeast domestication combined with thousands of generations of artificial selection. As the evolutionary constraints that were applied to the *S. cerevisiae* genome during these domestication events were ultimately dependent on the desired function of the yeast (e.g baking, brewing, wine or bioethanol production), these multitude of selective schemes have produced large numbers of *S. cerevisiae* strains, with highly specialized phenotypes that suit specific applications [1], [2]. As a result, the study of industrial strains of *S.*

This chapter was originally published under the Creative Commons Attribution License. Borneman AR, Desany BA, Riches D, Affourtit JP, Forgan AH, Pretorius IS, Egholm M, and Chambers PJ. Whole-Genome Comparison Reveals Novel Genetic Elements That Characterize the Genome of Industrial Strains of Saccharomyces cerevisiae. PLoS Genetics 7,2 (2011). doi:10.1371/journal.pgen.1001287 cerevisiae provides an excellent model of how reproductive isolation and divergent selective pressures can shape the genomic content of a species.

Despite their diverse roles, industrial yeast strains all share the general ability to grow and function under the concerted influences of a multitude of environmental stressors, which include low pH, poor nutrient availability, high ethanol concentrations and fluctuating temperatures. In comparison, non-industrial isolates such as laboratory strains, have been selected for rapid and consistent growth in nutrient rich laboratory media, thereby producing markedly different phenotypic outcomes when compared to their industrial relatives [3]. The outcomes of these very different selection pressures are therefore most evident when comparing industrial and non-industrial yeasts. As an example, laboratory strains of *S. cerevisiae*, such as S288c, are unable to grow in the low pH and high osmolarity of most grape juices and therefore cannot be used to make wine. This is a clear difference between industrial and non-industrial strains of *S. cerevisiae*, however there are numerous subtle differences not only between industrial strains, but also between strains used within the same industry [4], [5], highlighting the overall genetic diversity found in this species.

There have been several attempts to characterize the genomes of industrial strains of *S. cerevisiae* which have uncovered differences that included single nucleotide polymorphisms (SNPs), strain-specific ORFs and localized variations in genomic copy number [6]–[14]. However, the type and scope of genomic variation documented by these studies were limited either by technology constraints (e.g arrayCGH relying on the laboratory strain as a "reference" genome), or by the resources required for the production of high-quality genomic assemblies which has limited the scope and number of whole-genome sequences available for comparison. In addition, to limit genomic complexity to a manageable level, previously published whole-genome sequencing studies on industrial strains used haploid representations of diploid, and often heterozygous, commercial and environmental strains [9]–[13].

We sought to address these shortcomings by sequencing the genomes of four wine and two brewing strains of *S. cerevisiae* in their industrially-used forms. The industries of winemaking and brewing were targeted for this work as they have the longest association with *S. cerevisiae* (measured in the thousands of years) and each industry has accumulated large numbers of phenotypically distinct strains for which genetic comparisons can be made. This study demonstrates that industrial yeasts display significant genotypic heterogeneity both between strains, but also between alleles present within strains (i.e. heterozygosity). This variation was manifest as SNPs, small insertions and deletions, and as novel, strain and allele-specific ORFs, many of which had not been found previously in the *S. cerevisiae* genome and may provide the basis for novel phenotypic characteristics. Interestingly, several ORFs were shown to comprise a gene cluster that was present in multiple copies and at a variety of genomic loci in a subset of the strains examined. Furthermore, this cluster appears to have integrated into genomic locations by a novel circular intermediate, but without employing classical transposition or homologous recombination, which we believe represents the first time such an element has been characterized in *S. cerevisiae*.

Overall, this work suggests that, despite the scrutiny that has been directed at the yeast genome, there remains a significant reservoir of ORFs and novel modes of genetic transmission which may have significant phenotypic impact in this important model and industrial species.

13.2 RESULTS

Six industrial yeasts were chosen for genomic analysis, comprising four commercial wine strains and two brewing strains used for the production of ales (ale strains are primarily *S. cerevisiae*, while lager-style brewing strains are *S. pastorianus*, a hybrid of *S. cerevisiae* and *S. bayanus* [15], [16]). These six strains were sequenced to an average coverage of 20 fold with a combination of shotgun and paired-end methods using the GS FLX Titanium series chemistry [17], which resulted in six high quality genomic assemblies (Table 1).

13.2.1 LARGE CHROMOSOMAL VARIATIONS IN INDUSTRIAL YEAST STRAINS

Rather than being strictly diploid, many industrial yeast strains display chromosomal copy number variation (CNV) [18]. In order to catalogue

CNV in the industrial yeast genomes, the depth of sequencing coverage determined for each sequence contig were calculated such that areas of CNV could be detected as localized variations in that coverage (Figure 1). There were several large areas of increased copy number across the strains including six potential whole-chromosome amplifications (chrI of AWRI796, chrVIII of VL3, chrIII of FostersO and chrIII, V and XV of FostersB) and one potential reduction in chromosomal copy number (chrXIV of FostersO). There were also several partial chromosomal CNVs, including amplification of 200 kb of chrXIV in AWRI796, 600 kb of chrII and 200 kb of chrX in FostersO and a 400 kb reduction from chrVII of FostersO (Figure 1). However, while the ale strains had a higher number of large CNVs than wine strains, the overall fold change of these CNVs was generally reduced. This reduction can be most easily explained by the brewing strains having a polyploid genetic base while the wine strains are diploid, an observation which has been seen previously in these industrial yeasts [18].

TABLE 1: Strains sequenced in this study.

Strain	Industry	Supplier	Contigs[a]	N50[a] (kb)	Scaffolds[a]	Assembly size[a]	Genbank Accession[b]
Lalvin QA23	wine	Lallemand Inc.	96	185	39	11.6 Mb	ADVV00000000
AWRI796	wine	Maurivin	49	409	31	11.6 Mb	ADVS00000000
Vin13	wine	Anchor Bio-Technologies	80	308	29	11.5 Mb	ADXC00000000
FostersO	brewing (ale)	Fosters Group Ltd.	95	219	35	11.4 Mb	AEEZ00000000
FostersB	brewing (ale)	Fosters Group Ltd.	78	209	25	11.5 Mb	AEHH00000000
VL3	wine	Laffort	70	316	29	11.4 Mb	AEJS00000000

[a]*Excluding repetitive sequencing contigs such as sub-telomeric regions and Ty elements.*
[b]*These Whole Genome Shotgun projects have been deposited at DDBJ/EMBL/GenBank. The versions described in this paper are the first versions; ADVV01000000, ADVS01000000, ADXC01000000, AEEZ01000000, AEHH01000000, and AEJS01000000*

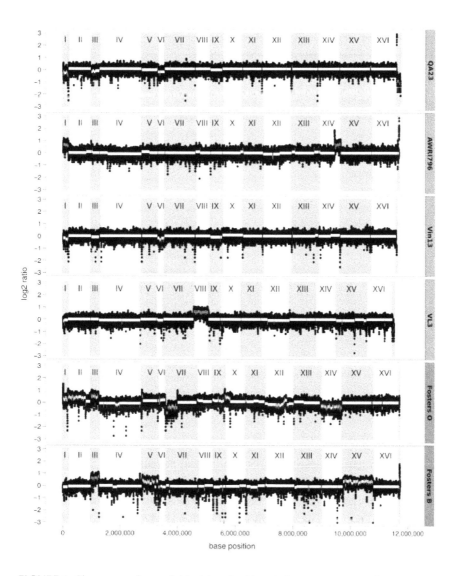

FIGURE 1. Chromosomal aneuploidy determined by whole-genome sequencing coverage. Sequencing coverage was determined for each contig using a sliding window of 1001 bp, with a 100 bp step frequency and plotted in chromosomal order (black circles). Regions of copy number variation were scored as either being greater than 1.25-fold (light lines; approximating either three or five copies in a tetraploid genome) or 1.5-fold (darker lines; one or three copies in a diploid genome) different to the median coverage for that strain. Strains are shaded according to their industry (wine, lighter; ale, darker).

TABLE 2: Heterozygosity in industrial *S. cerevisiae* strains.

Strain	Origin	Ploidy	Homozygous SNPs[a]	Heterozygous SNPs[a]
S288C	Lab	1n	41708	0
YJM789	Human isolate	1n	40675	0
JAY291	Bioethanol	1n	25648	0
RM11-1a	Vineyard isolate	1n	10825	0
EC1118	Wine	1n[b]	13241	0
AWRI1631	Wine	1n	9935	0
QA23	Wine	2n	4913	18861
AWRI796	Wine	2n	8996	1041
Vin13	Wine	2n	3544	15216
VL3	Wine	2n	5108	9904
FostersO	Ale	>2n[c]	25802	27215
FostersB	Ale	>2n[c]	23125	33071

[a]*SNPs were calculated relative to the most common base across all twelve strains at each position.*
b *EC1118 is a diploid commercial strain but the available sequence is a haploid representation of this genome*
c*As estimated from overall sequencing coverage.*

13.2.2 HETEROZYGOSITY IN INDUSTRIAL STRAINS

As existing published industrial yeast genome sequences were either generated from haploid derivatives of industrial strains [9]–[12] or had heterozygous regions discarded during analysis [13], the level of genome-wide heterozygosity present in industrial strains remains largely unknown. However, as the assemblies performed in this study retained genomic heterozygosity, it was possible to determine the level of allelic differences within each of these strains (Table 2). While every industrial strain contained heterozygous single nucleotide polymorphisms (SNPs), the proportion of these varied over thirty-fold between wine strain AWRI796 (1041 total heterozygous bp) and the brewing strain FostersB (33071 bp). Heterozygous insertions and deletions (InDels) were also present and ranged from single base pair variants to large InDels of up to 35.3 kb. Strains were also shown to contain heterozygous instances of Ty element insertion, although, due to the repetitive nature of these elements, their presence in the genome could generally only be estimated through paired-end information (data not shown).

13.2.3 NUCLEOTIDE VARIATION PRESENT IN S. CEREVISIAE

In addition to the intra-strain variation that was present between homologous chromosomes within individual strains, there was also significant nucleotide variation between strains. As seen for the allelic variation, both SNPs and InDels were found between strains, with inter-strain InDels of up to 45 kb being observed. Many of the smaller InDels (both heterozygous and homozygous) were located in regions comprising tandem repeats (Figure 2A, Table S1) and primarily in the expansion and contraction of di- and tri-nucleotide tandem repeats (Figure 2B). Indeed, when using chromosome XVI as an example, over 86% of the instances of di- and tri-nucleotide repeats displayed variable length in at least one of the strains. As the size of tandem repeats has been associated with differences in gene expression [19], this suggests that there are both strain and allele-specific differences in the expression of genes proximal to these repeat-associated InDel events.

SNP variation was also common throughout the strains with a total of 165,913 non-degenerate SNPs (unique points of nucleotide variation) that were present in at least one allele of the twelve strains investigated (~1.3% of the total genome length). However, given the influence of large, strain-specific InDels (which were filtered out of the SNP analysis) the apparent SNP density is much higher than 1.3%, such that these SNPs were shown to display a median inter-SNP distance of only 37 bp.

By using the number of SNPs separating any two isolates as an estimation of their relatedness (Figure 3A), we were able to show that industrial yeasts are distinct from both the laboratory and human pathogenic strains and were also found to group by industry. This was especially true of the brewing strains which displayed a high degree of genetic distance not only from the laboratory and human isolates, but also from the wine and bioethanol strains. The only exception to this pattern of grouping by industry or environment niche was with the 'natural' isolate RM11-1a which grouped closely with wine strains. However, given that it is descended from a strain

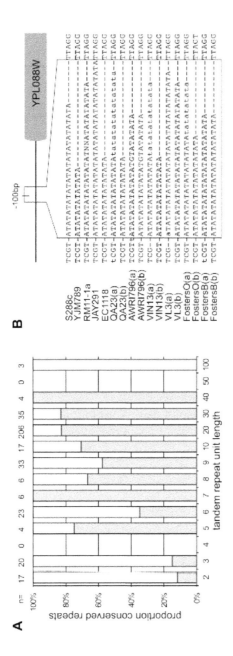

FIGURE 2. Nuclectide variation in *S. cerevisiae*. (A) InDels associated with tandem repeats. Histogram showing the proportion of tandem repeats of various sizes (repeated size indicated on x-axis) present on chrXVI that were either conserved in repeat length (blue) or contained strain-specific InDels (yellow). The total number of repeat loci present in each class is listed above the histogram. (B) An example of a strain- and allele-specific InDel in a tandem repeat in the promoter region of YPL088W.

sourced from a vineyard, RM11-1a may well share genetic origins with those strains used in winemaking.

In order to put the genetic variation observed in these genomic alignments in a larger population context, twelve strains were selected to represent each of the six main *S. cerevisiae* population groups as proposed by Liti et al [12] for further SNP comparison (Figure 3B). In this broader context, wine strains sequenced in this study were shown to also group tightly with the wine/European strains DBVPG1106 and DBVPG1373, showing that the data produced across these two studies are directly comparable. However, while the ale strains were still shown to be distinct from the wine isolates they were found to be far closer to the wine strains than isolates such as those used in sake production, which display the greatest level of nucleotide diversity when compared to the wine strains. Indeed, when the SNP data from these additional strains in included in the calculations of SNP density, the total number of non-degenerate SNPs increases to 216,207 (~1.7%) with a median inter-SNP distance of only 27 bp. However, despite comparisons to eighteen other diverse strains of *S. cerevisiae* 15,576 of these SNPs were found solely in this study (2,501 in more than one strain) and with the vast majority of these SNPs being present in a heterozygous form (only 1,864 novel SNPs were homozygous in at least one strain).

13.2.4 ORF CONSERVATION ACROSS S. CEREVISIAE

To determine how inter-specific variation at the nucleotide level translated into protein-coding differences, the predicted coding potential of each strain was compared. ORFs were predicted from each sequence (including the pre-existing whole genome sequences) using Glimmer [20] and compared using a combination of BLAST [21] homology matches and genomic synteny to differentiate instances of orthology from gene duplication (Table S2). When using the laboratory strain S288c as a reference, there was an average of 92% ORF coverage across the strains. The majority of S288c ORFs without a match in other strains were shown to be located in repetitive regions of the *S. cerevisiae* genome such as in the sub-telomeric zones or the numerous Ty retrotransposons that are present in S288c genome relative to other strains. Due to the repetitive nature of these regions it was often impossible to unambiguously position these sequences in the industrial yeast genome assemblies and they remain within repetitive, unmappable contigs in the various genome assemblies. It therefore appears that, due to its persistent propagation in the laboratory, the genome of S288c may represent a reduced genomic state as it does not appear to contain additional genes that provide unique metabolic or cellular potential outside of those present in other strains. It does however contain a far greater number of Ty transposons relative to all of the other strains suggesting that transposon proliferation occurred on at least one occasion during the development of this laboratory strain.

13.2.5 NOVEL ORFS

While the laboratory strain S288c is considered the reference for the genomic complement of *S. cerevisiae*, it is becoming apparent that it lacks a multitude of ORFs which exist in other strains of *S. cerevisiae* [9]–[13], [22], [23]. This is confirmed n the present study with between 36 (FostersB) and 110 (Lalvin QA23) ORFs lacking significant homology to the S288c genome but for which there were clear matches to sequences in other *S. cerevisiae* strains or microbial species (Table S2). Orthologs of 102 out of 218 of the non-degenerate set of these 'non-S288c' ORFs have been identified previously in *S. cerevisiae* strains, mainly through whole-genome sequencing of AWRI1631, EC1118 and RM11-1a and YJM789 [8], [9], [13] (Table S2). These include genes encoding proteins such as the Khr1 killer toxin [24] which is found in YJM789, EC1118, Vin13, VL3, FostersB and FostersO and orthologs of the MPR1 stress-resistance gene (which was originally identified in the Sigma 1278b strain[23]) in RM11-1a, EC1118, AWRI1631, JAY291, QA23 and VL3.

Interestingly, in addition to these ORFs there were at least three proteins present in the human pathogen YJM789 and the FostersB and FostersO ale strains but which were lacking from the wine, biofuel and laboratory strains (Figure 4C). These included the YJM-GNAT GCN5-related N-acetyltransferase [8] and a separate gene cluster which is predicted to

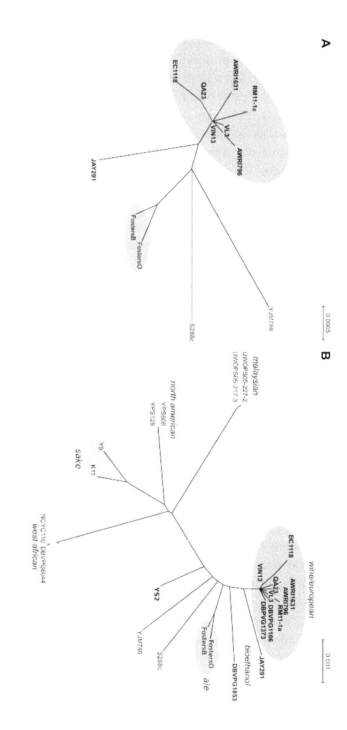

FIGURE 3: Nucleotide relationships between S. cerevisiae strains. (A) A neighbor joining tree representing the genetic distance between strains as calculated from the total SNP diversity present in whole genome alignments. (B) A neighbor joining tree representing the genetic distance between strains presented in part (A) and representative strains from several S. cerevisiae geographical populations [12]. Industrial strains are color-coded based upon their primary industry (wine/European, including RM11-1a, pink; a e, blue; bioethanol, green; sake, yellow). Strains that are predicted to contain the heterogeneous five-gene cluster are labeled in bold.

contain both *RTM1*, which was identified previously as a distillery-strain specific gene that provides resistance to an inhibitory substance found in molasses [22], and a large ORF of around 2.3 kb which, despite its large size and high-degree of conservation across the brewing and human pathogenic strains, lacks significant homology to any other protein sequences except for six isolates from the large *S. cerevisiae* population genomic screen which also appear to encode this protein [12] (Figure S1). In addition to these two conserved ORFs, in the ale strains this cluster also appears to encode an invertase that would be expected convert sucrose into the sugars glucose and fructose.

Despite the presence of at least two existing high-coverage wine strain sequences and at least an additional six low coverage genomes, the entire repertoire of ORFs present in wine strains of *S. cerevisiae*, let alone the species as a whole, is far from complete. In addition to expanding the strain range of previously identified non-S228c proteins, it was possible to identify at least eleven ORFs that lacked homology to existing proteins from *S. cerevisiae*, in addition to many new paralogs of existing *S. cerevisiae* genes. These novel ORFs often clustered in large InDels, the largest of which was a 45 kb fragment in the wine strain AWRI796. This novel genomic region is located adjacent to a large repetitive element present on chromosomes XIII, XV and XVI, which hampered initial efforts to assign this region to a specific chromosome. However, through the application of a 20 kb paired-end library, it was possible to bridge the repetitive region and position this novel region at the end of the right arm of chromosome XV. This fragment is predicted to encode nineteen ORFs (Figure 4A), three of which are predicted to encode aryl-alcohol dehydrogenases (AADs). AADs have been extensively characterized in filamentous fungi where they catalyze the reversible reduction of aldehydes and ketones to aromatic alcohols during lignin-degradation [25], [26]. These new AAD homologs are phylogenetically distinct from other AAD enzymes that have been identified, including the seven predicted AADs that are present in the S288c genome [27], [28] (Figure 4B).

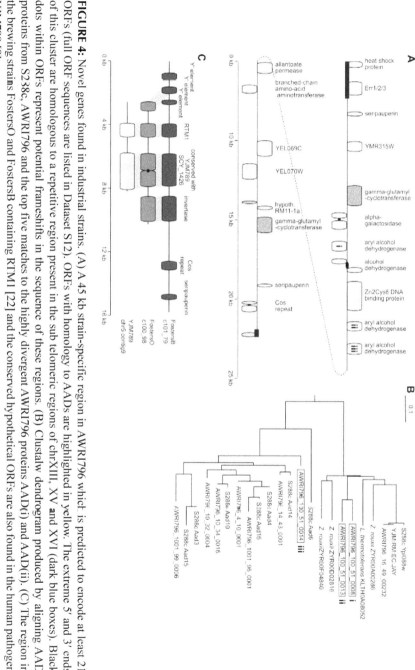

FIGURE 4: Novel genes found in industrial strains. (A) A 45 kb strain-specific region in AWRI796 which is predicted to encode at least 21 ORFs (full ORF sequences are listed in Dataset S12). ORFs with homology to AADs are highlighted in yellow. The extreme 5' and 3' ends of this cluster are homologous to a repetitive region present in the sub telomeric regions of chrXIII, XV and XVI (dark blue boxes). Black dots within ORFs represent potential frameshifts in the sequence of these regions. (B) Clustalw dendrogram produced by aligning AAD proteins from S238c, AWRI796 and the top five matches to the highly divergent AWRI796 proteins AAD(i) and AAD(ii). (C) The region in the brewing strains FostersO and FostersB containing RTM1 [22] and the conserved hypothetical ORFs are also found in the human pathogen YJM789 [8].

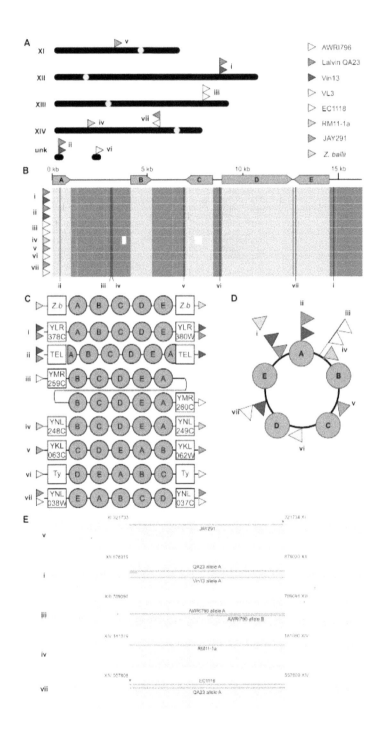

FIGURE 5: A divergent cluster of genes with a possible circular intermediate. (A) The location and orientation of the gene cluster throughout the genomes of the industrial yeasts. Upper case roman numerals refer to standard *S. cerevisiae* chromosomes (unk – location unknown) with individual loci labeled with lower case roman numerals. (B) Nucleotide conservation of the five-gene clusters. An alignment of the nucleotide sequence of all eleven clusters is shown below a schematic depiction of the five predicted ORFs present in this nucleotide sequence (A, zinc-cluster transcription factor; B, cell-surface flocculin; C, nicotinic acid permease; D, 5-oxo-L-prolinase; E, C6 transcription factor). In order to produce contiguous alignments, the sequence of each cluster was manually split to begin with the start codon of ORF A, with the position of each break indicated. Conserved bases are shaded blue (light blue for ORFs sequences). Insertions are highlighted in red and substitutions in green. (C) Differences in gene order within individual clusters. Each of the five genes are represented by filled circles (labeled as in partB), with the systematic name of the ORFs that border each insertion listed in open squares (Z.b, this cluster is present in Z. bailii (Accession number FN295481.1); Ty, transposon sequence; TEL, sub-telomeric repeat (COS) sequence). Colored arrows bordering each cluster indicate the strain(s) in which this insertion is present. (D) Each of the nine cluster locations and orders can be resolved through the use of a circular intermediate that integrates into the genome via breakage at locations indicated by each colored triangle. (E) Conservation of genomic sequences flanking individual cluster insertion events. Nucleotide alignments are shown for the 50 bp directly adjacent to either side of the five chromosomally-mapped insertion events (shaded when conserved) in addition to the first and last 50 bp of the each cluster (shaded according to partB). Insertions are shaded darker, substitutions lighter with both additionally highlighted by asterisks. Sequences used for the alignment are (from top to bottom) S228c, JAY291, RM11-1a, EC1118, AWRI1631, QA23 allele A, QA23 allele B, AWRI796 allele A, AWRI796 allele B, Vin13 allele A, Vin13 allele B, VL3 allele A, VL3 allele B, Fosters B allele A, Fosters B allele B, Fosters O allele A, Fosters O allele B. Nucleotide coordinates for the bases directly flanking the insertion are relative to the S288c genome.

13.2.6 CHARACTERIZATION OF A NOVEL, AND POTENTIALLY TRANSMISSIBLE, GENE CLUSTER

One particularly curious feature of many of the industrial yeast strains analyzed in this study, was a cluster of five conserved ORFs that was present in all of the wine strains, RM11-1a and the bioethanol strain JAY291, and potentially in at least four of the strains present in the Liti et al [12] study (Figure 3). This cluster is predicted to encode two potential transcription factors (one zinc-cluster, one C_6 type), a cell surface flocullin, a nicotinic acid permease and a 5-oxo-L-prolinase, and has been suggested to be horizontally acquired by *S. cerevisiae* from *Zygosacharomyces spp* [13]. In this study we have been able to show that while the sequences of the individual genes within this cluster are highly conserved between strains, the cluster itself is actually highly diverse with respect to copy number, genomic location and overall gene order (Figure 5, Table S3). The cluster was present in one to at least three copies across strains, with individual clusters being located in at least seven different genomic loci (Figure 5A). For example, wine strain Lalvin QA23 was shown to contain at least three copies of the cluster, found in three different genomic loci and with at least two copies being heterozygous. However, despite this diversity, the sequence of the ORFs and intergenic regions of the cluster were highly conserved, with only fifteen nucleotide substitutions (0.01%) recorded across the eleven known copies of the cluster (Figure 5B, Figure S2).

In addition to the differences in copy number and location, the exact order of the ORFs within the cluster differed in a location dependent manner (Figure 5B, 5C). However, all of these different ORF arrangements could be resolved into a syntenically-conserved order if the linear genomic copy of each cluster resulted from the differential resolution of a common circular intermediate, with a unique breakpoint in this circular arrangement being observed for each genomic location (Figure 5B–5D). However, despite the differential location of these clusters these integration events appear to select for functional conservation of the genes with the majority of the breakpoints being located within intergenic regions (Figure 5B). Of the two exceptions to this, one of these events occurs at the extreme 3' end (~100 bp from the predicted stop codon) of one ORF such that a functional protein is likely to still be produced from this gene.

Adding further interest to the mode of transfer of this cluster, its integration into the genome appears to occur without the production of the terminal repeated sequences that would be expected if integration of this element occurred by either homologous recombination or classical mobilization via a transposon-like mechanism. In fact, for at least three of the seven different integration events characterized in this study, integration of the cluster has occurred between two directly adjacent, conserved nucleotides, with a further two events showing only single nucleotide indels at the junction between the cluster and the flanking genomic sequences (Figure 5E).

13.3 DISCUSSION

While *S. cerevisiae* is one of the most intensively studied biological model organisms and economically-important industrial microorganisms, many characteristics of its genome remain unknown, especially in strains other than the laboratory reference S288c. Through the analysis of six industrial strains, it was possible to show that the industrial members of this species are distinct, with wine and brewing strains being almost as distantly related at the DNA level as they are to either the laboratory or human pathogenic strains. This suggests that despite their roles in performing industrial fermentations, the two groups comprise genetically separate *S. cerevisiae* lineages. While this is a situation similar to that proposed previously for wine and sake strains of *S. cerevisiae* [2], the wine and ale strains were much more closely related to each other than to strains with origins outside of Europe [12], and this may reflect a distant common European-type ancestor. The bioethanol strain JAY291 displays an intermediate level of sequence relatedness to the wine strains (compared to ale strains) and also contains the five-gene cluster, suggesting that this strain shares at least some of its genomic origins with the wine isolates. With the relatively recent development of the bioethanol industry, it is not entirely unexpected that yeasts used in this process may well have their origins in commercial strains used in established ethanologenic industries. Wine strains would therefore make a logical choice for this starting point given their highly efficient production of ethanol and relatively high tolerance to a variety inhibitory substances, such as ethanol or polyphenols, that also exist in bioethanol fermentations [29].

In addition to mapping the relationships between these strains, this study uncovered a number of genetic elements not previously identified in the *S. cerevisiae* genome, as well as expanding the range of several strain specific elements that had been identified previously. This highlights the fact that the genetic variation that underlies the phenotypic diversity of *S. cerevisiae* goes well beyond that of SNPs or small InDels and is similar to the situation observed with many bacterial species where the pan (species-wide) genome is larger than that observed in any single strain [30]. As for the situation observed with single nucleotide variation, several of these genetic elements link strains to specific industries (e.g. the *RTM1* cluster in the ale strains and the five-gene cluster in the wine strains). It would therefore be expected that these ORFs provide selective advantage within specific industries that have favored their retention. For some of these ORFs, such as the *RTM1* cluster, the phenotypic benefits that they have historically provided in one industry may be advantageous in modern incarnations of others. For example, modern wine production generally makes use of inoculated commercial strains (rather than the historical use of wild yeast), which are produced on a large scale using molasses as a feedstock. Genes such as the *RTM1* cluster may therefore provide advantages in the production of modern commercial wine yeast, but which are lacking from the genomic complement of this group of strains due to the historical practices of winemaking.

While other strain-specific ORFs were shown to have much narrower strain ranges (often single strains), it was possible to predict industrially-relevant roles for some of these genes. For example, the novel AAD proteins that were identified in the wine strain AWRI796 may have a direct impact on the range of volatile aromas produced during fermentation, as the aromatic alcohols produced through the action of the AAD enzymes can present very different aromas profiles to their corresponding aldehydes and ketones [31]. The presence of these AADs in specific industrial yeasts may therefore alter the profile of volatile aromas produced during winemaking or brewing, contributing to strain-specific aroma characteristics that are vitally important to many flavor and aroma-based industrial applications.

The role of ORFs such as those present in the wine yeast five-gene cluster are less clear but, given the potential regulatory role for at least two of these proteins, they could produce significant phenotypic effects. The generally similar characteristics of high sugar and ethanol tolerance of *Zygosacharomyces spp* and the wine and bioethanol strains of *S. cerevisiae* [29], [32], may provide a selective advantage for growth under these conditions. However, understanding the function of individual ORFs is overshadowed by questions regarding the origins of this novel cluster in addition to its effect on genome structure and dynamics. It was recently proposed that this cluster entered the *S. cerevisiae* genome from *Zygosacharomyces spp* [13]. Our data suggests that if this is the case, the transfer has either occurred on multiple occasions via a conserved circular intermediate that has integrated randomly into different genomic loci, or the fragment has entered the *S. cerevisiae* genome on a single occasion but has subsequently mobilized to new genomic locations via a circular intermediate (Figure S3). Alternatively, this cluster is a mobile feature of the *S. cerevisiae* genome that has been lost from many strains and was transferred to *Zygosacharomyces spp.* Regardless of the direction or precise mode of transfer it appears that this genetic cluster may mobilize throughout the genome via a method which has yet to be characterized in yeast and therefore provides an entirely new mechanism for the generation of variation in the *S. cerevisiae* genome.

A thorough understanding of the scope of plasticity of the yeast genome is a vital prerequisite for the systematic understanding of yeast biology or for the development of the next generation of yeasts for industrial applications. As more *S. cerevisiae* strains are sequenced, the suitability of S288c as a "reference" strain for this species is becoming less clear, especially as it appears to lack a large numbers of ORFs found in many other *S. cerevisiae* strains while containing an abnormally high number of Ty transposable elements [8], [9]. Given the ubiquitous nature of the S288c genome for the design of 'omics experiments, these novel elements have generally not been considered when studying strains other than S288c. Thus, little data exists regarding the functional contributions of these proteins. As such, they represent a significant knowledge gap with respect to cellular and metabolic modeling strategies. This is especially true for proteins such as the ORF located next to *RTM1* which is large (~800 amino acids) and highly conserved but has no significant homologs outside of a small subset of *S. cerevisiae* strains on which a function can be based. Fortunately, the continued development of next generation sequencing, such as that applied in this work, have provided the means to now characterize large numbers of yeast strains to provide this information and outline the true scope and variability of this species.

13.4 MATERIALS AND METHODS

13.4.1 YEAST STRAINS

Each commercial strain was obtained from the original mother cultures from the supplier. Genomic DNA was prepared by zymolase digestion and standard phenol-chloroform extraction.

13.4.2 SEQUENCING AND ASSEMBLY

Library construction and sequencing was performed at 454 Life Sciences, A Roche Company (Branford, CT) using a pre-release development version of the GS FLX Titanium series shotgun and 3 kb paired-end protocols. Sequences were assembled using MIRA (http://sourceforge.net/apps/mediawiki/mira-assembler/index.php?title=Main_Page) and manually-edited using Seqman Pro (DNAstar).

Regions of chromosomal CNV were determined by calculating the per-base sequencing coverage across each sequencing contig with median smoothing (1001 bp window, 100 bp step size). The ratio between the coverage at each genomic location and the overall median genomic coverage was the calculated to determine the level of over-representation for each location. Large-scale chromosomal aneuploidies were detected by screening for regions in which median ratio for a contiguous stretch of at least 101 individual segments differed from the overall genomic median by either 1.25 (5:4 ratio representing at least 1 extra genomic copy in a tetraploid) or 1.4 fold (3:2 ratio representing at least 1 extra genomic copy in a diploid).

13.4.3 SNP PREDICTION

Chromosomal scaffolds from each yeast strain were aligned using FSA [33]. Diploid sequences were assigned into two haploid alleles by converting any degenerate bases into their non-degenerate pairs. Heterozygous regions were divided into both an insertion and deletion allele. A chromosomal consensus was computed for the alignment based upon the most frequent allele at each position in the alignment. Nucleotides that varied from the consensus in each strain were scored as sequence variants and were subsequently divided into SNPs (nucleotide substitution) or InDels (nucleotide insertion or deletion). To enable the comparison to strains with low coverage sequences [12], SNPs that were calculated for each strain relative to S288c (imputed SNPs) were used to create synthetic S288c-based genome sequences that contain the SNPs present in these strains. The genetic relationship between the strains was calculated by editing and concatenating the nucleotide alignments of all sixteen chromosomes using Seaview [34] followed by calculating the distance tree using the NJ algorithm of Clustalw (ignoring gapped regions in the alignment). Tandem repeats were predicted from the chromosomal alignment of all twelve yeast strains using Tandem Repeats Finder [35] using default parameters (match weight, 2; mismatch, 7; indel, 7; pM, 0.80; pI, 0.10; minimum alignment score, 50; maximum period size, 500). Individual repeats were then scored as either being variable if the specific tandem repeat region contained strain- or allele- specific InDels.

13.4.4 ORF PREDICTION AND COMPARISON

ORFs were predicted using Glimmer [20] with the predicted ORFs of S288c being used to build the prediction model (See Datasets S1, S2, S3, S4, S5, S6, S7, S8, S9, S10, S11 for actual CDS sequences for each strain). Initial ORF designations were made by identifying the best sequence match for each ORF when compared to S288c using BLASTn [21]. Glimmer was also used to predict ORFs from the sequence of S288c (Accession numbers NC001133-NC001148) to correct for false-negatives in the predictions when compared to existing ORF designations in S288c. ORFs with no match to S288c were searched against the full list of non-redundant Genbank proteins to identify a closest existing homology match. ORFs from each strain were then arranged in syntenic order (Table S2 for a full list of ordered ORFs). For protein sequence comparisons, predicted protein sequences were aligned using Clustalw [36] (http://align.genome. jp).

REFERENCES

1. Querol A, Belloch C, Fernandez-Espinar MT, Barrio E (2003) Molecular evolution in yeast of biotechnological interest. Int Microbiol 6: 201–205. doi: 10.1007/s10123-003-0134-z.
2. Fay JC, Benavides JA (2005) Evidence for domesticated and wild populations of Saccharomyces cerevisiae. PLoS Genet 1: e5. doi:10.1371/journal.pgen.0010005.

3. Mortimer RK, Johnston JR (1986) Genealogy of principal strains of the yeast genetic stock center. Genetics 113: 35–43.

4. Lambrechts MG, Pretorius IS (2000) Yeast and its importance to wine aroma - a review. Sth Afr J Enol Vitic 21: 97–129.

5. Swiegers JH, Pretorius IS (2005) Yeast modulation of wine flavor. Adv Appl Microbiol 57: 131–175. doi: 10.1016/S0065-2164(05)57005-9.

6. Goffeau A, Barrell BG, Bussey H, Davis RW, Dujon B, et al. (1996) Life with 6000 genes. Science. 274. New York, , NY: pp. 546, 563–547.

7. Dunn B, Levine RP, Sherlock G (2005) Microarray karyotyping of commercial wine yeast strains reveals shared, as well as unique, genomic signatures. BMC Genomics 6: 53. doi: 10.1186/1471-2164-6-53.

8. Wei W, McCusker JH, Hyman RW, Jones T, Ning Y, et al. (2007) Genome sequencing and comparative analysis of Saccharomyces cerevisiae strain YJM789. Proc Natl Acad Sci USA 104: 12825–12830. doi: 10.1073/pnas.0701291104.

9. Borneman AR, Forgan AH, Pretorius IS, Chambers PJ (2008) Comparative genome analysis of a Saccharomyces cerevisiae wine strain. FEMS Yeast Res 8: 1185–1195. doi: 10.1111/j.1567-1364.2008.00434.x.

10. Doniger SW, Kim HS, Swain D, Corcuera D, Williams M, et al. (2008) A catalog of neutral and deleterious polymorphism in yeast. PLoS Genet 4: e1000183. doi:10.1371/journal.pgen.1000183.

11. Argueso JL, Carazzolle MF, Mieczkowski PA, Duarte FM, Netto OV, et al. (2009) Genome structure of a Saccharomyces cerevisiae strain widely used in bioethanol production. Genome Res 19: 2258–2270. doi: 10.1101/gr.091777.109.

12. Liti G, Carter DM, Moses AM, Warringer J, Parts L, et al. (2009) Population genomics of domestic and wild yeasts. Nature 458: 337–341. doi: 10.1038/nature07743.

13. Novo M, Bigey F, Beyne E, Galeote V, Gavory F, et al. (2009) Eukaryote-to-eukaryote gene transfer events revealed by the genome sequence of the wine yeast Saccharomyces cerevisiae EC1118. Proc Natl Acad Sci USA 106: 16333–16338. doi: 10.1073/pnas.0904673106.

14. Stambuk BU, Dunn B, Alves SL Jr, Duval EH, Sherlock G (2009) Industrial fuel ethanol yeasts contain adaptive copy number changes in genes involved in vitamin B1 and B6 biosynthesis. Genome Res 19: 2271–2278. doi: 10.1101/gr.094276.109.

15. Tamai Y, Momma T, Yoshimoto H, Kaneko Y (1998) Co-existence of two types of chromosome in the bottom fermenting yeast, Saccharomyces pastorianus. Yeast 14: 923–933. doi: 10.1002/(SICI)1097-0061(199807)14:10<923::AID-YEA298>3.0.CO;2-I.

16. Dunn B, Sherlock G (2008) Reconstruction of the genome origins and evolution of the hybrid lager yeast Saccharomyces pastorianus. Genome Res 18: 1610–1623. doi: 10.1101/gr.076075.108.

17. Margulies M, Egholm M, Altman WE, Attiya S, Bader JS, et al. (2005) Genome sequencing in microfabricated high-density picolitre reactors. Nature 437: 376–380. doi: 10.1038/nature03959.

18. Mortimer RK (2000) Evolution and variation of the yeast (Saccharomyces) genome. Genome Res 10: 403–409. doi: 10.1101/gr.10.4.403.

19. Vinces MD, Legendre M, Caldara M, Hagihara M, Verstrepen KJ (2009) Unstable tandem repeats in promoters confer transcriptional evolvability. Science 324: 1213–1216. doi: 10.1126/science.1170097.

20. Delcher AL, Bratke KA, Powers EC, Salzberg SL (2007) Identifying bacterial genes and endosymbiont DNA with Glimmer. Bioinformatics 23: 673–679. doi: 10.1093/bioinformatics/btm009.

21. Altschul SF, Madden TL, Schaffer AA, Zhang J, Zhang Z, et al. (1997) Gapped BLAST and PSI-BLAST: a new generation of protein database search programs. NAR 25: 3389–3402. doi: 10.1093/nar/25.17.3389.

22. Ness F, Aigle M (1995) RTM1: a member of a new family of telomeric repeated genes in yeast. Genetics 140: 945–956.

23. Takagi H, Shichiri M, Takemura M, Mohri M, Nakamori S (2000) Saccharomyces cerevisiae sigma 1278b has novel genes of the N-acetyltransferase gene superfamily required for L-proline analogue resistance. J Bacteriol 182: 4249–4256. doi: 10.1128/JB.182.15.4249-4256.2000.

24. Goto K, Iwase T, Kichise K, Kitano K, Totuka A, et al. (1990) Isolation and properties of a chromosome-dependent KHR killer toxin in Saccharomyces cerevisiae. Agric Biol Chem 54: 505–509. doi: 10.1271/bbb1961.54.505.

25. Constam D, Muheim A, Zimmermann W, Fiechter A (1991) Purification and Partial Characterization of an Intracellular NADH - Quinone Oxidoreductase from Phanerochaete chrysosporium. J Gen Microbiol 137: 2209–2214.

26. Reiser J, Muheim A, Hardegger M, Frank G, Fiechter A (1994) Aryl-alcohol dehydrogenase from the white-rot fungus Phanerochaete chrysosporium. Gene cloning, sequence analysis, expression, and purification of the recombinant enzyme. J Biol Chem 269: 28152–28159.

27. Delneri D, Gardner DC, Bruschi CV, Oliver SG (1999) Disruption of seven hypothetical aryl alcohol dehydrogenase genes from Saccharomyces cerevisiae and construction of a multiple knock-out strain. Yeast 15: 1681–1689. doi: 10.1002/(SICI)1097-0061(199911)15:15<1681::AID-YEA486>3.0.CO;2-A.

28. Delneri D, Gardner DC, Oliver SG (1999) Analysis of the seven-member AAD gene set demonstrates that genetic redundancy in yeast may be more apparent than real. Genetics 153: 1591–1600.

29. Pretorius IS (2000) Tailoring wine yeast for the new millennium: novel approaches to the ancient art of winemaking. Yeast 16: 675–729. doi: 10.1002/1097 0061(20000615)16:8<675::AID-YEA585>3.0.CO;2-B.

30. Lefebure T, Stanhope MJ (2007) Evolution of the core and pan-genome of Streptococcus: positive selection, recombination, and genome composition. Genome Biol 8: R71. doi: 10.1186/gb-2007-8-5-r71.

31. Ugliano M, Henschke PA (2009) Yeasts and Wine Flavour. In: Moreno-Arribas MV, Polo MC, editors. Wine Chemistry and Biochemistry. New York: Springer. pp. 313–392.

32. Sponholz W (1993) Wine spoilage by microorganisms. In: Fleet G, editor. Wine Microbiology and Biotechnology. London: Taylor and Francis. pp. 395–420.

33. Bradley RK, Roberts A, Smoot M, Juvekar S, Do J, et al. (2009) Fast statistical alignment. PLoS Comput Biol 5: e1000392. doi:10.1371/journal.pcbi.1000392.

34. Gouy M, Guindon S, Gascuel OSeaView version 4: A multiplatform graphical user interface for sequence alignment and phylogenetic tree building. Mol Biol Evol 27: 221–224. doi: 10.1093/molbev/msp259.

35. Benson G (1999) Tandem repeats finder: a program to analyze DNA sequences. NAR 27: 573–580. doi: 10.1093/nar/27.2.573.

36. Thompson JD, Higgins DG, Gibson TJ (1994) CLUSTAL W: improving the sensitivity of progressive multiple sequence alignment through sequence weighting, position-specific gap penalties and weight matrix choice. NAR 22: 4673–4680. doi: 10.1093/nar/22.22.4673.

This article has supplemental information that is not featured in this version of the text. To view these files, please visit the original version of the article as cited in the beginning of this chapter.

AUTHOR NOTES

CHAPTER 1

Funding
This work was supported by a National Institute of Health post-doctoral fellowship to JMS, grants R01 GM78153 and GM47238 to FRC, and grants R01 GM064813 and GM039978 to BF. The funders had no role in study design, data collection and analysis, decision to publish, or preparation of the manuscript.

Competing Interests
The authors have declared that no competing interests exist.

Acknowledgments
We thank Nicolas Buchler, Alejandro Colman-Lerner, Frank Luca, Mary Miller, Robert Singer, and David Stillman for strains and plasmids. We thank Eric Siggia for many critical discussions and advice on quantitative analysis, and Catherine Oikonomou and Jonathan Robbins for comments on the manuscript.

Author Contributions
The author(s) have made the following declarations about their contributions: Conceived and designed the experiments: SDT HW JMS BF FRC. Performed the experiments: SDT HW APR BF FRC. Analyzed the data: SDT HW JMS APR BF FRC. Contributed reagents/materials/analysis tools: SDT HW JMS APR BF FRC. Wrote the paper: SDT JMS BF FRC.

CHAPTER 2

Funding
The work was supported by NIH grants R01GM75309 (through the joint NSF/NIGMS initiative to support research in the area of mathematical biology) to QN and TMY, and P50GM76516 to QN and TMY.

Competing Interests

The authors have declared that no competing interests exist.

Author Contributions

Conceived and designed the experiments: CSC QN TMY. Performed the experiments: CSC. Analyzed the data: CSC QN TMY. Contributed reagents/materials/analysis tools: CSC QN. Wrote the paper: TMY.

CHAPTER 3

Funding

This work was supported by grant number R01GM057063 from the National Institutes of Health to RL. The funders had no role in study design, data collection and analysis, decision to publish, or preparation of the manuscript.

Competing Interests

The authors have declared that no competing interests exist.

Acknowledgments

We are grateful to R. Wedlich-Soldner and G. Rancati for their insightful discussions and technical suggestions. We are also grateful to C. Egile, I. M. Lister, and P. Marina for their advice regarding protein purification. We would like to thank E. Bi for generously providing the MBP-Cdc24 construct and an aliquot of the anti-Cdc24 antiserum, P. Crews for his gift of Latrunculin A, and K. Delventhal and E. Jessen for their much-appreciated assistance with site-directed mutagenesis.

Author Contributions

Conceived and designed the experiments: SCW SAG RL. Performed the experiments: SCW SAG. Analyzed the data: SCW SAG. Wrote the paper: SCW RL.

CHAPTER 4

Funding

This project is supported by the grant of "100 Talent Program" from Chinese Academy of Sciences to GH and National Natural Science Foundation of China (30873129) to LZ. The funders had no role in study design, data collection and analysis, decision to publish, or preparation of the manuscript.

Competing Interests

The authors have declared that no competing interests exist.

Acknowledgments

The authors are indebted to Drs. David Soll at the University of Iowa, Joachim Morschhäuser at the University of Wurzburg, Gemany, and Alexander Johnson at the University of California, San Francisco for generous gifts of plamids and strains.

Author Contributions

Conceived and designed the experiments: HD LZ GH. Performed the experiments: HD GG JX YS YT GH. Analyzed the data: HD GH. Contributed reagents/materials/analysis tools: HD GG JX YS YT GH. Wrote the paper: HD GH.

CHAPTER 5

Funding

This work was supported in part by the National Institutes of Health National Research Service Award (NRSA) Postdoctoral Fellowship GM076954 (to DWN), grant R01-NS065890 (to DJT), and grants from the Fundaçao para a Ciencia e Tecnologia (No. SFRH/BD/39389/2007 to LB-N and No. PTDC/BIAMIC/108747/2008 to CR-P). The funders had no role in study design, data collection and analysis, decision to publish, or preparation of the manuscript. PCCL contributed to this manuscript before employment at Incyte Corporation, which had no involvement in this study.

Acknowledgments

We thank Dr. David E. Levin for plasmid YEp351-Slt2-FLAG and members of the Thiele laboratory for critical reading of this manuscript.

Author Contributions

Conceived and designed the experiments: LB-N DWN PCCL DJT. Performed the experiments: LB-N DWN PCCL. Analyzed the data: LB-N DWN PCCL CR-P DT. Wrote the paper: DWN DJT.

CHAPTER 6

Funding

This work was supported by grants from the US National Institutes of Health (R01NS056086) and the US National Science Foundation (MCB 1122135) to LL. The funders had no role in study design, data collection and analysis, decision to publish, or preparation of the manuscript.

Competing Interests

The authors have declared that no competing interests exist.

Acknowledgments

We thank the members of the Li Laboratory for comments on this article. We regret that many related publications were not included due to space limitations.

CHAPTER 7

Funding

This work was supported by NIH grant to SWL (R01 GM 056350) and ILD (7 R01 GM 070934). The funders had no role in study design, data collection and analysis, decision to publish, or preparation of the manuscript.

Competing Interests

The authors have declared that no competing interests exist.

Acknowledgments

We thank A. Brestscher, L. Brodsky, Y. O. Chernoff, E. A. Craig, E. Herrero, H. Kurahashi, S. L. Lindquist, D. C. Masison, Y. Nakamura, and R. B. Wickner for strains, plasmids, and antibodies. We especially thank T. R. Serio for providing us with the Sup35-GFP yeast strains essential for this study, prior to publication. We thank R. Levine, T. Cahill, and S. Bagriantsev for help with experiments in the initial phase of this work. We also thank D. Kaganovich for helpful suggestions about this work. Finally, we thank D. Stone for helpful comments and for allowing Z.Yang to complete this work in his laboratory.

Author Contributions

Conceived and designed the experiments: ZY ILD SWL. Performed the experiments: ZY JYH. Analyzed the data: ZY SWL. Contributed reagents/materials/analysis tools: ZY JYH ILD SWL. Wrote the paper: ZY SWL

CHAPTER 8

Funding

This work was supported by a grant from the Kahn Fund for Systems Biology at the Weizmann Institute of Science, and by the Tauber Fund through the Foundations of Cognition Initiative.

Competing Interests

The authors have declared that no competing interests exist.

Acknowledgments

We thank S. Maizel for the help and advice with setting up the chemostat system. We thank B.Z. Shilo for helpful comments on the manuscript. We also thank I. Tirosh and A. Koren for helpful discussions and careful reading of the manuscript.

Author Contributions

Conceived and designed the experiments: NB JI SL GF. Performed the experiments: MC SL AW GF. Analyzed the data: NB JI SL. Contributed reagents/materials/analysis tools: JI. Wrote the paper: NB SL.

CHAPTER 9

Funding

This work was supported by grants from the Swiss National Science Foundation (3100A0-112235) and the Bonizzi-Theler Foundation to APG. NM was supported by a Commonwealth Split-Site fellowship. SCJ acknowledges financial support from the Medical Research Council (MRC) Laboratory of Molecular Biology and Cambridge Commonwealth Trust. The funders had no role in study design, data collection and analysis, decision to publish, or preparation of the manuscript.

Competing Interests

The authors have declared that no competing interests exist.

Acknowledgments

We thank members of the Gerber and Detmar labs for discussions. We thank Dr. K. Harshman and his team at the Center for Integrative Genomics (University of Lausanne, Switzerland) for the production of yeast oligo microarrays.

Author Contributions

Conceived and designed the experiments: TS SJC APG. Performed the experiments: TS NM. Analyzed the data: TS SJC APG. Contributed reagents/materials/analysis tools: SJC APG. Wrote the paper: TS APG.

CHAPTER 10

Funding

This work was supported by a grant from the KRIBB Research Initiative Program and by a grant from the National Research Foundation of Korea (2012019094). GKB and JKC were supported by the Agency for Science, Technology, and Research (A*STAR) of Singapore. JKC is a recipient of the TJ Park Science Fellowship. Computing facilities were supported by the CHUNG Moon Soul Center of KAIST and a grant from the National Research Foundation of Korea (2009-0086964). The funders had no role in study design, data collection and analysis, decision to publish, or preparation of the manuscript.

Competing Interests

The authors have declared that no competing interests exist.

Acknowledgments

We thank Jachwan Jahng and Dae Woong Jung for help with the data analysis, Dr. Leonid Kruglyak and Dr. Rachel Brem for providing the yeast strains, and Dr. Terry Furey for help with F-Seq.

Author Contributions

Conceived and designed the experiments: JKC BL. Performed the experiments: KL JS H-SL. Analyzed the data: SCK IJ KK. Contributed reagents/materials/analysis tools: GKB DK SL. Wrote the paper: JKC.

CHAPTER 11

Competing Interests

The authors declare that they have no competing interest.

Author Contributions

JW carried out the phenotyping experiments. SM made the sequencing libraries. CAM, FD-C and FAC performed the remaining experiments. MB generated the genome assembly. ANNB, AB, CCS, NM and SK performed computational analysis. ANNB generated the web tools. GL, JW, AMM, EJL and CAN designed and supervised the experiments and computational analysis. GL and CAN wrote the manuscript. All authors read and approved the final manuscript.

Acknowledgments

We thank Dr. Feng-Yan Bay for kindly providing S. arboricolus strains. GL is supported by CNRS, ATIP-AVENIR and ARC. JW is supported by the Royal Swedish Academy of Sciences and the Carl Trygger foundation. CAM, CCS and CAN are supported by the Biotechnology and Biological Sciences Research Council (grant numbers BB/E023754/1, BB/G001596/1); CAN is a David Phillips Fellow. ANNB is supported by a postgraduate scholarship from the Natural Sciences and Engineering Research Council of Canada.

CHAPTER 12

Competing Interests

The authors declare they have no competing interests.

Author Contributions

YF developed the LNS metric, calculated global LNS, and correlated it to sequence features. MJD and AAC analyzed LNS data and sequence similarity measures. YF calculated and clustered condition-specific LNS. All authors conceived the study, participated in interpretation of the results, and wrote the paper. All authors read and approved the final manuscript.

Acknowledgments

Thanks to Greg Lang for GPA1 data in advance of publication; David Botstein, Ian Ehrenreich, Justin Gerke, Greg Lang, and Dannie Durand for helpful comments; and Kevin Byrne for orthology mapping data. OGT is supported by the NIH grant R01 GM071966, NSF grant IIS-0513552 and NSF CAREER award DBI-0546275. All authors were supported by the NIGMS Center of Excellence P50 GM071508, and by donations from the A. V. Davis Foundation and Princeton University for funding of QCB301, Experimental Project Laboratory. MJD is supported in part by grants from the National Center for Research Resources (5P41RR011823-17) and the National Institute of General Medical Sciences (8 P41 GM103533-17) from the National Institutes of Health. MJD is a Rita Allen Foundation Scholar. OGT is supported by the NIH grants R01 GM071966 and R01HG005998, NSF grant IIS-0513552 and NSF CAREER award DBI-0546275. AAC is supported by the Canadian Institutes for Health Research.

CHAPTER 13

Funding

The AWRI, a member of the Wine Innovation Cluster in Adelaide, is supported by Australian grapegrowers and winemakers through their investment body, the Grape and Wine Research Development Corporation, with matching funds from the Australian Government. Systems Biology research at the AWRI is performed using resources provided as part of the National Collaborative Research Infrastructure Strategy, an initiative of the Australian Government, in addition to funds from the South Australian State Government. The funders of this work had no role in study design, data collection and analysis, decision to publish, or preparation of the manuscript.

Competing Interests

BAD, DR, JPA, and ME were employees of 454 Life Sciences, A Roche Company, at the time this work was performed.

Acknowledgments

The authors would like to thank the commercial yeast suppliers for access to original mother cultures of all strains used in this study and Paul Henschke (AWRI) for critical comments on the manuscript.

Author Contributions

Conceived and designed the experiments: ARB JPA ISP ME PJC. Performed the experiments: BAD DR AHF. Analyzed the data: ARB. Wrote the paper: ARB PJC.

INDEX

N

O